T0236847

AUTOMOTIVE ACCIDENT
RECONSTRUCTION
Practices and Principles

Ground Vehicle Engineering Series

Series Editor:

Dr. Vladimir V. Vantsevich

Professor and Director
Program of Master of Science in Mechatronic Systems Engineering
Lawrence Technological University, Michigan

Automotive Accident Reconstruction: Practices and Principles
Donald E. Struble

Dynamics of Wheel–Soil Systems: A Soil Stress and Deformation-Based Approach
Jaroslaw A. Pytka

Road Vehicle Dynamics: Fundamentals and Modeling
Georg Rill

Driveline Systems of Ground Vehicles: Theory and Design
Alexandr F. Andreev, Viachaslau Kabanau, Vladimir Vantsevich

AUTOMOTIVE ACCIDENT
RECONSTRUCTION
Practices and Principles

Donald E. Struble

CRC Press
Taylor & Francis Group
Boca Raton London New York

CRC Press is an imprint of the
Taylor & Francis Group, an **informa** business

CRC Press
Taylor & Francis Group
6000 Broken Sound Parkway NW, Suite 300
Boca Raton, FL 33487-2742

First issued in paperback 2017

Version Date: 20130729

ISBN 13: 978-1-138-07672-3 (pbk)
ISBN 13: 978-1-4665-8837-0 (hbk)

Library of Congress Cataloging-in-Publication Data

Struble, Donald E. (Donald Edward), 1942-
  Automotive accident reconstruction : practices and principles / Donald E. Struble.
    pages cm. -- (Ground vehicle engineering series)
  Includes bibliographical references and index.
  ISBN 978-1-4665-8837-0 (hardcover : alk. paper)
  1. Traffic accidents. 2. Motor vehicles. 3. Traffic accident investigation. I. Title.

HE5614.S84 2014
363.12'565--dc23   2013028514

Visit the Taylor & Francis Web site at
http://www.taylorandfrancis.com

and the CRC Press Web site at
http://www.crcpress.com

Contents

Preface...xiii
Acknowledgments xv
Author..xvii

1. General Principles 1
   An Exact Science?..1
   Units, Dimensions, Accuracy, Precision, and
   Significant Figures 2
      A Word About Mass 2
      Another Word About Inches 2
   Newton's Laws of Motion..4
   Coordinate Systems 5
   Accident Phases...7
   Conservation Laws 8
   Crush Zones..9
   Acceleration, Velocity, and Displacement 10
   Crash Severity Measures 12
   The Concept of Equivalence..14
   Objectives of Accident Reconstruction 16
   Forward-Looking Models (Simulations) 17
   Backward-Looking Methods..17
   References 18

2. Tire Models...19
   Rolling Resistance...19
   Longitudinal Force Generation...20
   Lateral Force Generation...29
   Longitudinal and Lateral Forces Together.......................................30
   The Backward-Looking Approach..32
   Effects of Crab Angle...32
   References 33

3. Subdividing Noncollision Trajectories with Splines...........................35
   Introduction..35
   Selecting an Independent Variable..36
   Finding a Smoothing Function 37
   Properties of Splines..38
   Example of Using a Spline for a Trajectory 40

4. A Program for Reverse Trajectory Calculation Using Splines............47
Introduction...47
Developing Velocity–Time Histories for Vehicle Run-Out Trajectories.....47
Other Variables at Play in Reverse Trajectory Calculations 49
Vehicle Headings and Yaw Rates 49
Example Reverse Trajectory Calculation..50
Yaw Rates 53
Secondary Impacts with Fixed Objects 53
Verifying Methods of Analyzing Post-Crash Trajectories.......................53
The RICSAC Crash Tests..54
Documenting the Run-Out Motions..55
Data Acquisition and Processing Issues..56
Separation Positions for the RICSAC Run-Out Trajectories 58
Side Slap Impacts 59
Secondary Impacts and Controlled Rest...59
Surface Friction 60
     Brake Factors..61
Sample Validation Run..62
Results of Reverse Trajectory Validation...65
References 68

5. Time–Distance Studies 69
Purpose 69
Perception and Reaction 70
Constant Acceleration 71
Example of Constant Acceleration Time–Distance Study 74
Variable Acceleration...77
References 80

6. Vehicle Data Sources for the Accident Reconstructionist..............81
Introduction..81
Nomenclature and Terminology 81
     SAE Standard Dimensions 90
Vehicle Identification Numbers 92
Vehicle Specifications and Market Data...95
Vehicle Inertial Properties...97
Production Change-Overs and Model Runs 98
Sisters and Clones...98
Other Information Sources...99
People Sizes..100
References 101

7. Accident Investigation..103
Introduction..103
Information Gathering...103

Scene Inspection.. 106
Vehicle Inspection.. 109
Crush Measurement 114
References 114

8. Getting Information from Photographs 115
Introduction.. 115
Photographic Analysis 116
Mathematical Basis of Photogrammetry 118
Two-Dimensional Photogrammetry 119
Camera Reverse Projection Methods... 122
Two-Photograph Camera Reverse Projection 127
Analytical Reverse Projection... 128
Three-Dimensional Multiple-Image Photogrammetry..................... 128
References 132

9. Filtering Impulse Data 135
Background and Theory... 135
Analog Filters 136
Filter Order 137
Bode Plots.. 137
Filter Types.. 139
Digital Filters 139
FIR Filters 140
IIR Filters 140
Use of the Z-transform.. 141
Example of Finding the Difference Equation
from the Transfer Function 143
Bilinear Transforms 144
References 144

10. Digital Filters for Airbag Applications 147
Introduction.. 147
Example of Digital Filter in Airbag Sensor 147
References 151

11. Obtaining NHTSA Crash Test Data 153
Contemplating Vehicle Crashes.. 153
The Crush Zone... 153
Accelerometer Mount Strategy 154
Other Measurement Parameters and Transducers 156
Sign Conventions and Coordinate Systems 157
Processing NHTSA Crash Test Accelerometer Data.......................... 157
Summary of the Process 158
Downloading Data from NHTSA's Web Site 158

Identifying the Accelerometer Channels to be Downloaded 159
Downloading the Desired Channels 162
Parsing the Data File 163
Filtering the Data 164
References 165

12. Processing NHTSA Crash Test Acceleration Data 167
Background 167
Integrating the Accelerations 167
Filtering the Data 169
Filter(j) Subroutine 170
Parsing the Data File 171
NHTFiltr.bas Program Output 172
Averaging Two Acceleration Channels 172
Using the NHTSA Signal Browser 174
References 175

13. Analyzing Crash Pulse Data 177
Data from NHTSA 177
Repeatability of Digitizing Hardcopy Plots 181
Effects of Plotted Curve Quality 182
Accuracy of the Integration Process 183
Accuracy of the Filtering Process 184
Effects of Filtering on Acceleration and Velocity Data 185
Effect of Accelerometer Location on the Crash Pulse 188
Conclusions 190
Reference 190

14. Downloading and Analyzing NHTSA Load Cell Barrier Data 191
The Load Cell Barrier Face 191
Downloading NHTSA Load Cell Barrier Data 192
Crash Test Data Files 194
Grouping Load Cell Data Channels 194
Computational Burden of Load Cell Data Analysis 195
Aliasing 195
Example of Load Cell Barrier Data Analysis 198
Using the NHTSA Load Cell Analysis Software 201
References 201

15. Rollover Forensics 203
Introduction 203
Measurements of Severity 204
Evidence on the Vehicle 207
Evidence at the Scene 218
References 224

16. Rollover Analysis 227
Introduction 227
Use of an Overall Drag Factor 228
Laying Out the Rollover Trajectory 229
Setting Up a Reverse Trajectory Spreadsheet 231
Examining the Yaw and Roll Rates 236
Scratch Angle Directions 239
Soil and Curb Trips 243
References 244

17. Vehicle Structure Crash Mechanics 247
Introduction 247
Load Paths 247
Load–Deflection Curves 249
Energy Absorption 252
Restitution 254
Structural Dynamics 257
Restitution Revisited 261
Small Car Barrier Crashes 263
Large Car Barrier Crashes 263
Small Car/Large Car Comparisons 264
Narrow Fixed Object Collisions 264
Vehicle-to-Vehicle Collisions 265
Large Car Hits Small Car 270
Barrier Equivalence 272
Load–Deflection Curves from Crash Tests 273
Measures of Crash Severity 274
References 275

18. Impact Mechanics 277
Crash Phase Duration 277
Degrees of Freedom 278
Mass, Moment of Inertia, Impulse, and Momentum 279
General Principles of Impulse–Momentum-Based
Impact Mechanics 280
Eccentric Collisions and Effective Mass 282
Using Particle Mass Analysis for Eccentric Collisions 285
Momentum Conservation Using Each Body as a System 286
The Planar Impact Mechanics Approach 287
The Collision Safety Engineering Approach 288
Methods Utilizing the Conservation of Energy 289
References 290

19. Uniaxial Collisions 291
Introduction 291

Conservation of Momentum 291
Conservation of Energy 295

20. **Momentum Conservation for Central Collisions** 299
    Reference 301

21. **Assessing the Crush Energy** 303
    Introduction 303
    Constant-Stiffness Models 303
    Sample Form Factor Calculation: Half-Sine Wave Crush Profile 308
    Sample Form Factor Calculation: Half-Sine Wave Squared
    Crush Profile 309
    Form Factors for Piecewise-Linear Crush Profiles 311
    Sample Form Factor Calculation: Triangular Crush Profile 314
    Constant-Stiffness Crash Plots 315
    Example Constant-Stiffness Crash Plot 316
    Constant-Stiffness Crash Plots for Uniaxial Impacts by Rigid
    Moving Barriers 320
    Segment-by-Segment Analysis of Accident Vehicle Crush
    Profiles 322
    Constant-Stiffness Crash Plots for Repeated Impacts 324
    Constant Stiffness with Force Saturation 326
    Constant Stiffness Model with Force Saturation, Using Piecewise
    Linear Crush Profiles 329
    Constant-Force Model 333
    Constant-Force Model with Piecewise Linear Crush Profiles 335
    Structural Stiffness Parameters: Make or Buy? 337
    References 343

22. **Measuring Vehicle Crush** 345
    Introduction 345
    NASS Protocol 348
    Full-Scale Mapping 351
    Total Station Method 356
    Loose Parts 357
    Other Crush Measurement Issues in Coplanar Crashes 357
    Rollover Roof Deformation Measurements 360
    References 360

23. **Reconstructing Coplanar Collisions, Including
    Energy Dissipation** 363
    General Approach 363
    Development of the Governing Equations 366
    The Physical Meaning of Two Roots 370
    Extra Information 370

Sample Reconstruction 371
References 373

24. Checking the Results in Coplanar Collision Analysis 375
Introduction.. 375
Sample Spreadsheet Calculations... 375
Choice of Roots... 376
Crash Duration... 376
Selecting Which Vehicle Is Number 1 378
Yaw Rate Degradation.. 378
Yaw Rates at Impact.. 379
Trajectory Data 379
Vehicle Center of Mass Positions 379
Impact Configuration Estimate... 380
Vehicle Headings at Impact.. 381
Crab Angles at Impact... 381
Approach Angles 382
Restitution Coefficient.. 383
Principal Directions of Force.. 384
Energy Conservation... 385
Momentum Conservation... 386
Direction of Momentum Vector... 387
Momentum, Crush Energy, Closing Velocity, and
Impact Velocities 387
Angular Momentum 388
Force Balance.. 389
Vehicle Inputs.. 390
Final Remarks... 390
References 390

25. Narrow Fixed-Object Collisions.. 393
Introduction.. 393
Wooden Utility Poles.. 394
Poles that Move 397
Crush Profiles and Vehicle Crush Energy... 398
Maximum Crush and Impact Speed... 402
Side Impacts.. 403
References 404

26. Underride/Override Collisions... 407
Introduction.. 407
NHTSA Underride Guard Crash Testing... 408
Synectics Bumper Underride Crash Tests 409
Analyzing Crush in Full-Width and Offset Override Tests 409
The NHTSA Tests Revisited 410

More Taurus Underride Tests 411
Using Load Cell Barrier Information.. 412
Shear Energy in Underride Crashes... 413
Reconstructing Ford Taurus Underride Crashes 415
Reconstructing Honda Accord Underride Crashes 417
Reconstructing the Plymouth Reliant Underride Crash......... 419
Conclusions.. 419
References 419

27. Simulations and Other Computer Programs 421
Introduction... 421
CRASH Family of Programs.. 422
   CRASH 422
   Crash3 and EDCRASH.. 423
   WinSMASH... 426
SMAC Family of Programs.. 429
   SMAC... 429
   EDSMAC 430
   EDSMAC4... 433
PC-CRASH... 435
Noncollision Simulations.. 437
   HVOSM 437
   EDSVS (Engineering Dynamics Single Vehicle Simulator) 438
   EDVTS (Engineering Dynamics Vehicle–Trailer Simulator) 439
   Phase 4 440
   EDVDS.. 441
Occupant Models 443
References 443

Index.. 447

Preface

Before entering the field of motor automotive safety, I specialized in aircraft and missile structures all the way through graduate school and into a university teaching career. However, the fascinating field of automotive crashworthiness beckoned, and before long I was focused on car crashes, energy-absorbing structures, and occupant protection. Eventually, I went into accident reconstruction.

Becoming a reconstructionist meant that there was much to learn; like so many other newbies, I needed mentoring. Don Friedman, Bob Cromack, and Chuck Warner were instrumental in that regard. Nevertheless, there was no single textbook, no single source for the requisite theory, and no road map showing how to acquire the data and information I needed. So, I did what so many others have probably done—worked out the theory and the practical techniques myself, aided by technical papers written by those who had gone before, and spurred on by invigorating discussions with my colleagues at Minicars, Inc. and Collision Safety Engineering. The theory gradually wounded up in the form of pen and paper calculations, computer programs, and written documents that were distributed to the younger engineers and referred to constantly as time went on. Some of this material found its way into technical papers. At the same time, experience gradually revealed investigation and measurement techniques—some of which worked better than others—and how the necessary data and information could be obtained.

To the younger engineers who both received and helped develop this knowledge base, much credit is due—particularly Kevin Welsh and John Struble—who went through the written materials and participated in many discussions, asking probing questions, challenging assumptions, and working out procedures. This book is an attempt to gather, in one place, the material that other young engineers will need to master in order to investigate and reconstruct crashes. My aim has been to make it an authoritative source they can consult throughout their careers, enabling them to base their work on the stoutest possible foundation.

The material consists of practical matters, like where to find the technical information one needs, how to acquire and analyze publicly available data, and how to interpret evidence, for example, as well as more theoretical subjects such as how to apply the principles of mechanics so as to analyze crashes. Of course, the book does not cover everything; journal articles and even other books will always be important, particularly as the field evolves. The discerning reader will notice, for example, that crashes involving heavy trucks and other articulated vehicles are not covered. Simulation models are discussed only in an introductory manner. But one has to stop somewhere, lest the tome becomes unwieldy. Most of the investigative techniques and all of the fundamental principles and resulting equations will still apply. If this book does its job, it will serve as a valuable resource for reconstructionists as they build their own careers.

Acknowledgments

This book would not have been undertaken except for the prompting by my wife Lonny and my son John, and could not have come to fruition without the continuous and enthusiastic support by my entire family. My deepest thanks go to all of them, particularly to John, who read and reviewed various chapters and—as usual—provided valuable comments and insights, and pointed out errors. Those that remain are due to me and no one else.

Author

Donald E. Struble holds BS, MS, and PhD from California Polytechnic State University (San Luis Obispo), Stanford University, and Georgia Institute of Technology, respectively, all in engineering with an emphasis on structural mechanics. Dr. Struble was assistant professor of aeronautical engineering at Cal Poly, manager of the Research Safety Vehicle program and senior vice president of Engineering and Research at Minicars, Inc., and president of Dynamic Science in Phoenix, Arizona. He has worked in automotive safety since 1972, including occupant crash protection and crashworthiness at the highest crash severities (50 mph barrier crashes, for example), and has been reconstructing accidents since 1983.

Dr. Struble has published numerous papers in these areas. A 2001 paper, "SAE 2001-01-0122" (SAE International, Warrendale, PA), received the Arch T. Colwell Merit Award, for papers based on "their value as contributions to existing knowledge of mobility engineering, and primarily with respect to their value as an original contribution to the subject matter." He has had three papers selected for inclusion in *SAE Transactions*, "judged by a distinguished panel of engineering experts to be among the most outstanding SAE technical papers." Three other papers were selected for inclusion in SAE technical compendia on air bags and accident reconstruction. Dr. Struble was editor of *Advances in Side Airbag Systems*, published by SAE International, 2005.

Dr. Struble has delivered invited presentations at two ESV (Experimental Safety Vehicle) conferences, and is co-holder of a patent on a side impact airbag. Dr. Struble has been an SAE TopTec instructor on the following topics: air bags, high-speed rear impacts, and accident reconstruction. He is a member of SAE, AAAM, and Sigma Xi, the Scientific Research Society. Formerly senior engineer at Collision Safety Engineering in Phoenix, Arizona, and president of Struble–Welsh Engineering in San Luis Obispo, California; he is now retired.

1

General Principles

An Exact Science?

Science is the endeavor of examining the world around us, developing hypotheses that may explain its behavior, testing those hypotheses, and thereby obtaining a deeper understanding of how that world works. Notice that the word "exact" was not used in this description of science. Science seeks exactness, but there are always limits. In fact, Werner Heisenberg pointed out that, in the limit, the very act of observing one property degrades our knowledge of another.[1] The best science can do is approach exactness as closely as possible, within the limits of time, money, and practicality.

Engineering is different from science in that it seeks to apply the knowledge of science in the design, development, testing, and manufacture of new things. Certainly motor vehicles, roads, and roadside appurtenances are engineered things that must be understood by the reconstructionist. Motor vehicle crashes are events out of the ordinary that occur outside of the laboratory (and outside the presence of the reconstructionist), without many (or even all) of the measurement and observation tools available to the scientist. Very often important information is entirely missing.

So reconstruction is neither exact; nor is it a science. It is partly engineering, in that it deals with engineered things. It is also an art, significantly shaped by experience and intuition. It is not the purpose of this book to emphasize this latter aspect, since that is covered more thoroughly elsewhere, although certain practices and observations from the author's experience will be introduced where they may be helpful. Rather, it is hoped that fundamentals essential to reconstruction will be set forth, and illustrative examples included, so that the reconstructionist can put numbers on things and ensure that his opinions are consistent with the physical evidence and the laws of physics, and are therefore as close to the truth as he or she can make them. After all, it was Sir William Thompson, Lord Kelvin, who said, "When you can measure what you are speaking about, and express it in numbers, you know something about it; but when you cannot measure it, when you cannot express it in numbers, your knowledge of it is of a meager and unsatisfactory kind; it may be the beginning of knowledge, but you have scarcely, in your thoughts, advanced it to the state of science"[2] (p. 73). So, in this book, we will be concerned more about the quantitative than the qualitative measure.

Units, Dimensions, Accuracy, Precision, and Significant Figures

To put numbers on things, we must speak a common language. It is the case that the Système International d-Unitès (International System of Units), or metric system, abbreviated SI, has been adopted and used throughout the world. It is also the case that a notable exception is the American public, which populates the majority of jury boxes, judges' benches, and law practices around the world. These are the people with whom the reconstructionist must communicate in the US legal system. Indeed, the Technical English System of Units is the language spoken by most reconstructionists.

The base units used in this book are force (pounds, abbreviated lb), length (feet, abbreviated ft), and time (seconds, abbreviated sec). Metric equivalents will be provided on occasion. An example would be barrier forces in Newtons, even though this author has yet to encounter a bathroom scale that reads in such units. In vehicle crashes, times are often discussed in milliseconds (thousandths of a second, abbreviated msec). Derived units are obtained from the base units. For example, area is a measurement derived from length and is reported in square feet (abbreviated ft²). Velocity is derived from length and time and is measured in feet per second (abbreviated ft/sec), as is acceleration, measured in feet per second per second (abbreviated ft/sec²).

A Word About Mass

It is a measure of the amount of substance—that which resists acceleration. It is a derived unit; namely, the amount of mass which would require the application of one pound of force to achieve an acceleration of one foot per second per second. This amount of mass, called a slug, would weigh about 32.2 lb on the surface of the earth. (But on the moon, one slug would weigh about 1/6 as much, because the moon's gravity is about 1/6 Earth's.) By Newton's Second Law, we see that $m = F/a$, and so one slug equals one lb·sec²/ft. Since lay persons usually have no concept of a slug mass, it has been this author's practice to speak only of weight (units of force), and reserve the slug for computation only. The concept of pound mass does not relate to base units, is easily confused with pound force, and is not used herein.

Another Word About Inches

Generally, length quantities for vehicles are reported in inches (abbreviated in.). This includes the all-important (to reconstructionists) measurement of crush and stiffness. This practice is retained herein, but calculations regarding the laws of physics are carried out in length length units of feet. For example, energy is expressed in foot-pounds (abbreviated ft-lb), and moment of inertia is calculated in slug-inches squared, but converted to slug-feet squared when used in physics computations. For consistency, physics calculations are carried out in feet, even though inputs and outputs relating to vehicle dimensions are expressed in inches to maintain familiarity for the user and the consumer of the results. For example, vehicle crush is expressed in inches and crush stiffness in lb/in. or lb/in². In the technical literature, this author has yet to encounter a stiffness value in proper metric units, such as: Newtons per centimeter, or kiloNewtons per square meter.

Finally, lay persons generally understand feet per second when applied to velocity. However, the speedometers in their vehicles read in miles per hour, so it has been this author's practice to use miles per hour (abbreviated mph) when communicating about speed. Of course, conversions to ft/sec are used for computational purposes. Similarly, lay persons have some understanding of angle measurements in degrees (abbreviated deg), but radians (abbreviated rad) are mostly unknown to them. Therefore, angles are communicated in degrees, and angle rates (such as roll rate and yaw rate) are communicated in degrees per second (abbreviated deg/sec).

The ability to detect small changes of a property is known as precision, and is often related to the resolution (degree of fineness) with which an instrument can measure. A set of scales may report a weight of 165.76 lb, but if those scales cannot detect a difference between 165 and 166 lb, it is misleading to report weights to 0.01 lb when the precision of the instrument is only 1 lb. Using two decimal places implies more knowledge than is actually present. This effect is seen when examining computer files for crash barrier load cells, which may show multiple readings that are identical to six decimal places! Close examination of the data may reveal an actual precision of about 1.5 lb, which is understandable for a device intended for measurements up to 100,000 lb.

Precision is not to be confused with accuracy, which reflects the degree of certainty inherent in any measurement. Uncertainty means that the true value is never known. The best we can do is to make an estimate of the true value, using an instrument that has been calibrated against a standard (whose value is known with some published precision).

The precision of calculations in the computer is a function of the hardware and software in the computer. Excel 2007, running under Windows XP Professional, claims a number precision of 15 digits, for example. This is far more than needed to avoid round-off error during calculations, and it is hardly representative of the precision of inputs, measurements, and so on. The precision—or better yet, the imprecision—of the results should be conveyed by limiting the number of significant figures in the output. In determining the number of significant figures, the leftmost nonzero digit is called the most significant digit, and the rightmost nonzero digit is the least significant digit. They and the digits between them are significant figures. If a decimal point appears, the rightmost digit is the least-significant digit, even if it is zero. For example, the following numbers all have six significant figures:

3.00000

3.14159

314159

3141590

0.314159

0.00314159

0.00314160

Even though digitized load cell values in National Highway Traffic Safety Administration (NHTSA), computers appear to have more than seven significant figures that is not representative of the actual precision of measurement. It would be more appropriate to report load cell values to the nearest pound, based on our earlier discussion. As another example, it has been the author's practice to use a value of 32.2 ft/sec² (only three significant figures) for the value of the acceleration of Earth's gravity at its surface (1 G).

Newton's Laws of Motion

Vehicle collisions are physical events. Fortunately, they occur at speeds far below that of light, so the laws of motion set forth by Sir Isaac Newton[3] provide an accurate framework for describing (and quantifying) vehicle crashes. They may be summarized as follows:

1. Every body persists in its state of being at rest or of moving uniformly straightforward, except insofar as it is compelled to change its state by force impressed. In other words, it remains at a constant velocity (including zero) in an inertial (i.e., unaccelerated) reference frame, unless an unbalanced force acts on it. This law, originally formulated by Galileo, may be considered a special case of the Second Law, expressed below.

2. The net force on a particle is equal to the time rate of change of its linear momentum (the product of mass and velocity), when measured in an inertial, or Newtonian, reference frame. (When there is no force, there is no momentum change, and hence no velocity change.) Since the mass is constant, Newton's Second Law can be written as

$$F = \frac{dp}{dt} = \frac{d(mV)}{dt} = m\frac{dV}{dt} = ma \tag{1.1}$$

as referred to above. F is the force, p is the particle's momentum, m is the mass, V is the velocity, t is the time, and a is the acceleration. Integrating the above equation over a finite time yields

$$\Delta p = m\Delta V \tag{1.2}$$

which is closer to Newton's original wording.

3. To every action, there is always an equal and opposite reaction: or the forces of two bodies on each other are always equal and directed oppositely.

It is helpful to define a system before applying Newton's Laws to it. That way, one avoids mistaking external forces (those acting from outside the system) with internal forces (those internal to the system). According to Newton's Third Law, internal forces are self-equilibrating and have no net effect on the system as a whole.

Coordinate Systems

One can never get too far into the discussion of dynamics without bringing in the notion of a reference frame. See the previous section, for example. A reference frame is used to describe the position of any body or particle; that position is quantified by making distance and/or angle measurements from a set of lines or surfaces affixed to the reference frame. When we say that the moon is 240,000 mi. away and is low in the western sky, it is understood that the reference frame is the Mother Earth, which itself is spinning on its axis and hurtling through space. It may not be a satisfactory reference for inter-planetary travel, but for getting from Detroit to Toledo it is just fine.

When we speak of an inertial reference frame, we mean one in which Newton's Laws hold. This may sound like a circular definition, but what it really means is that the reference frame must be unaccelerated (i.e., moving at a constant speed and in a constant direction). Rotating reference frames do not qualify, since there is centripetal acceleration even if the reference frame is spinning at a constant speed on a stationary axis. Thus, the Mother Earth is not exactly an inertial reference frame, but the centripetal acceleration at its surface is negligibly small compared to everything else we will be discussing. So, we often use the Earth as reference, we attach coordinates to it, and we call them "global coordinates." Coordinates affixed to cars crashing on its surface are not global; they are usually called "local coordinates." The distinction must be kept in mind when one is dealing with data collected (e.g., accelerations) in a non-Newtonian reference frame.

The reference frame lines or planes from which measurements are made constitute a coordinate system. The most familiar system is described by Cartesian coordinates, which will be used almost exclusively in this book. In three dimensions, coordinate axes lie at the intersection of mutually perpendicular planes; therefore, the lines themselves are mutually perpendicular. In two dimensions, all the action takes place in a plane defined by the intersection of two mutually perpendicular coordinate axes. In any case, coordinate naming convention will follow the right-hand rule, in which a curled right hand shows the direction of rotation from the X-axis to the Y-axis, and the thumb points in the direction of the Z-axis.

The right-handed coordinate system familiar to most beginning students of mathematics and engineering is shown in Figure 1.1. Here, the X-axis is horizontal, the Y-axis is vertical, the Z-axis is out of the plane (up), and angles ψ are measured counterclockwise, in accordance with the right-hand rule. If a distinction must be made between global and local coordinates (such as those attached to a vehicle), then the former coordinates are in upper case and the latter coordinates are in lower case.

Notwithstanding all the above, the SAE (formerly, Society of Automotive Engineers; now, SAE International) has developed its own coordinate system, shown in Figure 1.2. It has come to be used by almost all automotive engineers, at least those in the United States.

At first glance, the SAE coordinate system might appear to be backwards, and possibly left handed. However, the X-axis points in the direction of travel, which is straight ahead. And when we look at a map, the North arrow is usually up, so we are used to measuring angles from that direction. In fact, that direction is like all clocks (and hence is called "clockwise"). Therefore, the Y-axis points to the right. A local y-axis would point to the right also, which is the passenger side in many countries, including the United States. Since the SAE coordinates are right handed, the Z-axis is into the plane (down). In fact, one can go from one coordinate system to the other by simply rotating it through an angle of 180° about the X-axis.

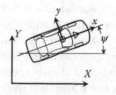

FIGURE 1.1
Mathematical coordinate system.

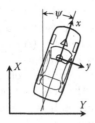

FIGURE 1.2
SAE coordinate system.

It has been this author's practice for many years to work in a mathematical coordinate system, since that was used by the computerized drawing software AutoCAD®. In this book, mathematical derivations will occur in a mathematical coordinate system, as do mathematical derivations everywhere. However, it must be kept in mind that both mathematical and SAE coordinates are right handed. Therefore, the equations have equal applicability in each system. As a result, it has also been this author's practice, when interpreting test data, using software that assumes SAE coordinates, or analyzing data from others, to stay with the coordinate system that has already been utilized.

If it is necessary to convert from one coordinate system to another (while leaving the X-axis alone), it is helpful to remember the coordinate rotation about the X-axis, mentioned above. Values for X and for rotations about the X-axis (roll angle) are unaffected by the switch. On the other hand, values for Y and for rotations about the Y-axis (pitch angle) will have their signs reversed, as will values for Z and for rotations about the Z-axis (yaw angle).

Finally, it should be noted that in dealing with vehicle motions, the origin of vehicle coordinates is located at the vehicle center of mass, commonly called the CG. However, the origin is almost always located elsewhere—usually on an easily identifiable landmark—in test measurements and vehicle design documents. One must keep that in mind while preparing inputs for reconstruction calculations.

Accident Phases

Vehicle crashes can be thought of as having three separate phases: pre-crash, crash, and post-crash. Credit for this paradigm generally goes to William H. Haddon, the director of the National Highway Safety Bureau, the predecessor of the National Highway Traffic Safety Administration (NHTSA). The Safety Standard numbering system reflects this point of view.

During all three phases, each vehicle moves on the ground, and dissipates energy while doing so. During the pre- and post-crash phases, ground forces and elevation changes dominate. (Aerodynamic forces are generally ignored.) While they may act over fairly long periods of time and distance, ground forces are generally less than the vehicle weight, due to the limitations of friction. In other words, the accelerations are <1 G.

The crash phase is much different. It is defined by the existence of contact forces between the vehicles. The crash phase begins when the vehicles first come into contact, and ends when they separate. In a high-speed crash, a vehicle may experience a peak acceleration of 30–80 times the acceleration of gravity (30–80 G), which means that the force acting on the vehicle as a whole may reach 30–80 times its weight. The average acceleration is the area under the acceleration-versus-time curve (crash pulse), divided by the duration. It is easily seen that if the crash pulse can be roughly characterized as a triangle, the average acceleration will be half the peak. Even the average force far exceeds the ground forces in significance. For this reason, analyses of the crash phase often assume that ground forces are negligible.

Exceptions to this observation may occur in low-speed collisions, where the collision forces are much reduced in magnitude. In such cases, one should not apply analytical methods designed for high-speed crashes unless one can be satisfied that errors due to ignoring ground forces are still small.

Conservation Laws

A system is a collection of matter that, for purposes of analysis, is analyzed as a whole. If a system is isolated from its environment, then certain mechanical properties of the system cannot change. Such properties are said to be "conserved." Even if the isolation is not total, it is often sufficient to allow an assumption of conservation. For example, an artillery projectile is known to interact with the surrounding atmosphere; yet ignoring aerodynamic drag can permit fairly simple calculations to predict its trajectory with reasonable accuracy.

The first great conservation law is that if a system can be defined so as to be isolated from external forces, then its momentum is conserved. This follows immediately from Newton's Second Law, as long as the system is defined such that external forces are absent, or at least negligible. In the absence of such forces, its momentum is conserved. A pertinent example was alluded to above. It is often helpful to define a system containing two (or more) vehicles engaged in a collision. If tire forces can be ignored, then this system's momentum is conserved, even though both vehicles are moving and interacting within it. In fact, momentum conservation implies that the system center of mass moves at a constant speed and in a constant direction throughout the crash. Thus, an observer who is stationary with respect to the system center of mass will see the colliding vehicles come to rest, even though they may have come together at very different speeds and angles.

The second great conservation law is a rotational analog of the first: if a system can be defined so as to be isolated from external torques, then its angular momentum (defined as the moment of momentum) is conserved. The law may have a significant effect on the analysis of vehicle crashes where eccentricities and rotations are present.

The third great conservation law is that if a system, such as two colliding vehicles, can be defined so that no work is done on it, its total energy remains constant. Of course, energy comes in many forms, and energy can be converted from one form to another. Forms of energy that are most interesting to reconstructionists are kinetic energy (the energy of motion), potential energy (the energy of position), and crush energy (the energy of deformation).

Sometimes it is helpful to define a system on which work is done, such as a vehicle sliding or rolling on a surface. As alluded to above, the analysis of pre-crash and post-crash vehicle motions may be accomplished by calculating the change in kinetic energy due to changes in potential energy, and due to work being done by braking or sliding on the ground. In such cases, the work being done by friction forces causes a conversion of some of the energy to heat. We say that such energy that cannot be recovered is "dissipated." Crush energy falls into that category.

Crush Zones

A motor vehicle is a collection of many parts that are connected, but distributed in three-dimensional space. This can be appreciated by watching vehicles being put together in any vehicle assembly plant. Crashworthiness, a term derived from "seaworthiness" and "airworthiness," deals with the ability of those assembled parts to protect an occupant during a crash. The intention is to have some of those parts retain their positions in a crash until high force levels are reached; other parts are intended to move and deform, thereby dissipating some of the energy that went into the crash.

Thus, the crashworthiness engineer distinguishes between sections exhibiting little or no relative deformation (the zone of negligible crush), and those having measurable relative deformation (the crush zone). In practice, these are idealizations; the crush zone is different from one crash mode to another, and there is no dotted line or color-coding system to differentiate them. In practice, accelerations vary throughout time and space (location in the vehicle), and the further one gets away from the crush zone, the more similar the acceleration histories (crash pulses) become.

Since it is hoped that the occupant is outside the crush zone, it is the accelerations measured away from the crush zone that provide the clearest assessment of the risk of occupant injury. Usually, in frontal crashes, accelerometers are placed on stiff structure near the base of the B-pillar (the middle pillar), or in the rear seat area for just this purpose. In side impacts, accelerometers are usually placed on the rocker panel across from the impact. In rear impacts, there may be accelerometers near the base of the A-pillar or on the front occupants' toe boards.

It may be desirable to place accelerometers and other instruments at or near the center of gravity, particularly if the vehicle rotates during or after the crash. However, there may not be sufficiently stiff structure in the area to provide an adequate accelerometer mount. Accelerometer signals from such areas may be subject to "ringing," which means that high-frequency vibrations may be present on top of the signal being monitored. As a result, accelerometer data from "CG" locations should be used with caution. Of course, it is desirable to view any accelerometer trace with a discerning eye.

Acceleration, Velocity, and Displacement

Acceleration, velocity, and displacement are all measurable with transducers (devices which convert the mechanical quantity into an electrical signal). Fifth wheels and laser guns can be (and are) used to measure vehicle velocity during braking and handling maneuvers. String pots or laser devices are used to record positions and displacement during quasi-static or static evaluations. However, high accelerations, high speeds, and high displacements may render all such devices useless. Therefore, it is nearly universal practice to use acceleration-measuring transducers (accelerometers) in crash tests. This forces the investigator to derive velocity and displacement information from acceleration traces.

A typical history of acceleration versus time (or acceleration trace, or crash pulse) is shown in Figure 1.3. Immediately apparent is the jaggedness of the curve, with high peaks and low valleys. Filtering (a topic taken up later) is often used to suppress the "noise" and reveal the underlying phenomena. Consider, for example, the same information as presented in the crash test report, shown in Figure 1.4.

As indicated, the data have been filtered to Channel Frequency Class (CFC) 60, a topic to be taken up in Chapters 9 through 13.

If one assumes that, in this test, the vehicle rotations can be ignored (an assumption that should be examined), one can integrate this trace and interpret the results as velocity and displacement that are associated with this channel (i.e., measured at the same location and in the same direction). This is done by utilizing their definitions, as follows:

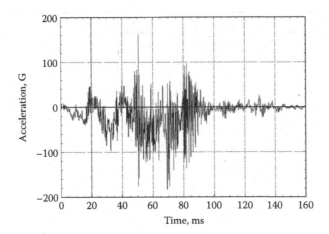

FIGURE 1.3
Raw data, target vehicle CG acceleration, test DOT 5683.

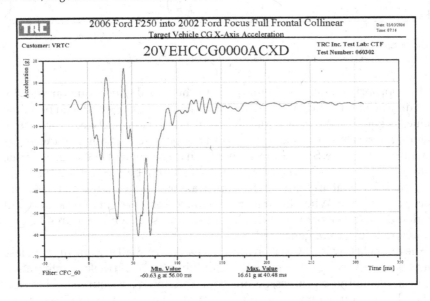

FIGURE 1.4
(**See color insert.**) CG X-axis acceleration, target vehicle, test DOT 5683.

$$V = \int a \, dt \tag{1.3}$$

and

$$d = \int V \, dt = \iint a \, dt \tag{1.4}$$

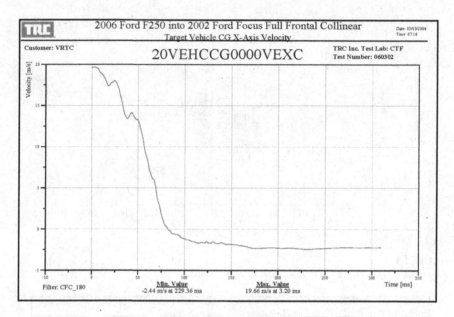

FIGURE 1.5
(See color insert.) CG X-axis velocity, target vehicle, test DOT 5683.

with appropriate attention being paid to the initial conditions on velocity and displacement. Thus, we see that the change in velocity between two time values is equal to the area under the acceleration curve between those time values; the acceleration at any point is the slope (rate of change) of the velocity trace at that time. Similarly, the change in displacement between two time values is equal to the area under the velocity curve between those time values; the velocity at any time is the slope (rate of change) of the displacement curve at that time. When the integration processes were applied to the acceleration data, the velocity and displacement traces shown in Figures 1.5 and 1.6, respectively, were generated.

The filter class was CFC 180 on both traces. As we shall see, the higher class number indicates a greater smoothing of the data. Nevertheless, integration alone is a smoothing process, because ups and downs in a curve tend to cancel each other out when the area under the curve is computed. By contrast, differentiating (measuring the slope of a curve) amplifies any noise present, because ups and downs produce positive and negative slopes, respectively.

Crash Severity Measures

Suppose a 2000-lb vehicle crashes head-on into a rigid, fixed, infinitely massive barrier at 30 mph, and bounces away from the wall at 5 mph. The system

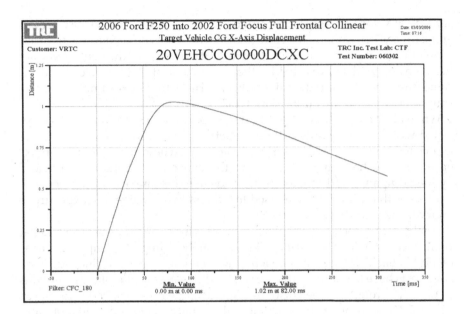

FIGURE 1.6
(**See color insert.**) CG X-axis displacement, target vehicle, test DOT 5683.

center of mass is stationary in this impact (because of the infinite mass of the barrier). The car's speed at impact is 30 mph, its closing speed at impact is 30 mph, its separation speed is 5 mph, and its change in velocity as a result of the crash (final velocity minus original velocity) is −5 − (+30) = −35 mph. (The negative sign indicates that the change was opposite to the direction of travel.)

Now suppose a second crash involves two more of these vehicles colliding head-on, with each vehicle traveling 30 mph at impact, and rebounding away from each other at 5 mph each. Because of symmetry, the system center of mass for the two vehicles has a velocity of zero (before and after). For each of the vehicles, the speed at impact is 30 mph, the closing speed (at impact) is 60 mph, the separation speed is 10 mph, and the change in velocity is −35 mph.

Crash Number Three involves two more identical vehicles colliding head-on, with one of the vehicles traveling 20 mph, and the other one 40 mph. After the crash, the faster car moves forward at 5 mph, and the slower car bounces backwards at 15 mph. In this case, the momenta are not equal; the system center of mass moves at a speed of 10 mph, in the direction of the faster car. The speeds at impact are 20 and 40 mph, their closing speed is 60 mph, their separation speed is 10 mph, and each car undergoes a velocity change of −35 mph.

Closing speed is a scalar quantity; closing velocity is a vector that is simply the vector difference between the two impact velocity vectors. Similarly, separation velocity is the vector difference between the two velocity vectors at the time of separation. The restitution coefficient is defined as the negative ratio of the component of separation velocity normal to the contact surface between the vehicles, divided by the component of closing velocity in the same direction, measured on the contact surface. In these examples, the restitution coefficients are all 10/60, or 0.167.

While all three crashes have different impact and separation speeds associated with them, the only difference between them is the reference frame associated with the system center of mass. We would not expect the reference frame to have any influence on the exposure to injury experienced by the occupants, and indeed in these idealized examples, it does not. The parameter that is most widely used to express this sameness of exposure to injury is the velocity change, or ΔV.

Finally, let us consider a fourth crash, in which a 2000-lb car traveling at 40 mph collides head-on with a 4000-lb car traveling at 20 mph. Each car rebounds backward at 5 mph. Here, the momenta are equal (indeed, some crash tests have been set up this way), and the system center of mass remains stationary throughout the crash. In this crash, the impact speeds are 20 and 40 mph. The closing speed is 60 mph, the separation speed is 10 mph, and the restitution coefficient is 0.167, as before. However, due to the mass differences, the heavier car has a ΔV of −25 mph, and the lighter car has a ΔV of −45 mph. Here, the crash severities and injury exposures are definitely different, as seen in the ΔV comparisons.

These examples are illustrative of why ΔV has become almost universally accepted as the measure of crash severity. The ΔVs of the accident (particularly for the vehicle of interest) have become the most important questions asked of the reconstructionist. Of course, the speeds at impact (and the travel speeds, if they are different) are also important, particularly if traffic law violations are at issue. However, ΔV is the variable of most interest to vehicle and restraint designers, and medical personnel.

The Concept of Equivalence

The previous examples have suggested a possible "equivalence" between a barrier crash and a vehicle-to-vehicle impact. From a crashworthiness point of view, it would seem that a barrier crash would have to produce the same ΔV as a vehicle-to-vehicle crash to even be considered "equivalent." A look at the time histories (such as shown in Figures 1.3 through 1.6) would suggest even more stringent requirements. The crash pulse itself (e.g., Figures 1.3 and 1.4) has a variation in it, even after filtering to CFC class 60. Moreover, successive crashes of nominally identical vehicles under nominally identical conditions produce noticeable variations in the crash pulses. There seems to be some inherent randomness built into crash pulses, due both to measurement uncertainties and to the complexity of vehicles as a whole. However, the occupant is not necessarily affected by the high-frequency components of the crash pulse, as the occupant is not coupled to the vehicle rigidly, if at all. So perhaps those variations do not matter too much as far as occupant protection is concerned.

From a practical view, the velocity–time trace represents a more fully attainable target in terms of equivalence. It is a less irregular-looking curve. At the very minimum, if the starting and ending velocities are achieved, then the difference between them (the ΔV) will also be achieved. Moreover, if one matches the time at which the final velocity is attained (i.e., the crash duration is the same), then the velocity–time curve will have the right average slope (i.e., the right average acceleration). If these three parameters are within acceptable limits, then the crash can be said to be "equivalent," within acceptable limits.

But in the crash duration lies the rub. A vehicle-to-vehicle crash involves two structures connected in series, which combination is inherently less stiff than the single structure involved in a rigid barrier crash. Therefore, the pulse duration in a barrier impact is less than in a vehicle-to-vehicle crash, unless the barrier face is fitted with an energy-absorbing element of the appropriate stiffness. This subtlety is generally overlooked when the term "barrier equivalent velocity" is used. It is preferable, in this author's view, to avoid the terms "barrier equivalent velocity" and its abbreviation, BEV, altogether.

As the above examples have indicated, the restitution coefficient enters into the calculation of ΔV. So if a barrier test is run so as to duplicate the ΔV of a vehicle-to-vehicle crash, the barrier impact speed will need to be lower than the ΔV, because the rebound velocity will add to the ΔV of the barrier crash. Generally, the vehicle crush energy will not be matched because the overall structural stiffness of the two crash partners is different, as explained above.

On the other hand, if crush energy is matched, what then? We can define an equivalent energy speed, or *EES*, which is the speed at which a certain mass m achieves a specified kinetic energy *KE*. By definition, we can write

$$KE = \frac{1}{2}m(EES)^2 \tag{1.5}$$

from which it follows that

$$EES = \sqrt{\frac{2(KE)}{m}} = \sqrt{\frac{2g(KE)}{W}} \tag{1.6}$$

In the context of a crash, *KE* is the amount of kinetic energy converted to crush energy; that is, $KE = CE$.

However, some of the initial kinetic energy in a crash is recovered due to restitution. When restitution is present, CE is not the initial kinetic energy, and *EES* is not the speed at impact. One needs to think instead of the change in kinetic energy, or ΔKE. Thus, $\Delta KE = CE$. To decide at what speed a crash test must be run in order to achieve a certain *EES*, one must estimate what the restitution coefficient will be in the barrier crash. One can then find the impact speed such that the initial kinetic energy of the barrier impact, minus the energy recovered in rebound, equals the crush energy (and hence ΔKE) that one is attempting to duplicate in the test.

This subject will be explored further in Chapter 17.

Objectives of Accident Reconstruction

Central to the concept of accident reconstruction is the idea that evidence is left behind after almost every event, and that almost every observable feature has a causation. Of course, there are myriad events, and piles of evidence that are not accident related. So the task of the reconstructionist would seem to be one of culling out the nonaccident-related evidence, and then connecting accident events with the evidence.

Of course, it is more complicated than that. Often, important evidence is overlooked, not documented, inadvertently altered during investigations, destroyed in the treatment of crash victims, deteriorated due to time, temperature, wind, moisture, and so on, or simply lost. Hardly ever does the reconstructionist actually witness the accident, and never are there laboratory instruments to measure or record data. Event data recorders, or "crash recorders," however, can supply significant information in actual crashes.

So the reconstructionist is called in after the fact (often years after the fact) and is asked, often in the absence of important facts, "What happened?" From limited inputs must come limited outputs, and thus limited opinions. Often, the most important ones for co-planar crashes involve speeds—particularly velocity changes, as discussed previously. Travel speeds and impact speeds often come into the picture as well. The direction of the velocity change vector, and by implication, the direction of the impulse vector, is also important because biomechanics experts often rely on it for their opinions on occupant kinematics and injury causation.

For rollovers, questions tend to focus on the distance from roll initiation to rest, number of rolls, roll direction (driver-side or passenger-side leading), roll rates, ejection portals (if any), and ejection point (if ejection occurs).

This book focuses on how to obtain quantitative answers to such questions. There may be many other questions posed to the accident reconstructionist, particularly with regard to the interpretation and meaning of physical evidence. A noteworthy example is whether the belt restraints were in use during the crash, or not. Such topics have been treated adequately elsewhere, and will not be repeated here, except where the author's experience may shed new light on a subject.

Forward-Looking Models (Simulations)

One approach to the analysis of a crash is to start with a free-body diagram, enumerating the forces at work on the system. Newton's Second Law may then be used to derive the differential equations of motion. Solving them is termed an initial value problem, because one sets certain initial conditions—such as velocities—and then proceeds with time moving forward, much like aiming an artillery piece and then firing it. The equations of motion may be solved—perhaps in closed form, but most often numerically—by starting the clock at the beginning of the event, and incrementing it in small time steps. The solution proceeds from the initial values to the final values. Unfortunately, the initial values are unknown, and the results of the calculations (final values) are already known. Therefore, new and improved estimates of the initial values must be made in an iterative fashion until reasonable convergence is achieved with the (known) position and velocities at the end of the event.

Backward-Looking Methods

It is fair to say that reconstructionists work backwards in time, even when they use forward-looking methods. That is, they start with the vehicles at rest or some other known condition post-crash, and work backwards through the post-crash, or run-out, phase, then further backwards through the crash phase, and then (sometimes) backwards through the pre-crash phase, arriving at the beginning of the accident. This is also a process of working from the knowns, captured in post-accident measurements, photographs, physical evidence, and so on, and toward the unknowns, which are the speeds and other conditions present at the beginning of the event.

While forward-looking simulation models will be discussed, they will not be treated in detail. That has been done elsewhere. Instead, the focus of the book is on backward-looking methods. The discussion is organized in a similar way. It will start with the analysis of post-crash vehicle movement, and will then proceed to a discussion of the various ways of analyzing the crash phase.

References

1. Heisenberg, W., *The Physical Principles of the Quantum Theory*, University of Chicago Press, Chicago, IL.
2. Kelvin, Lord William Thomson, Electrical Units of Measurement, lecture delivered to the Institution of Civil Engineers on May 3, 1883, in *Popular Lectures and Addresses*, Vol. 1, London: McMillan, 1889, p. 73.
3. Newton, Sir I., *Philosophiae Naturalis Principia Mathematica*, Royal Society of London, London, UK, 1687.

2

Tire Models

Rolling Resistance

While it seems so obvious that it need not be said, tires are deformable, and it is this deformation that is required to generate the lateral and longitudinal forces on the roadway that allows the vehicle to accelerate, brake, and maneuver. This same property of deforming under load creates rolling resistance. When a vehicle is set down on the ground, the tires must flatten out where they contact the roadway, until the so-called contact patches are large enough that the contact pressures can equilibrate the vehicle's weight.

When the tire is rolling, the side walls and tread must change their shapes every time they pass near or through the contact patch. This continual flexing back and forth absorbs energy, because the rubber has a property known as hysteresis (the stress–strain curves are different in loading and unloading). In other words, it takes work to flatten out the tire, and not quite all of the strain energy in the rubber is recovered when the tire returns to its regular shape. That is why tires heat up when in use, even on a cold day.

This energy dissipation in a rolling tire is manifested as rolling resistance. In other words, it takes some force to push the vehicle along a smooth, level road. In the absence of such an external force on a level surface, the vehicle will slow down; the retarding force, divided by its weight, will equal its deceleration (negative acceleration) as a multiple of the Earth's gravity. Thus, we can speak of rolling resistance in terms of Gs. Friction works in similar ways; we can measure a sliding friction coefficient of a body on a surface by measuring the force required to cause one level surface to slide on another, and dividing by its weight. The coefficient of friction, a dimensionless quantity, can also be spoken of in terms of Gs.

One can envision a test plan in which a car is brought up to speed, the throttle is closed, the transmission is taken out of gear, and the car is then allowed to coast down without braking while its speed (and its derivative, acceleration) is recorded. Since rolling resistance is small, it produces small decelerations, and is mixed in with friction in the bearings, resistance from the air (aerodynamic drag), and even with losses in the drive train. These effects, deliberately kept small in the interest of fuel economy, tend to be lumped together under the term of "rolling resistance."

One can also envision leaving the transmission in gear during such tests. Then the engine would be forced to turn, compression losses would kick in, and decelerations would be higher, depending on the gear used. One could even envision extending the tests to include braking, coasting out of gear, and even acceleration, such are used in normal driving. Such tests have been reported by J. Stannard Baker.[1] The actual data, the test procedures used, and the vehicles tested are not indicated.

Longitudinal Force Generation

It is clear from these considerations that the various wheels have differing contributions to the overall rolling resistance. For example, wheels that are driven (i.e., connected to the drive train) would generate different longitudinal forces ("engine drag") from those that are "free" to roll. The distribution of braking force from the brake system also has an influence during braking, but for purposes of our analysis, braking force in passenger cars will be assumed to be uniformly distributed among all the wheels. The vertical forces reacting to the vehicle's weight also have an effect on the contribution to rolling resistance (although the effects of longitudinal and lateral weight transfer due to braking and cornering will not be treated by the analysis; that will be left to three-dimensional simulation models). Thus, each wheel is generally considered separately, even though in our method the computed longitudinal forces will eventually be combined into a single number.

The effect of vertical loading leads to the interest in front/rear weight distribution, including those of the vehicles tested by Baker. (Left/right weight distribution is assumed to be 50/50 for simplicity, though any asymmetry could easily be accounted for, as seen in Chapter 4.) Given the time frame of Baker's experiments, it is safe to say that the vehicles were rear-wheel drive and probably less front-biased than many present-day front-wheel drive cars. For purposes of analysis, we assume a 50/50 distribution for Baker's data. Therefore, modeling Baker's tests by assuming no wheels bound up, no braking demand from the driver, properly inflated tires, and engine drag on the rear wheels only, will not lead to Baker's coast-down accelerations unless the analysis also reflects a 50/50 front/rear weight distribution.

Tire pressure is another factor affecting rolling resistance, which can change significantly if the tire is flat (which often happens in crashes). Also, it is not unusual to find one or more wheels bound up in the wreckage.

Braking is an often-studied topic which could easily demand a chapter of its own, just for a discussion of tire forces. Suffice it to say that for an individual tire, braking is discussed by first defining slip, which is the ratio of the wheel's tangential velocity at the tire patch to its forward speed at the hub. In other words,

$$s = \frac{V_F}{V_H} = \frac{V_H - R\omega}{V_H} = 1 - \frac{R\omega}{V_H} \tag{2.1}$$

where s is the slip, V_F is the velocity of the tire at the tire patch, V_H is the velocity of the wheel hub, ω is the angular velocity of the wheel, and R is the rolling radius (which, because of the flatness of the tire at the tire patch, is not equal to the measurement one would obtain elsewhere on the tire). If the tire patch is stationary, there is no slip, and the wheel is behaving like a pinion gear on a rack. When the tire is not rotating, the wheel is said to lock up, and the slip is 1.0 or 100%. Thus, $0.0 \le s \le 1.0$.

Of course, the slip of a specific tire depends on both the tire and the road surface. If one attaches a section of tire tread to a drag sled and measures the force required to drag the sled along the road, one obtains the sliding coefficient of friction, commonly abbreviated μ. Coulomb's friction law would then predict the sliding force as μF_Z, where F_Z is the normal force (or the weight of the drag sled, if it is towed properly). Tests of a tire operating on a particular roadway would then reveal how the tire drag F_X would vary with slip. A typical curve would look like Figure 2.1.

We see that the tire generates more resistive force at some low slip value, say about 20%, than it does in pure sliding (slip $s = 1.0$). This corresponds to the difference between static friction and sliding friction. The curve also shows the benefit of antilock braking: more deceleration capability can be achieved if the tire is kept to low values of slip. Of course, keeping slip low

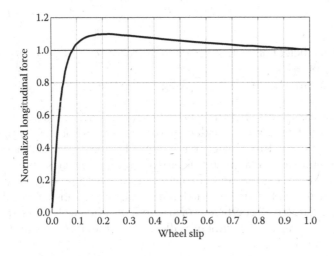

FIGURE 2.1
Longitudinal tire slip characteristic.

means that the tire rotational speed and acceleration have to be continuously monitored, and the brakes have to be modulated accordingly.

A tire's rotation can be increased due to positive driveline torque, or—as described previously—decreased by these factors:

- Braking demand from the driver
- Wheel being partly or wholly bound up in the crash damage
- Engine drag
- Tire deflation

Positive driveline torque (traction) has the opposite effect of these negative factors, but will not be discussed further. It is seldom encountered during accident avoidance maneuvers or run-outs from crashes. The effects of varying tire pressure will also be ignored, except for those due to tire deflation, since vehicle control, fuel economy, and so on are not being considered.

The acceleration of a vehicle on a straight, level path is given by

$$a = \frac{\sum F_j}{m} \tag{2.2}$$

where
F_j is the longitudinal drag force at wheel j
m is the total vehicle mass.

The longitudinal drag force at wheel j is subject to the relationship

$$F_j \leq \mu_j N_j \tag{2.3}$$

where
μ_j is the coefficient of friction between the roadway and the tire, at the location of tire j.
N_j is the normal (vertical) force acting on the tire.

From our earlier discussion, it is apparent that the coefficient of friction typically varies during the accident sequence because of the change in longitudinal slip of the tire. However, it is typical reconstruction practice to use a constant value of μ_j in a given trajectory segment; in fact, segments may be defined by the very assumption of a constant μ_j over a portion of the vehicle run-out. Since the objective is to calculate vehicle velocities, typical practice is to use an "effective" coefficient of friction; in other words, one that when applied over a trajectory segment will produce the appropriate dissipation of kinetic energy.

Many vehicle skid tests have been carried out to determine appropriate coefficients of friction for various road surfaces, tire conditions, and so on.

Results that reconstructionists have widely relied upon were reported by Baker.[2] Sometimes, police perform skid testing in or near the accident scene (with the road blocked off, presumably!), or sliding friction tests may be performed using a drag sled. In those cases, the use of actual data at the actual scene produces the most credible results, as long as the reported friction values seem reasonable. Otherwise, it is common practice to use a value of about 0.7 for a well-traveled stretch of pavement.

When a vehicle is sitting stationary on a flat, level surface, the normal force N_j is simply due to the tire carrying its portion of the vehicle weight. In turning or braking maneuvers, there is "weight transfer" due to additional normal forces being required to resist the pitch and roll moments emanating from the fact that the vehicle center of mass is above the ground (i.e., not in the same plane as the tire/roadway friction forces). For two-dimensional (coplanar) analysis of vehicle motions, such weight transfer effects are necessarily ignored, and will not be discussed further here.

It is important for the reconstructionist to evaluate each tire/wheel combination separately, because their conditions may vary, and they may be located on different surfaces. It is useful, therefore, to employ the concept of "brake factor," which is defined for wheel j as

$$bf_j = \frac{T_j/r_j}{N_j} \tag{2.4}$$

where
> T_j is the retardation torque on the wheel.
> r_j is the rolling radius of the wheel (distance from the spin axis to the ground).
> N_j is the normal force, as before.

It is seen that T_j/r_j is the retardation force due the following:

- Internal friction (rolling resistance) as discussed previously, including aerodynamic drag, expressed as coasting deceleration on a level surface.
- Whether the wheel is on a driven axle, and thus subject to resistance from the engine and transmission.
- The level of brake application demand, expressed as desired deceleration in Gs, which, because it comes from driver inputs, would be the same for all four wheels in any particular trajectory segment. The resulting brake force is also assumed to be evenly distributed.
- Whether the tire is deflated.
- The degree to which a wheel is bound up in the wreckage (expressed as a decimal fraction).

The first two effects can be treated as either/or (either it is not driven, and subject only to tire and bearing losses, or there is also resistance from the drive line). For an all-wheel drive vehicle, both axles would be specified as driven. On the other hand, a transmission in neutral would be modeled by identifying neither of the axles as being driven. In fact, a transmission in neutral is the way the first effect is measured.

The third effect comes from the brakes and is additive. There is often no indication in the vehicle from brake application, but the brake pedal surface or the bottom of a shoe have been known to leave behind distinctive evidence.

The fourth effect reduces the rolling radius and increases the tire patch size, but more importantly, greatly increases the tire losses. It is usually measured and reported in combination with the first effect, but is in fact additive. Therefore, in the analysis it is treated as a factor which is added. It is important to note that at-scene photographs should be examined for tire deflation, since wrecker personnel or others sometimes inflate one or more tires so that the vehicle can be moved more easily.

The fifth effect is very important and needs to be handled with care. If the vehicle is lifted off the ground for inspection purposes, and a wheel is still found to be bound up, then it is safe to apply a 1.00 factor for this effect. (On the other hand, if the vehicle structure bears directly on the ground, a friction coefficient reflecting metal on pavement may be necessary.) In other cases, the wheel may appear to be bound up when the vehicle is on the ground, but may not be so when the vehicle is lifted. A bound-up factor less than unity may then be indicated. In that case, the tire should be examined for sheet metal cuts in the tread or sidewalls, or for evidence of sliding on the ground.

The cumulative effect of these various brake factors can overwhelm the available friction at the tire/road interface. Therefore, the limiting effect of available friction must be included in the analysis. After all, a wheel that does not lock on a dry paved surface may well do so when there is ice present.

The basic rolling resistance factor depends on whether the vehicle is disengaged from the engine ("free to roll"), or whether it is driven. To this can be added a factor that comes into play if the tire is deflated.

Data for these effects must come from testing, preferably on an actual vehicle. Two sources are the experiments conducted by Robinette et al.,[3] and those conducted by Cliff and Bowler.[4] A summary of Cliff and Bowler's data is shown in Table 2.1.

Also reported were brake factors for uninflated (flat) tires, whose averages ranged from 0.13 for a drive wheel in a passenger car, and 0.18 for a non-driven wheel in an SUV. The vintages of the vehicles tested were all in the 1990s, except for a 1984 Toyota Tercel.

These data illustrate the importance of knowing what gear an accident vehicle was in. An inspection of the vehicle is of little value (except immediately post-accident), because the transmission is often placed in neutral to facilitate clearing the vehicle from the scene. The best opportunity to ascertain the gear probably lies in careful examination of any at-scene photographs

TABLE 2.1

Brake Factor Measurements Reported by Cliff and Bowler

Automatic Transmission Drive Wheels	Gear				
	1st	2nd	3rd	Overdrive	Neutral
26.8 mph					
Passenger cars					
Average	0.219	0.095	0.060	0.048	0.020
Range	0.144–0.284	0.082–0.116	0.034–0.082	0.027–0.070	0.018–0.022
Trucks and SUVs					
Average	0.268	0.111	0.074	0.020	0.025
Range	0.258–0.277	0.088–0.134	0.074–0.075	—	0.024–0.026
4.5 mph					
Passenger cars					
Average	0.015	0.015	0.017	0.014	0.011
Range	0.013–0.016	0.011–0.022	0.013–0.023	0.013–0.015	0.007–0.015
Trucks and SUVs					
Average	0.018	0.008	0.020	0.008	0.017
Range	0.015–0.021	0.004–0.011	0.015–0.025	—	0.013–0.021

Manual Transmission Drive Wheels	Gear					
	1st	2nd	3rd	4th	5th	Neutral
26.8 mph						
Passenger cars						
Average	0.258	0.131	0.082	0.062	0.049	0.021
Range	0.217–0.364	0.102–0.200	0.064–0.116	0.051–0.090	0.035–0.059	0.017–0.025
4.5 mph						
Passenger cars						
Average	0.114	0.073	0.051	0.054	0.056	0.010
Range	0.097–0.172	0.059–0.108	0.042–0.065	0.040–0.071	0.036–0.064	0.008–0.013

that may have been taken (and were almost always taken to show something other than the gear shift lever), perhaps comparing the picture(s) to an exemplar vehicle. Even then, the gear selector position may be difficult to determine with certainty, particularly with a manual transmission.

Nondriven wheel brake factors were not separately listed because any vehicle has at least two driven wheels. The best approximation for nondriven wheels comes from tests with the transmission placed in neutral. Such brake factors are typically small but nonzero, as can be seen from Table 2.1.

All that remains is to add in any brake factor contributions from brake demand and/or the wheel being bound up in the wreckage. The brake demand

and the fraction of wheel binding could conceivably be large. However, any values higher than the friction coefficient would have no interest from a practical point of view, inasmuch as the friction coefficient of the tire/road interface would govern, and the wheel would remain fully locked up. On the other hand, the wheel could only be partially bound up. This author has seen instances where the wheel could still turn, but where sheet metal edges had gouged or cut the tire, or the tire had rubbed on the suspension or the wheel house. Obviously, the vehicle should be carefully examined for such evidence.

To keep track of and calculate the various factors, a spread sheet can be set up, with separate calculations for each tire. We start by calculating (or specifying, in the case of crash tests) the normal (vertical) force on the tire. Then we set about calculating its brake factor. The rolling resistance factor is taken from test data, as explained above. Factors are added in for the fraction the wheel is bound up, and for whether the tire is deflated (also from test data). Then any demand from the brakes (assumed to be the same for all four wheels) is added in. The result is compared to the coefficient of friction applicable to that tire in the particular segment of the trajectory. If the sum of the brake factors exceeds the friction coefficient, the brake factor is truncated accordingly.

In an Excel spread sheet, the IF() function* may be used to assign the rolling resistance factor for the left front wheel, for example, as follows:

$$\text{Roll\_R\_LF} = \text{IF[UPPER(FADriven)= "Y ", BF\_Driven, BF\_Non\_Driven]} \quad (2.5)$$

where the UPPER function capitalizes the user input, FADriven (which is "Y" or "N," depending on whether the front axle is driven), and where BF_Driven and BF_Non_Driven are brake factors that apply to the driven or non-driven wheels, respectively. Thus, FA_Driven could be entered by the user in either upper or lower case. The additional brake factor due to tire deflation can be handled with similar logic:

$$\text{Addition} = \text{IF[UPPER(LFDeflated) "Y", BF\_Deflated, 0]} \quad (2.6)$$

Then the Excel MIN() can be used to truncate the brake factors as follows:

$$\text{BF\_LF} = \text{MIN[Friction\_LF, (Sum of the various brake factor components)]}$$

$$(2.7)$$

where Friction_LF is the friction coefficient at the left front tire, for that segment of the trajectory. The brake factor, multiplied by the normal force for the wheel, results in the drag force acting at the wheel in question. The sum of all the drag forces equals the total retardation force on the vehicle, for the particular segment. The retardation force may be divided by the total vehicle weight to obtain the overall longitudinal brake factor.

---

* Variable = IF(condition, value if condition is true, value if condition is false).

The overall segment friction coefficient is the weighted average (literally!) of the friction coefficients at the various wheels. It is the fractional weight at each wheel, multiplied by the friction coefficient at that location, and summed over the various wheels.

Beyond academic interest, the reason for calculating the overall segment friction coefficient is to find how much of the available friction force the vehicle would be using if it were traveling longitudinally over that particular trajectory segment. (Travel at angles other than straight ahead is considered in the next section.)

The question of how much of the available friction the vehicle is using is encapsulated in a single parameter: the "lock fraction." We want a number which, when multiplied by the overall friction coefficient, will relate to the drag force on the vehicle as a whole as it traverses the trajectory segment. Although the calculations are somewhat involved, the lock fraction and overall friction coefficient permit the use of fairly simple calculations later on, in the analysis of vehicle run-outs. This is true even if the wheels are on various surfaces, having various friction coefficients for the various tires.

In the spreadsheet that follows, Figure 2.2, the input variables are highlighted with bold type, and their cells are shaded. All other quantities are calculated. Front weight distribution is input as a decimal fraction, and the axle weights are computed therefrom. If one knows the axle weights but not the front distribution, Excel's "Goal Seek" feature can be used to find it.

The Y/N cells are not case sensitive. Anything other than a "Y" (or "y") is treated as a "no" entry, including a blank in that cell.

In Figure 2.2, the friction coefficients vary from segment to segment, but are the same for each wheel. Segment 6 has different friction coefficients at each wheel. The first five segments show the effect of decreasing the friction coefficient; various tires lock up according to how the friction coefficient compares to other parameters.

In segment 1, comparing the front tires shows the additive effect of tire deflation. The adhesion limits are 840 and 560 lb for each of the front and rear wheels, respectively. The brake factor for the left front wheel is 0.395, and the drag force at that wheel is $0.395 \times 1200 = 474$ lb, which is within the adhesion limits, even though the wheel is partly bound up (0.2). The right front wheel is completely bound up, so it is locked (skidding); the force at that wheel is the adhesion limit of $0.7 \times 1200 = 840$ lb. The left rear wheel is not deflated and is not driven, so its brake factor is 0.022, in accordance with our inputs of test data. The right rear is deflated and is also nondriven, so its brake factor is calculated at $0.022 + 0.108$, or 0.130.

In this example, one of the wheels is locked and three are not, so the overall lock fraction LF is 0.513. That number, when multiplied by the segment friction coefficient of 0.700, yields the overall longitudinal brake factor, 0.359.

In segment 2, the friction coefficient has dropped to 0.5 for all wheels. Now the braking demand on all wheels of 0.4 exceeds the adhesion limit on three of them, so those three are now locked up, with a segment brake factor equal

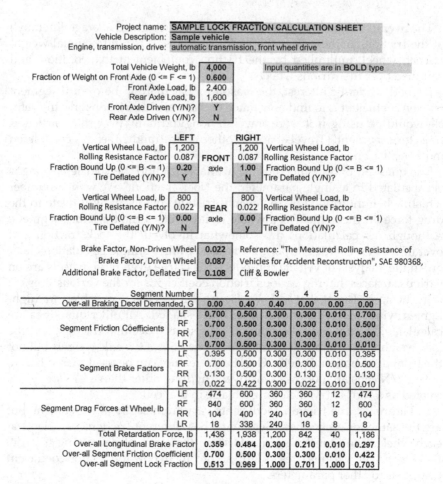

FIGURE 2.2

Sample calculation of lock fraction for various trajectory segments.

to the friction coefficient of 0.5. The left rear wheel is not bound up in the wreckage, not deflated, and not driven, so even though there is some brake force application, it is not quite enough to lock the wheel. The retardation force for the left rear wheel is (0.40 + 0.022 = 0.422) times the left rear vertical load of 800 lb, or 338 lb. On the locked wheels, the retardation force is simply the friction coefficient times the vertical load. With three wheels locked, the overall lock fraction LF climbs to 0.969.

As the friction coefficient drops to 0.3, the braking demand of 0.40 G is sufficient to lock all four wheels (segment 3). If the brakes are released as in segment 4 (0.00 brake demand), the front wheels are still skidding (because of tire deflation or being bound up). However, the rear wheels are within adhesion limits, so the roadway friction does not dominate unless the friction coefficient becomes low enough. In our example, if it were to drop to 0.01

(segment 5), even the so-called "free-to-roll" left rear wheel locks up due to rolling resistance, and the overall lock fraction LF is 1.000.

In segment 6, the tires have individual friction coefficients, as could be the case if the vehicle is crossing a terrain boundary, with some wheels on one side and the other wheels on the other side. In this case, the overall segment friction coefficient is different from any of the individual values. Only the right front wheel is skidding.

Lateral Force Generation

When the vehicle is traveling a straight path, the tire can generate longitudinal acceleration or deceleration forces because it deforms and serves as the interface between the moving vehicle and the stationary road. This is also true when the tire is at an angle to the path of travel, only now the tire deforms laterally, and lateral forces occur at the contact path. One might expect that the magnitude of such forces depends on just how much angle and how much normal force there is, and testing would confirm that expectation. (There may also be dependence on other factors, such as velocity and temperature, but those are generally considered to be second-order small for reconstruction purposes.) The angle between the tire's longitudinal direction (heading) and its path of travel is known as the "slip angle" α; that is,

$$\alpha = Arc\tan\left(\frac{V_{\text{lat}}}{V_{\text{long}}}\right) \qquad (2.8)$$

where V_{long} and V_{lat} are the longitudinal and lateral components of the tire's velocity vector, respectively. Note that these were not labeled V_X and V_Y, because the X–Y axis system is tied to the vehicle, not the tire. In other words, the tire may be at an angle to the vehicle, due to crash damage or steering input, for example.

When the tire is at right angles to its path of travel, the tire has ceased to roll; it is sliding sideways. Coulomb's friction law predicts its lateral force is simply the sliding coefficient of friction μ, multiplied by the normal force F_Z. In theory, this coefficient of friction could be different than that for longitudinal sliding, but in reconstruction practice, no distinction is usually made. The relationship between lateral force and slip angle for a given tire has the general appearance of Figure 2.3.

The first part of the curve is fairly linear. Because this is the regime in which a tire operates when a vehicle is under control, very often vehicle-handling models treat the lateral force as a linear function of the slip angle. (Reconstructionists often deal with vehicles out of control, however.) The

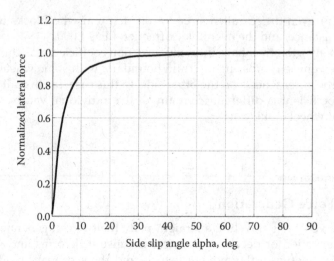

FIGURE 2.3
Typical tire cornering characteristic.

slope of the curve of lateral force versus slip angle, measured in pounds per degree, or pounds per radian, is known as the "cornering stiffness" or the "cornering coefficient." For an inflated tire, a typical value of cornering coefficient is about $F_Z/6$ pounds per degree, and for a flat tire, about $F_Z/30$ pounds per degree.[5]

Longitudinal and Lateral Forces Together

Longitudinal force and lateral force components combine to produce a resultant tire force. When a tire that has been generating longitudinal forces locks up, the wheel stops rotating (by definition), and the slip is unity. The longitudinal force is the product of the normal force and the coefficient of sliding friction, as we have seen. If there were a lateral force under such conditions, it would combine with the longitudinal component to produce a resultant force greater than μF_Z. This is impossible; μF_Z is the maximum force available. Therefore, the lateral force must go to zero when the tire locks up. The tire is said to be "saturated" under such conditions. (In fact, the resultant force for a locked tire opposes the velocity vector of the tire on the road, and is equal to μF_Z.)

Similar logic holds when the tire is at right angles to the direction of travel. In that case, the resultant force is μF_Z, is at right angles to the tire, and opposes the velocity vector. There is no force longitudinal to the tire.

For intermediate angles, the resultant tire force cannot exceed μF_Z; the lateral force capability of a tire decreases, and becomes zero when the tire locks

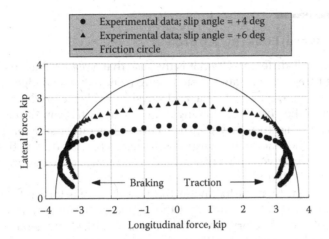

FIGURE 2.4
Tire forces for two slip angles,

up. The resultant tire force must stay within a locus of points that define a circle, called the "friction circle." (Technically, it would be a friction ellipse if the coefficients of sliding friction were different in the longitudinal and lateral directions.) Figure 2.4 shows how tire forces stay within the friction circle.[6]

Just like any other sliding object, a tire that is skidding on the ground generates a friction force that is opposed to its velocity vector, whatever direction it may be pointing. A tire can only do otherwise if it is rolling. It can be generating longitudinal forces due to braking or traction, and can be simultaneously generating lateral (cornering) forces, as long as it is not locked up. As it approaches lock-up (its rotation rate approaches zero), the available lateral forces decrease to zero. Thus, in a bootleg steer maneuver, where the parking brake is used to lock up the rear tires, their ability to resist lateral vehicle motions goes to zero, and the rear end can swing around violently.

A good feel for the complexity of analyzing tire forces can be had by studying the source code in the Simulation Model of Automobile Collisions (SMAC)[7] or EDSMAC (Engineering Dynamics, Inc.'s version of SMAC).[8] Even so, these programs also do not take into account weight transfer due to tire forces; the vehicles exist entirely in the plane. The programs undertake a separate analysis for each of the four tires. The ability to correctly predict the tire forces in run-out is key to predicting the trajectory the vehicle will take in response to those forces. For example, the force magnitude and direction at each of the four tires must be known so that both linear and angular (yaw) accelerations at the vehicle's center of mass can be calculated at each time value in the simulation process. The analyst can spend the greater part of time and effort dealing with the complexities of the run-out phase of the crash, when in fact it is the crash phase that is usually more important (since that is where the injuries typically occur).

The Backward-Looking Approach

Our approach is different. Often it is the case that a reasonable knowledge of the run-out trajectory is gained, stemming from our knowledge of the final vehicle position, the impact area, and physical evidence on the ground. We are not put in the position of having to guess the initial conditions and then find out what trajectory they produce. Our task is rather simpler: to reasonably evaluate the amount of energy dissipated in friction during the run-out, so that the vehicle velocities (linear and rotational) at the beginning of the trajectory can be found.

The approach, then, is to divide the (known) trajectory into a number of straight-line segments, such that the segments reasonably represent the geometry of the trajectory, and such that the vehicle deceleration in each segment can be calculated with reasonable accuracy. In each segment, we only have to find the velocity at the beginning, based on what it was at the end, the nature of the segment, and the amount of energy scrubbed off by the vehicle as it traverses the segment.

To do this, we define the vehicle's "crab angle" as the difference between the vehicle's heading ψ and the direction θ of the trajectory segment. It is the analog of the slip angles at the individual tires, but is given a different name to remind us that it does not necessarily represent the actual (individual) slip angles at any of the four wheels. (No individual steer angles are considered, as it is not our intent to predict the motion of the car due to driver inputs; steer angles could be accounted for by placing the appropriate offset into the crab angle computation, but this author has never found it necessary.) The vehicle yaw torques, due to the tire forces acting at distances from the center of mass, are not calculated. Rather, yaw rates are calculated from the nature of the trajectory itself, as we will see later. A single, simple "tire model" predicts the deceleration of the vehicle as a whole, which was the original intention anyway.

Effects of Crab Angle

Of course, the lock fraction is not the whole story when it comes to determining vehicle deceleration, unless the vehicle's heading happens to coincide with its direction of travel. Generally, this does not have to be the case, as post-crash vehicles are almost always out of control, so we define the crab angle as the difference between the vehicle's heading and the direction of its velocity vector. It could be thought of as the slip angle of the entire vehicle, but is called the crab angle to remind us that it does not necessarily represent the actual (individual) slip angles at any of the four wheels. If the vehicle is sideways, its crab angle is 90° or 270°. In this situation, it matters not what the lock fraction is; the force generated along the vehicle's direction of travel will equal the vehicle weight times the friction coefficient. On the other hand, if the crab angle is 0° or 180°, the vehicle is tracking, and the force along its direction of travel will equal the vehicle weight times the friction coefficient times the lock fraction.

To reflect a continuous variation between these extremes, a pseudo-ellipse is used to calculate the vehicle drag factor, as follows:

$$DragF_{j-1} = \mu_{j-1}\sqrt{[LF_{j-1}\cos(Crab_{j-1})]^2 + \sin^2(Crab_{j-1})} \tag{2.9}$$

where $DragF_{j-1}$ is the average drag factor in the segment $j - 1$, which is bounded by the positions j and $j - 1$, μ_{j-1} is the (weighted average) roadway friction in that segment, LF_{j-1} is the lock fraction, and $Crab_{j-1}$ is the crab angle.

The crab angle is calculated as an average over the trajectory segment, based on the vehicle heading angles at the ends of the segment, as compared to the direction of the segment itself:

$$Crab_{j-1} = \frac{1}{2}(\psi_j + \psi_{j-1}) - \theta_{j-1} \tag{2.10}$$

where ψ is the vehicle heading, and θ is the path direction in that particular trajectory segment. It is important to use a number of trajectory segments sufficient to represent the variation of the crab angle throughout. Generally, experience has shown that reasonable crab angle calculations result if the vehicle rotation in any given segment is no more than 90°.

The above consideration leads to a desire to sub-divide the trajectory into smaller segments prior to computing the crab angles and calculating the vehicle deceleration. A method has been developed to this end, and will be discussed in subsequent chapters.

References

1. Baker, J.S., *Traffic Accident Investigation Manual*, Northwestern University, Evanston, IL, 1975.
2. Baker, J.S., *Traffic Accident Investigation Manual*, Northwestern University, Evanston, IL, 1975, p. 210.
3. Robinette, R., Deering, D., and Fay, R., *Drag and Steering Effects of Under Inflated and Deflated Tires*, SAE Paper 970954, SAE International, 1997.
4. Cliff, W.E. and Bowler, J.J., *The Measured Rolling Resistance of Vehicles for Accident Reconstruction*, SAE Paper 980368, SAE International, 1998.

5. McHenry, R.R., Jones, I.S., and Lynch, J.P., *Mathematical Reconstruction of Highway Accidents: Scene Measurement and Data Processing System*, prepared for the National Highway Traffic Safety Administration, Calspan Corporation, 1974.
6. Statement of work, truck tire characterization: Phase 1, Part 2, Cooperative Research, Funding by NHTSA Contract No. DTNH22-92-C-17189, SAE International, 1995.
7. Baker, J.S., *Traffic Accident Investigation Manual*, Northwestern University, Evanston, IL, 1975.
8. *EDSMAC: Simulation Model of Automobile Collisions*, Version 4, Fifth Edition, Engineering Dynamics Corporation, Beaverton, OR, 2006.

3

Subdividing Noncollision Trajectories with Splines

Introduction

Before we can get to the reconstruction of the collision phase of the crash, we first have to reconstruct the post-collision phase—the run-out trajectory or trajectories of the vehicles involved. By these terms, we mean the path taken be the center of mass, or CG, of each of the involved vehicles. It is our task to describe each trajectory in enough detail to permit the calculation of the tire forces, as described in Chapter 2, so that we can find out how quickly each vehicle slowed down.

Of course, one way to obtain this detail is to run a computer simulation, which marches forward in time, and let the simulation predict where the vehicle will go, by solving the differential equations of motion for the system in question. This is most often done using numerical methods. Such methods involve subdividing time into small increments (0.010 sec is typical for trajectory simulations) and then integrating the differential equations of motion step by step. The increments are kept small so as to limit the errors caused by the approximate nature of the methods. Tire forces are calculated at each time, as they are part and parcel of the governing differential equations.

But why go to all that trouble to predict each vehicle's path, when we already have the physical evidence as to where the vehicles went?

What we really need to do is to place each vehicle on its trajectory, calculate the tire forces and decelerations, and find out directly what the vehicle velocities were when they exited the crash. In other words, we need to work backward from the known conditions at the end of the trajectories (which often entail the vehicle stationary—at rest—and documented by police measurements, photographs, etc.), through the trajectories in the reverse direction, until we arrive at each vehicle's exit, or separation, from the collision phase.

If we can do this, we can keep the reconstruction focused on the important issues—what happened during the collision phase, which is when the injuries occurred. To rely on a computer simulation, we risk shifting our focus to the problem of getting the simulation to correctly predict the vehicle's rest position, in effect taking our eye off the ball so we can tweak the tire/ground interaction.

Of course, a trajectory is not known precisely at any given instant of time. What we do have is physical evidence—tire marks, gouges in the pavement, furrows in the dirt, knocked over sign posts, and so on—that allows the vehicle position to be located at certain key points along the path. By "vehicle position," we mean the coordinates in Cartesian space (X, Y, and Z) of its CG, plus the angles required to describe its attitude (heading angle ψ for most accidents, and roll angle φ for rollovers). Thus, a reverse trajectory analysis starts with the construction of a curve that connects all the key points or key positions, which are described by the key position coordinate values, sometimes referred to as the key values. If the key positions are chosen carefully, the fitted curve is a good representation of the actual path followed by the vehicle CG during its run-out.

One might ask "Why is a fitted curve necessary? Why not just approximate the trajectory by drawing straight lines between the key points, and calculate the deceleration along each line?" The answer is found in Chapter 2, where we saw that the deceleration is a function of the crab angle, among other variables. We know that, in general, the crab angle varies continuously throughout the trajectory, and so does the deceleration as well. So the answer is simply that not only does drawing such straight lines fail to provide a reasonable representation of the path. It is also the case that reverse trajectory calculations depend on certain variables, such as crab angle, being held constant. Thus, reverse trajectory methods also involve approximations, as do simulations, but now it is physical space, rather than time, that has to be subdivided.

Selecting an Independent Variable

To do this subdivision, we need a suitable independent variable. After all, the notion of a "curve" implies that one or more variables depend on an independent variable. Consider, for example, computer animations, in which the independent variable is the image frame. Frames are shown at a constant rate—usually, 30 frames/sec, and are anchored to key frames (i.e., they fill in between the key frames so that the motion is continuous). Thus, in animations, time is a sort of hidden independent variable. In our case, however, time is one of the unknowns, so we have to find something else, preferably something related to the spatial coordinates of the trajectory.

For example, consider the ballistic formula for a particle—a parabola in which Z is a quadratic function of X. X is not the independent variable, though. Both X and Z are functions of time, which is the true independent variable; t fails to appear in the relationship only because it has been eliminated between X and Z.

However, time is not available to the reconstructionist. (There is never enough of it!) So consider another example: the formula for a circle in the XY plane. From the formula, it is obvious that neither X nor Y is the independent variable. To plot a circle, we set up a parameter θ, which is the angle from the X-axis. Then, we write the equations of a circle in parametric form:

$$X = R\cos(\theta)$$
$$Y = R\sin(\theta)$$

(3.1)

and let θ vary between 0 and 2π. The graphing software will draw the curve as a series of straight lines between the points, but if θ is subdivided into sufficiently small segments, the plot will bear a suitable resemblance to a circle.

What we need for reverse trajectory analysis is something like θ in the above example. It is a fairly obvious leap to settle on trajectory arc length S as the independent variable. As the trajectory is traversed, S increases monotonically. Therefore, all the dependent variables—X, Y, Z, ψ, φ, and so on—will be single-valued, because S never doubles back on itself. This doubling back could be a problem if any other spatial variable were chosen as the independent variable.

But we have another problem with S. How can we measure arc length along the trajectory if we do not know what the trajectory is? What we can do, though, is to measure the distance along the straight lines connecting the key points. That is not exactly the arc length of the derived trajectory curve, but it is a monotonically increasing parameter that is uniquely associated with every point on the trajectory, and that is all we need. If we need the actual arc length for the trajectory, we can take, as an approximation, the cumulative lengths of the small segments into which the trajectory is divided.

Finding a Smoothing Function

Now, the next trick is to find a mathematical function (so that it can be subdivided) that passes through an arbitrary number of key points distributed arbitrarily in space. Again we can refer to animation software, which fills in the frames, at 1/30 sec intervals, between the key frames. Programs such as WaveFront[*] use mathematical smoothing functions called "splines" to achieve this end. Also, the reader may be familiar with splines being used in AutoCAD[†] to draw smooth lines between points. There are other types of functions that could be used for connecting key points with a smooth line,

---

[*] Alias|Wavefront, Toronto, Canada.
[†] Autodesk, Inc.

and there are various types of splines; all we need is some kind of analytical function, available in some widely used software package, that can use our pseudo-arc length coordinate S as an independent variable to create intermediate values of the dependent variables.

Electronic spreadsheets fit the description of widely used software. The best known of these is Excel.[*] There are a great many functions—mathematical and otherwise—embedded within Excel. A search of its functions reveals nothing resembling a smoothing function or spline. However, a search of the Internet shows that additional functions have been developed for incorporation within Excel. One such add-in package is XlXtrFun,[†] which can be downloaded from the web site http://www.xlxtrfun.com. Available at no cost, "XlXtrFun contains functions that look intrinsic to Excel. These functions are primarily interpolating and curve fitting, and include both two-dimensional (2D) and three-dimensional (3D) interpolating."[‡]

Properties of Splines

For our purposes, the function we need is called, naturally enough, Spline, which is a natural cubic spline. It is a "third degree piecewise polynomial curve that (1) goes through all the [key points], (2) at each [key point] the first derivatives of the two curves that meet there are equal, (3) at each [key point], the second derivatives of the two curves that meet there are equal, and (4) at the two end [key points], the second derivatives of each curve equal zero."[§] This spline function takes as inputs the key points, described by one array containing the independent variable, and a second array (of the same size, of course) containing the dependent variable. The given value of the independent variable is also an input parameter, as is a switch that indicates whether extrapolation is to be allowed or not. The function returns the value of the dependent variable that lies on the natural cubic spline curve at the given value of the dependent variable.

The syntax of the function is

$$= \text{Spline (ArrayofS, ArrayofD, GivenS, Extrapolate?)} \qquad (3.2)$$

where ArrayofS is a "one-dimensional array of numbers, or a contiguous group of cells containing numbers arranged either in a row or a column. The values must be either constantly increasing or constantly decreasing."

---

[*] Microsoft Corporation, Redmond, WA.

[†] Advanced Systems Design and Development, Red Lion, PA.

[‡] Advanced System Designs and Development, http://www.xlxtrfun.com/XlXtrFun/ReadMeXlXtrFunAndSurfGen.htm.

[§] Advanced Systems Design and Development, "XlXtrFun Excel Extra Functions Help," 2005.

In other words, the independent variable cannot double back on itself, which would result in a multi-valued function.

ArrayofD is a "one-dimensional array of numbers, or a contiguous group of cells containing numbers arranged in a row or column." These are the values of the dependent variable, and correspond to the array of independent variable values cited above.

GivenS is a number, which is the given value of the independent variable. The spline function returns the value of the dependent variable at this given value of the independent variable, such that the resulting point lies on the natural cubic spline curve.

Extrapolate? is a switch that is either TRUE of FALSE, signifying whether to allow extrapolation. The default value is FALSE, but in our case the independent variable will have end values exactly at the first and last key points. Therefore, the switch should be set to TRUE for our purposes.

Of course, the spline will extend over a range of the dependent variable, and will be calculated within a range of cells. Therefore, the appropriate spline formula must be copied into a cell range, while keeping the references to the key values constant.

Suppose, for example, that there are six key points that describe the relationship between the independent variable S and the dependent variable Z. Suppose also that the key S values have been copied into column C, starting at row 2 and ending at row 7. The key Z values have been copied into column D, but for some strange reason the key values start at row 4 and end at row 9. It is desired to have the splined curve have its S values in column E, and the Z values in column F. There will be 26 points on the curve, starting at row 25 and ending at row 52.

The first step is to enter 28 monotonically increasing or monotonically decreasing values of the independent variable S into cells E25 through E52. They do not have to be evenly spaced. The dependent value Z of the first point on the spline curve goes in cell F25. The value of that cell is given by

$$= \text{Spline}(\$C\$2{:}\$C\$7,\ D\$4{:}D\$9,\ E25,\ TRUE) \tag{3.3}$$

The absolute cell reference indicators ($) are used because the intent is to copy the formula without changing the references to the key values of the independent variable. Then one simply copies the contents of cell F25 into cells F26 through F52. Of course, the reference to cell E25 changes automatically.

If one uses the "define name" feature of Excel to name the key value arrays, then the formula for cell F25 could look like

$$= \text{Spline}(Key\_S,\ Key\_Z,\ E25,\ TRUE) \tag{3.4}$$

which is much easier to read. However, if one copies the formula to another spreadsheet having a different size or location of the key value arrays, one needs to remember to use Excel's "name manager" feature to reflect the changes (i.e., have the array names point to the proper cells).

Example of Using a Spline for a Trajectory

Most post-crash trajectories are fairly simple curves. For the purposes of example, however, suppose that a vehicle is thought to have had an S-shaped trajectory, on which four key points have been identified. The coordinates of these points, as entered into an Excel spreadsheet, appear in Figure 3.1.

A graph of these points, represented by the hollow circles, is shown in Figure 3.2. A dotted line is drawn between the key points, but unless we are looking at the trajectory of a pin ball, drawing straight lines between the key points does not appear to be representative of the actual trajectory of an out-of-control car. So we will see what a spline curve between these key points looks like.

We can see right away that neither X nor Y would make a very good independent variable. For one thing, why should one axis be preferred over the other? For another, trying to use either one as an independent variable would produce a multi-valued function in this example. So we construct a pseudo arc length variable between the key points, and we will call it Key S. To establish that variable, we calculate the distances between the key points using the Pythagorean theorem, and then do a cumulative sum. Note in Figure 3.2 that while there are four key points, there are only three lines connecting them. Therefore, the columns in Figure 3.1 that relate to the key points (A, B, C, and E) have four entries, while column D, Delta Key S, relates to the lines between them, and therefore only has three entries. Cell D2 is blank. The length of the line between key points is shown in the same row as the key point at which the line ends. That cell in row D is where the Pythagorean theorem calculation is made.

	A	B	C	D	E
	Key Point Number	Key X	Key Y	Delta Key S	Key S
1					
2	1	1.0	2.0		0
3	2	10.0	2.0	9.00	9.00
4	3	8.0	9.0	7.28	16.28
5	4	11.0	9.0	3.00	19.28

FIGURE 3.1
Data on key points in Excel.

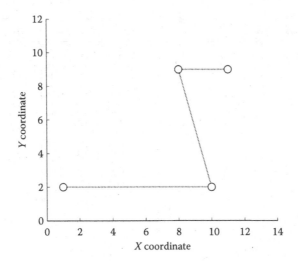

FIGURE 3.2
Plot of example key points.

Column E, Key S, is the cumulative sum of the numbers in column D. The values in columns D and E are shown to two decimal places to remind us that they are calculated, rather than input.

Key S appears in column E of Figure 3.1. It is the distance traveled along the piecewise linear path sketched in between the key points, so in that sense it is an arc length coordinate. However, it is not a measure of arc length measured along the trajectory we are going to draw with a spline curve. It is simply a parameter from which the variables X and Y for the spline curve will be calculated. It will increase monotonically from zero at the beginning of the piecewise linear path, to a value of 19.28 at the end.

The next step is to create a set of given values by subdividing S. Let us subdivide it into 19 segments—a purely arbitrary choice that will lead to 20 points, one at the beginning, one at the end, and 18 in between. To do the subdivision, we will fill 20 cells of column G with values that come from uniformly subdividing a line that is 19.28 units long into 19 segments. This produces an array of evenly spaced values, although it is not necessary that they be evenly spaced, as we shall see later. We simply need 20 monotonically increasing values. The results are the variable S, shown in column G of Figure 3.3, which will serve as the independent variable of the spline curve.

Now, we set up the spline that will produce the (dependent variable) X values. Into cell H2, we can type the first spline formula. It is shown in Expression (3.5).

$$= \text{Spline}(\$E\$2:\$E\$6, \$B\$3:\$B5, H2, \text{TRUE}) \tag{3.5}$$

	A	B	C	D	E	F	G	H	I		
1	Key Point Number	Key X	Key Y	Delta Key S	Key S	Spline Pt Number	S	Spline X	Spline Y	Seg Len	Cum S
2	1	1.0	2.0		0	1	0.00	1.00	2.00		0
3	2	10.0	2.0	9.00	9.00	2	1.01	2.53	1.61	1.58	1.58
4	3	8.0	9.0	7.28	16.28	3	2.03	4.03	1.25	1.54	3.12
5	4	11.0	9.0	3.00	19.28	4	3.04	5.44	0.95	1.45	4.56
6						5	4.06	6.73	0.74	1.31	5.88
7						6	5.07	7.87	0.65	1.14	7.01
8						7	6.09	8.80	0.71	0.93	7.95
9						8	7.10	9.49	0.95	0.73	8.68
10						9	8.12	9.90	1.41	0.61	9.29
11						10	9.13	9.99	2.11	0.70	10.00
12						11	10.15	9.76	3.05	0.97	10.96
13						12	11.16	9.31	4.15	1.19	12.16
14						13	12.18	8.76	5.33	1.30	13.46
15						14	13.19	8.23	6.50	1.28	14.74
16						15	14.21	7.84	7.55	1.13	15.87
17						16	15.22	7.72	8.41	0.87	16.73
18						17	16.24	7.98	8.98	0.63	17.36
19						18	17.25	8.70	9.21	0.75	18.11
20						19	18.27	9.76	9.17	1.07	19.18
21						20	19.28	11.00	9.00	1.25	20.43

FIGURE 3.3
Sample spline operation in Excel.

As mentioned previously, the absolute cell reference indicators ($) are used to keep the key value arrays Key S, Key X, and Key Y located in fixed places on the spreadsheet while we copy the cell formulas. In similar fashion, the first of the Y values is given by

$$= \text{Spline}(\$E\$2:\$E\$6, \$C3:\$C5, I2, \text{TRUE}) \tag{3.6}$$

Now, the contents of cells H2 and I2 can be copied onto the clipboard and pasted into the appropriate columns of rows 3 through 21. The results are shown in Figure 3.3.

Note that because the spline started at the first key point and ended at the last one, the X and Y values for the spline end points are seen to match exactly the coordinates of those two key points. Actually, the curve passes exactly through the intermediate key points as well, as can be seen in Figure 3.4. However, the coordinates of those points do not show up in the table representing the spline curve because when the independent variable S was created, there were no values (9.00 and 16.28) corresponding to those points.

The reader can verify that while the subdivided segments are approximately one unit long, they in fact vary in length from 0.61 to 1.58 units. It is also the case that the total length of the segments is 20.43 units—not 19.28. This is because the spline curve is slightly longer than the straight lines drawn between the key points.

Because Figure 3.4 was created by software that draws a straight line between each of the spline points, the figure actually shows a piecewise linear function that represents the trajectory subdivided into straight segments. That was the intent of the subdivision. If the intent was to show the underlying spline itself, which is a continuous curve, smaller segments could be

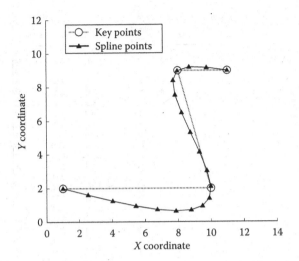

FIGURE 3.4
Segmented curve fitted through key points.

used to plot the underlying spline for comparison. Then, the two curves would be virtually indistinguishable.

Note that as the trajectories of moving bodies abhor sharp corners (except when there is a collision), so do splines. Thus, there tends to be some over-shoot or undershoot in the vicinity of sharp corners or even tight turns. This is the "price," if one wishes to look at it that way, of exactly matching the key points.

	A	B	C	D	E	F	G	H	I		
1	Key Point Number	Key X	Key Y	Delta Key S	Key S	Spline Pt Number	S	Spline X	Spline Y	Seg Len	Cum S
2	1	1.0	2.0	0	0	1	0.00	1.00	2.00		0
3	2	10.0	2.0	9.00	9.00	2	1.01	2.53	1.61	1.58	1.58
4	3	8.0	9.0	7.28	16.28	3	2.03	4.03	1.25	1.54	3.12
5	4	11.0	9.0	3.00	19.28	4	3.04	5.44	0.95	1.45	4.56
6						5	4.06	6.73	0.74	1.31	5.88
7						6	5.07	7.87	0.65	1.14	7.01
8						7	6.09	8.80	0.71	0.93	7.95
9						8	7.10	9.49	0.95	0.73	8.68
10						9	8.12	9.90	1.41	0.61	9.29
11						10	9.00	10.00	2.00	0.60	9.89
12						11	10.15	9.76	3.05	1.07	10.97
13						12	11.16	9.31	4.15	1.19	12.16
14						13	12.18	8.76	5.33	1.30	13.46
15						14	13.19	8.23	6.50	1.28	14.74
16						15	14.21	7.84	7.55	1.13	15.87
17						16	15.22	7.72	8.41	0.87	16.73
18						17	16.28	8.00	9.00	0.65	17.39
19						18	17.25	8.70	9.21	0.73	18.11
20						19	18.27	9.76	9.17	1.07	19.18
21						20	19.28	11.00	9.00	1.25	20.43

FIGURE 3.5
Independent variable Key S values represented in spline.

	A	B	C	D	E	F	G	H	I		
1	Key Point Number	Key X	Key Y	Delta Key S	Key S	Spline Pt Number	S	Spline X	Spline Y	Seg Len	Cum S
2	1	1.0	2.0		0	1	0.00	1.00	2.00		0
3	1a	8.0	1.5	7.02	7.02	2	1.02	1.97	1.84	0.99	0.99
4	2	10.0	2.0	2.06	9.08	3	2.04	2.95	1.68	0.99	1.98
5	3	8.0	9.0	7.28	16.36	4	3.06	3.94	1.55	1.00	2.98
6	4	11.0	9.0	3.00	19.36	5	4.08	4.95	1.46	1.01	3.99
7						6	5.09	5.98	1.42	1.03	5.02
8						7	6.11	7.03	1.43	1.06	6.07
9						8	7.02	8.00	1.50	0.97	7.04
10						9	8.15	9.20	1.69	1.22	8.26
11						10	9.08	10.00	2.00	0.86	9.11
12						11	10.19	10.55	2.60	0.81	9.92
13						12	11.21	10.68	3.35	0.77	10.69
14						13	12.23	10.51	4.27	0.93	11.62
15						14	13.25	10.10	5.32	1.12	12.75
16						15	14.26	9.52	6.47	1.29	14.04
17						16	15.28	8.81	7.68	1.40	15.44
18						17	16.36	8.00	9.00	1.55	16.99
19						18	17.32	7.27	10.18	1.39	18.37
20						19	18.34	6.56	11.40	1.41	19.79
21						20	19.36	5.96	12.56	1.30	21.09

FIGURE 3.6
Spreadsheet with additional key point.

Speaking of which, it may be desirable to have the key points represented in the spline arrays, especially for the calculation of velocity and other variables. That will leave a "paper trail" of calculations for all key points. To do that, and show exactly where the key points lie on the spline curve, one can simply copy (by value) the Key S cell into the cell containing the S coordinate

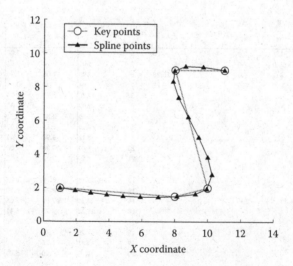

FIGURE 3.7
Trajectory with additional key points.

of the spline point nearest the key point. Then the spreadsheet shown in Figure 3.3 is changed (automatically) to that shown in Figure 3.5. See cells G11 and G18, and the spline values that result from those changes. Again, one can see that the spline curve passes exactly through the key points.

Suppose now that physical evidence indicates that the actual trajectory did not dip as low as the spline goes between key points 1 and 2. To produce a shallower spline in that area, insert an additional key point at the appropriate point, say, (8, 1.5). Then the spreadsheet of Figure 3.1 would be replaced with the one shown in Figure 3.6.

The ensuing spline operation produces the segmented curve shown in Figure 3.7. Note the additional key point. The reader can verify that the line connecting the key points has slightly increased in length, from 19.28 units to 19.36, while the segmented curve based on the spline has actually decreased in length, from 20.43 units to 19.73. Thus, with the addition of the key point, and a closer representation of a smooth curve, the length of the line connecting the key points is closer to the length of the segmented curve, as one might expect.

4

A Program for Reverse Trajectory Calculation Using Splines

Introduction

In Chapter 3, we saw how to flesh in the run-out trajectory, by drawing a spline curve between the key points, and then how to subdivide the trajectory into piecewise linear segments, such that the end points on the segments were positioned on the spline curve. We even saw how to adjust certain of the end points such that they would coincide with the key points, so that calculations pertaining to such points would also pertain to the key points. In this chapter, we shall see how to perform those calculations, so that we can discover the vehicle's motion as it exits the crash. The analysis can also be applied to the vehicle's pre-crash motions, once we know the conditions at impact.

Developing Velocity–Time Histories for Vehicle Run-Out Trajectories

This is not such a tall order as it may appear. In a sense, we have already done the heavy lifting. The main thing we need to do now is to establish a time base to go with the spatial curve we have already developed. To do this, we go back to the discussion in Chapter 2, showing how the vehicle deceleration could be calculated, once the friction coefficients, the various factors influencing the tire forces, and the crab angle were known. We will return to that subject shortly, but for now, let us assume that those parameters are in hand. Then in each segment, the work done on the vehicle in that particular segment can be calculated as the friction force multiplied by the distance traveled. This work done, or dissipation of energy, must equal the loss of kinetic energy plus any loss of potential energy.

Suppose there are NPts points on the subdivided trajectory. The total number of segments is then NPts-1. Point i is one of the points. Between that and the subsequent point $i + 1$ lies the segment i, the length of which is ΔS_i. In that segment, we can then write loss in kinetic energy = gain in potential energy + work done, or

$$\frac{1}{2}mV_{i+1}^2 - \frac{1}{2}mV_i^2 = W(Z_{i+1} - Z_i) + W(DragF_i)(\Delta S_i) \qquad (4.1)$$

where V_i is the velocity at point i, m is the vehicle mass, W is its weight, Z_i is the elevation (vertical coordinate) of point i from some reference point, $DragF_i$ is the vehicle's drag factor in segment i, assumed to be constant, and ΔS_i is the length of the segment. The above equation can be solved for V_i, as follows:

$$V_i = \sqrt{V_{i+1}^2 + 2g(DragF_i\Delta S_i + Z_{i+1} - Z_i)} \qquad (4.2)$$

Note that any loss in rotational energy has been left out of the calculation. If V is in mph and S and Z are feet, the above equation can be written

$$V_i = \sqrt{V_{i+1}^2 + 29.938(DragF_i\Delta S_i + Z_{i+1} - Z_i)} \qquad (4.3)$$

The constant is often rounded to 30 for simplicity.

It should be noted that considerations of potential energy lead to the inclusion of terrain topology in the computations. While many accident scenes are fairly flat, some are not. This formulation of the analysis allows for the inclusion of elevation changes without the necessity of artificially adjusting the friction coefficient, as is often done.

Since the distance traveled is the average velocity in the segment, multiplied by the time to traverse the segment, we can write

$$\Delta t_i = \frac{2\Delta S_i}{(V_i + V_{i+1})} \qquad (4.4)$$

where $\Delta t_i = t_{i+1} - t_i$. If we know the velocity at the end of the segment V_{i+1}, then both the beginning velocity and the time are known. This knowledge both motivates and enables the backward-looking analysis of vehicle run-outs (their motions in the absence of collision forces). If one starts with a known velocity (often zero) at the end of the trajectory, one can proceed backwards through the trajectory step by step, finding values for both velocity and time along the way.

Once all of the Δts are known, they can be added in cumulative fashion, in the forward direction. Thus, a forward-directed time line can be established, even though it is calculated in the reverse direction.

Other Variables at Play in Reverse Trajectory Calculations

So, the only roadblock to performing a segment-by-segment analysis of each trajectory segment, in reverse order, is knowing all the variables in Equation 4.2. Of course, Z is known from the splining operation, ΔS_i was calculated in the spreadsheet after the splining was done, and V_{i+1} is known from the previous calculation. But what about the drag factor $DragF_i$? We know from Equation 2.10 that the drag factor depends on the crab angle, the lock fraction, and the friction coefficient. From Equation 2.11, the crab angle $Crab_i$, in turn, depends on the vehicle headings ψ at the ends of the trajectory segment, and the direction θ of the segment. So, it is apparent that we will have to specify the heading ψ as well as X, Y, and Z at every key point, and a spline curve will have to be calculated for ψ in the same way it was for the other trajectory variables.

That leaves us with the lock fraction and the friction coefficient. Those could be splined, as well. However, it is more common to think of those variables as being constant over the trajectory segment, if not over the entire trajectory. It is common practice to change the friction coefficient when the vehicle passes over a terrain boundary (edge of pavement, for example). This can be neatly accomplished by a key position at that point. On either side of that key point, the friction coefficient is constant (perhaps at two distinct values); there is a step change when the vehicle crosses the key point. Thus, it is not recommended to spline the friction coefficient (and hence the lock fraction, which depends on the friction coefficient).

In summary, the CG position variables X, Y, and Z, and the vehicle heading ψ are splined. The friction coefficient μ and the lock fraction LF are held constant between key frames.

Vehicle Headings and Yaw Rates

Now that the time required to traverse each subdivided trajectory segment is known, and the vehicle headings are known at the ends of each segment, it is a simple matter to calculate the average yaw rate in the segment. Recalling that segment i is bounded by points i–1 and i, we can write

$$\omega_i = \frac{\Delta \psi_i}{\Delta t_i} = \frac{\psi_{i+1} - \psi_i}{\Delta t_i} \tag{4.5}$$

If ψ is in degrees, then ω will be in deg/sec; if ψ is in radians, then ω is in rad/sec.

One word of caution: the heading values must progress in a manner that makes sense as the vehicle traverses its path. For example, a heading change from 3° to 355° could be −8°, or it could be +352°. The computer will calculate the change as the latter; if the former is intended, the headings should be specified as 3° and −5°. Needless to say, the yaw rate calculations will produce bizarre results if headings are specified incorrectly.

Example Reverse Trajectory Calculation

Now, we are ready to put together the various elements of a reverse trajectory calculation: the tire analysis, friction coefficient, and lock fraction calculation from Chapter 2 (see Table 2.2), the determination of crab angle and drag factor from Chapter 2 (see Equation 2.10), the establishment of key positions and a spline curve running between them from Chapter 3 (see Figure 3.5), and the determination of velocity, time values, and yaw rate, as discussed above.

We start with a reverse trajectory reconstruction of the run-out of one of the vehicles involved in a staged crash test. In this instance, the run-out trajectory was divided into three segments, which required four key points: separation from the crash, two intermediate points, and the rest position. The portion of the Excel spreadsheet representing the key points is shown in Figure 4.1.

The cells containing direct user input are in bold type on shaded backgrounds. The other parameters are calculated quantities, as we have seen in Chapter 3.

Key Trajectory Points								
Key Pt #	X feet	Y feet	Z feet	ψ deg	Frict. Coeff.	Lock Frac	Δ Key S feet	Key S feet
1	6.00	-0.20	0.00	1.2				0.000
2	31.60	3.00	0.00	18.9	0.700	0.330	25.799	25.799
3	50.30	18.00	0.00	60.1	0.700	0.330	23.973	49.772
4	42.80	54.50	0.00	137.5	0.650	0.335	37.263	87.034

FIGURE 4.1
Key point and segment information for sample trajectory.

In general, the choice of key points is made for two basic reasons: (1) a certain number and locations of points are needed to provide definition to the spline curve, as we got a brief glimpse of in the previous chapter, or (2) as the vehicle traverses the point, there is a change in the vehicle behavior or in the surface characteristics that influences the drag factor. Where the surface characteristics change, some simulations provide for what is known as a "terrain boundary," which allows those changes to enter the calculations. In this case, there was no such change applicable; the key point locations were chosen on geometric considerations alone. However, to show how to handle a terrain boundary, let us pretend that the friction coefficient was 0.650 in the last part of the trajectory, instead of 0.700, and that key point number 3 was chosen for the purpose of locating that change.

Note that the segment information, such as friction coefficient, is shown at the end of the segment. For example, the first straight line between key points ends at key point 2; therefore, the parameters for the first straight line appear across from Key Point 2. Since there is no prior segment, the cells opposite Key Point 1 are blank.

The result of breaking the trajectory into 19 segments is shown in Figure 4.2. For this case, we have changed the Arc len S values of the two segment ends closest to the Key S values for the intermediate key points (25.799 and 49.722). This was discussed in Chapter 3. For situations where there

Spline Pt #	Arc len S feet	Spline X feet	Spline Y feet	Spline Z feet	Spline ψ deg	Seg Len feet	Cum S feet
1	0.000	6.000	-0.200	0.000	1.200		0.000
2	4.581	10.570	-0.154	0.000	3.181	4.570	4.570
3	9.162	15.135	-0.005	0.000	5.390	4.567	9.137
4	13.742	19.690	0.347	0.000	8.052	4.569	13.706
5	18.323	24.231	1.005	0.000	11.394	4.589	18.295
6	22.904	28.754	2.070	0.000	15.643	4.646	22.941
7	25.799	31.600	3.000	0.000	18.900	2.995	25.935
8	32.065	37.644	5.762	0.000	27.580	6.646	32.581
9	36.646	41.756	8.370	0.000	35.112	4.869	37.450
10	41.227	45.399	11.405	0.000	43.396	4.742	42.192
11	45.808	48.387	14.808	0.000	52.212	4.528	46.720
12	49.772	50.300	18.000	0.000	60.100	3.722	50.442
13	54.969	51.725	22.472	0.000	70.604	4.693	55.135
14	59.550	52.054	26.646	0.000	79.971	4.187	59.322
15	64.131	51.643	31.005	0.000	89.425	4.379	63.700
16	68.711	50.614	35.518	0.000	98.953	4.629	68.330
17	73.292	49.093	40.156	0.000	108.539	4.881	73.210
18	77.873	47.201	44.886	0.000	118.168	5.094	78.305
19	82.454	45.062	49.677	0.000	127.827	5.247	83.552
20	87.034	42.800	54.500	0.000	137.500	5.327	88.879

FIGURE 4.2
Spline results for sample trajectory.

Direc deg	Crab Ang deg	Mu	LockF	DragF	Veloc mph	Slope	Δ Time sec	Yaw Rate deg/sec	Cum time sec
					25.04				0
0.6	1.61	0.700	0.330	0.232	24.40	0.000	0.126	15.72	0.126
1.9	2.42	0.700	0.330	0.233	23.74	0.000	0.129	17.07	0.255
4.4	2.30	0.700	0.330	0.233	23.06	0.000	0.133	19.99	0.389
8.2	1.48	0.700	0.330	0.232	22.35	0.000	0.138	24.26	0.526
13.3	0.26	0.700	0.330	0.231	21.62	0.000	0.144	29.50	0.670
18.1	-0.82	0.700	0.330	0.231	21.14	0.000	0.095	34.10	0.766
24.6	-1.32	0.700	0.330	0.232	20.02	0.000	0.220	39.42	0.986
32.4	-1.04	0.700	0.330	0.231	19.16	0.000	0.169	44.44	1.156
39.8	-0.55	0.700	0.330	0.231	18.28	0.000	0.173	47.96	1.328
48.7	-0.91	0.700	0.330	0.231	17.40	0.000	0.173	50.95	1.501
59.1	-2.92	0.700	0.330	0.234	16.64	0.000	0.149	52.90	1.651
72.3	-6.97	0.650	0.335	0.230	15.63	0.000	0.198	52.96	1.849
85.5	-10.21	0.650	0.335	0.244	14.62	0.000	0.189	49.65	2.038
95.4	-10.69	0.650	0.335	0.246	13.48	0.000	0.212	44.50	2.250
102.8	-8.65	0.650	0.335	0.237	12.20	0.000	0.246	38.75	2.496
108.2	-4.42	0.650	0.335	0.223	10.78	0.000	0.290	33.10	2.785
111.8	1.55	0.650	0.335.	0.219	9.11	0.000	0.349	27.57	3.135
114.1	8.95	0.650	0.335	0.238	6.75	0.000	0.451	21.40	3.586
115.1	17.53	0.650	0.335	0.286	0.00	0.000	1.076	8.99	4.662

FIGURE 4.3
Remaining calculations for sample trajectory.

are no terrain boundaries, adjustment of selected points to match certain key points has no effect on the calculations, other than to provide a "paper trail." In this example, though, the friction coefficient changes at the third key point. Therefore, to maintain the proper spatial location of the friction coefficient change, it is important to reflect that change in the trajectory segments.

To assist in this process, the trajectory portions between key frames were alternately shaded in gray. One can see, in Figure 4.3, that shading was continued into the remaining calculations. In particular, the friction coefficients and lock fractions were copied by value into the cell representing the first segment following each key point. That cell was copied into the remaining cells until the next key point. That way, any changes in the calculations of friction coefficient or lock fraction are automatically reflected in the appropriate cells.

Now the reverse trajectory calculation of this vehicle's run-out may be considered complete. Not only do we have the velocity and yaw rate as the vehicle exited the collision phase, but we have the time history of those quantities during the run-out as well. A table of time versus all the position variables is called a "motion table," which is the technical foundation upon which an animation can be constructed.

Yaw Rates

It is a good idea to perform a reasonableness check on the yaw rates. As has been pointed out by Marquard,[1] yaw rates will tend to decrease more rapidly when the vehicle is longitudinal than when it is sideways. See Chapter 27 for further details. However, the reverse trajectory process does not attempt to calculate yaw torques. It simply takes account of the physical evidence (tire marks on the pavement, for example), which may or may not reveal the presence of Marquard's effect.

Having said that, we can generally expect yaw rates to diminish (not necessarily monotonically) as the vehicle bleeds off kinetic energy (unless, of course, the vehicle encounters curbs or other objects along the way), with the decrease being steeper if the initial yaw rate was high. Results may not conform to that expectation. It is often possible to obtain a more reasonable degradation of yaw rate by making fairly small changes in vehicle heading, within the limits of the physical evidence, at one or more of the vehicle key positions.

Secondary Impacts with Fixed Objects

It is sometimes the case that the reconstructionist needs to consider a secondary impact with a fixed object contacted by the vehicle on its way to rest. Such an event can be approximated and analyzed by inserting one to three small segments in the trajectory at the appropriate point. Probably the lock fraction would be set to unity for those segments. The "friction coefficient" is set to a value to mimic the accelerations experienced there (typically greater than unity). This may be done by estimating the energy dissipated in the secondary impact, and adjusting the accelerations in the segments until the energy dissipated in those segments provides an approximate match. Of course, one would not want to analyze a primary collision this way.

Verifying Methods of Analyzing Post-Crash Trajectories

There are very limited means of validating a method of analyzing post-crash vehicle trajectories. Crashed vehicles are by definition out of control, so handling or braking tests on a skid pad are not representative of a post-collision environment. Unless something unintended happens, the vehicle in a braking or handling test remains under control, so that kind of testing is not fertile ground for seeking validation data. As for crash tests, by definition the focus is on performance during the crash, and often the run-out distances are quite limited. In any case, the documentation of run-out trajectories is generally poor or nonexistent. Yet, the need has existed for decades to validate methods of accident reconstruction.

The RICSAC Crash Tests

In the late 1970s, this need was addressed by a research project funded by the National Highway Traffic Administration, titled "Research Input for Computer Simulation of Automobile Collisions" (RICSAC). The objective of the project was "to further evaluate the validity and accuracy of both the Simulation Model of Automobile Collision (SMAC) and the Calspan Reconstruction of Accident Speeds on the Highway (CRASH) computer programs"[2] (p. 1). To this end, a test matrix of 12 crashes was constructed. The test matrix contained two categories of vehicle sizes, five impact configurations, and two values of ΔV for each configuration. Each crash test involved an intermediate and a subcompact size vehicle, except for one test that involved two intermediate vehicles. In the front-to-rear collisions, the car struck in the rear was stopped; in all other tests, both cars were moving. The impact configurations are shown in Figure 4.4. Impact speeds were chosen so that the ΔV values for the struck vehicle were approximately 15 and 30 mph for the two exposures in each collision configuration.

Quantities measured typically included tri-axial accelerations (x, y, z) in the left front corner, left rear corner, right rear corner, rear deck, and firewall. Longitudinal accelerations were measured on some bumpers, and lateral accelerations on some doors. Also measured were pitch, roll, and yaw angles, and yaw angle rate, plus steer angle, and wheel angular velocity. A device was also designed and installed for the purpose of recording the vehicle's trajectory on the paved surface. Instrumented dummies, some restrained and some unrestrained, were also placed in the vehicles. Standard accident investigation protocol, including the documentation of vehicle damage and tire condition post-test, was employed. A report for each of the first five tests is contained in Volume II; Tests 6 through 12 are described in Volume III.[3] A discussion of the CRASH and SMAC reconstruction efforts is contained in Volume IV.

In short, these tests were designed with accident reconstruction techniques foremost in mind, which cannot be said for the motivation behind most tests. In particular, these tests afford a unique opportunity to test the validity of post-crash trajectory analysis methods, subject to this question: Can the crash separation conditions and the subsequent run-out be ascertained from the information that is available to us?

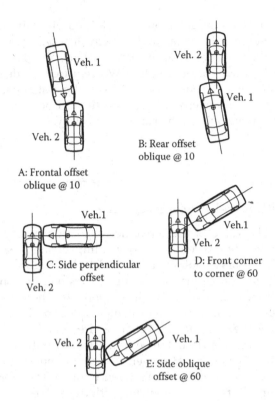

FIGURE 4.4
Impact configurations used in the RICSAC program.

Documenting the Run-Out Motions

When the vehicle comes to rest post-test, its translational velocity components and rotational velocity, or yaw rate, are known to be zero. Its position
and heading angle are also known, because they can be readily measured
and recorded, which was done in the RICSAC tests.

As in an actual crash, physical evidence in the form of tire marks, and
possibly gouges, is created on the pavement. In the RICSAC tests, such evidence was recorded in the form of lines on a scale drawing of the crash site.
Furthermore, the pavement was marked with colored water, showing the
paths of two points on the vehicle. These marks are also shown on the scale
drawings. Thus, every attempt was made to document the run-out path of
certain points on the vehicle. On the drawing, a vehicle schematic at the same
scale can be positioned on the marks at key points, thus allowing the vehicle
heading and CG location to be ascertained at those positions. However, there
is no way to identify the specific vehicle CG position at separation (certainly
an important key point), based on the physical evidence.

From there, things only get more complicated. Translational velocities were not, and could not be, measured directly. It was, and is, necessary to derive the velocities by analyzing the accelerations. Yet, the primary objective of the RICSAC tests—measuring the ΔV—required the determination of the velocity components at separation, as well as those at impact. These same separation velocity components, along with the CG position at separation, are also vital to understanding the run-out trajectory.

Data Acquisition and Processing Issues

Unfortunately, "the value of the separation velocity was contaminated by the effect of rotation of the vehicles between impact and separation,"[4] (p. 3) according to the authors of Volume IV of the RICSAC report.

The source of the contamination can be understood by considering that the separation velocity is a vector, with components. These components are obtained by integrating the acceleration components. Specifically, the (global) X component of acceleration does not change direction (by definition), so the X component of velocity can be obtained from it by integration; the same thing can be said for the Y direction. However, the accelerometers are mounted on the vehicle—a local (non-Newtonian) reference frame. Therefore, the vehicle acceleration components will not coincide with the global components if the vehicle is rotating; rather, the accelerometers will record some (unknown) linear combination of the global X and Y accelerations. Integration of the vehicle acceleration traces will produce something that looks like velocities, but the results are bound to be questionable if significant rotations are involved. The same remarks apply to displacement calculations.

The way to deal with this problem is to accurately measure the heading angle (yaw) at each time step of the integration, and resolve the accelerations into the appropriate components before they are integrated. However, in the RICSAC tests, the yaw angle was recorded graphically on a paper chart over a period exceeding 2 sec, which is far too coarse for analyzing a crash of 100 msec or so.

Another issue was related to the accelerometer locations. Primary reconstruction results deal with motions of the vehicle CG. However, the CG is not guaranteed to be located on stout structure, or even on structure at all. If there is an attempt to measure CG accelerations, test engineers seek to locate a triaxial accelerometer pack as close to the CG as possible, but they are constrained by the need to attach accelerometers directly to sturdy structure. Also, the location should be outside the crush zone, as discussed in Chapter 1. Often, a compromise is required. The RICSAC tests were no exception; accelerometer locations away from the CG were typically used, as can be seen in the description of the instrumentation employed.

In addition to the translational and rotational accelerations experienced at the CG, other locations in a rotating body will also experience a radially directed component of acceleration—the centripetal acceleration—along with the tangential acceleration. The magnitude of these effects will depend on the location on the vehicle. To the accelerometer, the effects will look just like—and be sensed as—linear accelerations. Therefore, to determine the CG accelerations from off-CG data, these additional effects will have to be calculated, and subtracted out of the data trace. Evidently, the data reduction software in use at Calspan at the time did not have this capability. In any case, the difficulties were compounded when yaw angle data were lost for both vehicles in Test 4 and for Car 2 in Test 5.

Owing to such data processing issues, the usefulness of the data has been called into question.[5-9] However, corrected data would be valuable, and efforts to resolve the issues have been made.[10,11] In particular, the McHenry study of 1997 and the Brach study of 2001 employed entirely different correction methods, while arriving at slightly different results for ΔV. They also obtained differing values for the X and Y components of separation velocity. Only the magnitudes of the resultant separation velocities (separation speeds) were compared between studies, however, in order to simplify the comparison (see Figure 4.5).

The dashed diagonal line represents exact agreement; the solid line is a regression line. In the square next to each data point, the number to the left of the dash is the test number; the number to the right is the vehicle number. While the two data interpretations are different, the overall correlation

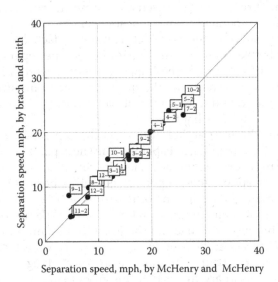

FIGURE 4.5
Comparison of separation speeds obtained by Brach and McHenry.

between the results was about 96%, so one may conclude that at last the (adjusted) RICSAC results are reliable for validation purposes. Only the usual—and inevitable—measurement uncertainties associated with any experiments, particularly experiments as complicated as these, remain.

Separation Positions for the RICSAC Run-Out Trajectories

To use the adjusted RICSAC results for validation, one needs to know where the vehicles were at separation (X_1, Y_1, and ψ_1) and where they came to rest (X_R, Y_R, and ψ_R). These define the end points of the trajectory. Translational velocity components and yaw rates during the run-out are not available. However, the Calspan personnel did make scale drawings of trajectory evidence (such as tire scuffs and trajectory paint trails left by the test vehicles). The copy quality in the reports is not good. However, vehicle schematics can be positioned on the representations of the evidence, producing some approximation of intermediate CG locations and vehicle headings. An example will be presented later.

The collision configuration was quantified by reporting the coordinates of the vehicle CGs at impact. It is noteworthy that this information was *not* used for the run-out trajectories. After all, during the collision phase, there are collision forces acting on the vehicles, and it is our purpose to study the motions of the vehicles post-crash, in the absence of those very forces.

(It is a theme throughout our discussion of reconstruction methods that the crash phase is separated from the post-crash phase. Many authors lump them together when considering vehicle run-out. In fact, there is finite vehicle movement between impact and separation, so the separation positions are necessarily different from the impact positions. Moreover, the applied forces during the crash phase and the run-outs come from different sources, and they are of a quite different magnitude, except for low-speed crashes. Refer to the discussion in Chapter 1.)

Unfortunately, the only authors to report the separation positions were McHenry and McHenry, in SAE Paper 970961. The separation positions were no doubt inferred from the impact velocities, the ΔVs, and the crash duration. With one exception, their reported separation positions were used in the present numerical descriptions of the run-out trajectories. It should be noted, though, that their impact durations were markedly different from those utilized by Jones and Baum. (The Jones and Baum durations ranged from 200 to 275 msec, in 25 msec increments.) According to the McHenry paper, the Jones and Baum durations were much longer than typically seen in crash tests—an argument that this author supports. Obviously, the assumed impact duration has an effect on where the vehicles are calculated to be at the time of separation.

In any case, the McHenry and McHenry separation positions were relied upon, except for Test 5. In that case, the McHenry X positions for both vehicles show opposite directions of motion between the crash phase and the run-out phase. This is nonsensical. The two separation X values were estimated by analyzing the ΔVs and the (known) impact velocities. The X value at rest for vehicle 1 was also recalculated, since the reported value was inconsistent with the trajectory drawing.

Side Slap Impacts

Two other tests were noteworthy. Tests 9 and 10 involved 90° side impacts, where the struck vehicle (the "target" vehicle, in test parlance) was impacted well away from the mass center, in the area of the front fender. These "L" collisions often produce a "side-slap," wherein the two vehicles rotate together and contact for a second time. When an investigation indicates an "L" collision, the sides of the vehicles should be examined carefully for evidence of that second contact. A side-slap makes the reconstructionist's life a lot more difficult, not only because two impacts and two separations are involved. It is also the case that the entire struck vehicle may be bent laterally in a mode known to civil engineers as "side-sway," which tends to affect the way the vehicles come together for the second time. No simulation model represents this behavior, so if one uses such an approach, one has to be prepared to accept (and explain) the discrepancies between the physical evidence and the model's simulation of the side-slap.

In any case, Test 10 did produce a side-slap, which is apparent in studying the post-test photos and acceleration time histories, and is noted by McHenry and McHenry. The side-sway deformation in the target vehicle (Car No. 1) is also seen in the investigator's notes and in the post-crash photographs, in addition to the contact damage on the sides of the vehicles.

Secondary Impacts and Controlled Rest

From our point of view, however, the most problematical aspect of Test 10 is the fact that the trajectory of vehicle 2, a Ford Torino, "extended beyond the paved surface, onto a gravel shoulder, and finally a grass field. Near the point at which the Torino exited the pavement (approximately 100 feet from the point of impact), it contacted a metal light pole transformer base. After this impact, the Torino's left front wheel mounted the transformer base, causing the vehicle to pitch and roll. Shortly thereafter, the instrumentation cable reached full extension and the Torino's motion was halted [probably by using the abort system]." This secondary impact is not atypical of what happens in the real world. However, there is no documentation of it beyond the above description. Moreover, the vehicle came to a "controlled" rest, which is investigator parlance to indicate that driver-type inputs were employed. Knowing nothing about how the vehicle was stopped, it was decided not to attempt to reconstruct this vehicle's run-out trajectory.

Test No. 9 was also interesting (in the sense of the Chinese blessing about living in interesting times). No side-slap is noted per se, and the reproductions of the post-test photographs are typically poor for that time period, so much so that a definitive statement cannot be made. However, the investigator's depiction of the damage to the side of Car 2, another Ford Torino, is suspicious. Most tellingly, the trajectory of Car 1, a 1975 Honda Civic, makes about a 90° turn before exiting the collision, as is clearly seen in the test site diagram and in the post-test photos. Moreover, the Civic experienced the highest yaw rate of any vehicle in the program—about 255°/sec. The peak value occurred just beyond 100 msec, but continued for about another 600 msec. Certainly, if there were data processing errors due to the effects of rotation, they would show up in this test. The run-out trajectories of both vehicles were reconstructed, but with misgivings.

Test No. 2 is missing the x-direction accelerometer trace for the firewall of Car 1, so Brach and Smith performed no analysis of that test. The missing data seems not to have bothered the McHenrys, who report results for Test 2. Except for relying on the separation positions reported by McHenry and McHenry for Car 2 in that test, the present analysis makes no reference to the accelerometer traces, so the trajectories for both vehicles were reconstructed. However, the post-test photographs appear to show the vehicles in contact with each other at rest, so the reconstruction could be in some error if a second impact occurred but was not accounted for (which it was not).

Post-test photographs for Test No. 3 also suggest that the vehicles may have been touching at rest. However, the contact is much narrower than it was on Test No. 2, causing less concern about the reconstruction.

Surface Friction

Each RICSAC test report has what appears to be a standard statement in the narrative "the roadway was dry with skid resistance value of 87." This would imply a friction coefficient of 0.87. There is no evidence that the friction coefficient was actually measured, however. While a friction coefficient measurement taken at the accident site should be used in the analysis if the measurement appears to have been properly conducted, in the absence of such measurements a value of 0.7 is often used for "dry pavement." This is usually justified by referring to the Traffic Accident Investigation Manual by J. Stannard Baker,[12] who quotes a range of friction coefficient values for Asphalt or Tar, traveled, and Portland Cement (traveled) of 0.60–0.80. In his entire career, this author has never used a value as high as 0.87 for tire-pavement friction. Other authors have commented on the inadvisability of blindly relying on skid number for accident reconstructions.[13] However, 0.87 is the friction coefficient used in both the CRASH and SMAC runs reported in Jones and Baum. In our reconstructions, a friction of 0.7 was used, in conformity with this author's usual practice.

Brake Factors

The test reports contain the weight measured at each wheel, so those numbers were entered directly into lock fraction calculations. Refer to the discussion in Chapter 2. The post-test investigator observations for the vehicles include a code for whether the wheels were bound up or not. As discussed in Chapter 2, such observations can be tricky, so post-test photos were also examined for evidence of binding in the wreckage, or deflation (which is often part and parcel of being bound up in the wreckage). No evidence of deflation was seen, although again the published copies of the photographs did not provide absolute certainty. The coding in the test reports for being bound up was applied to the lock fraction calculations, using in all cases a value of 1.00 (100%, or fully locked), which may well be an overestimate. One exception to the coding in the reports occurred with Test 9. The report shows a wheel bound up for each vehicle, but when Jones and Baum analyzed that test by reconstructing it with SMAC, they used braking percentages that, though arbitrary, indicated no wheel binding for either vehicle. Again, the published post-test photograph copies were inconclusive. Since Jones and Baum were at Calspan at the time and may have seen the vehicles, no binding was applied to our reconstruction, either. Indeed, the results suggest strongly that no wheels were bound up.

In Jones and Baum, the brake factors used in the CRASH runs seem arbitrarily chosen. In SMAC, actual brake forces are specified at each wheel. The SMAC brake forces in Jones and Baum do not match the brake factors used in CRASH. In the present reconstructions, for an automatic transmission vehicle, the individual brake factor for a nondriven wheel was 0.022; the brake factor for a driven wheel was 0.087; and the additional brake factor for a deflated tire was 0.108, as derived from Cliff and Bowler.[14] As can be seen from referring to Chapter 2, these values are on the high side, chosen because the RICSAC vehicle vintages were 1974 and 1975, whereas Cliff and Bowler tested vehicles of the 1990s (after many efficiency improvements had been incorporated for better fuel economy). For manual transmission vehicles, the brake factors for individual nondriven and driven tires were 0.025 and 0.059, respectively; the additional brake factor for a deflated tire was 0.105. The values reflect the report narrative indicating the vehicles were placed in gear or in drive, with the engines off. However, the actual gear is not noted.

Sample Validation Run

An overview of the trajectory reconstruction procedure may be gained by considering one of the RICSAC tests in which the curvature of both vehicle trajectories produced numerous scuff marks and necessitated the use of intermediate positions in order to get a good mathematical representation of the paths. Test No. 4 involved a 1974 Ford Torino impacted in an offset manner into the rear of a 1974 Ford Pinto, stopped at an oblique angle. Positions at impact and rest were shown on the original drawing (which was done to scale), but the coordinates of those positions were taken from McHenry and McHenry. As mentioned previously, McHenry and McHenry also reported the separation positions, of which this author added drawings. Figure 4.6 is an annotated diagram of the run-out trajectories shown in Jones and Baum.

It is noteworthy that the test report authors used SAE coordinates, as is so often the case in automotive publications. As discussed in Chapter 1, SAE coordinates were retained for our reconstruction.

A sight familiar to most traffic accident investigators is the north arrow that appears in the lower left corner. This is an indication that the original drawing was prepared using a template sold by the Traffic Institute (now the Institute for Public Safety) of Northwestern University. That template has cut-outs for vehicles of various sizes in two scales: 1 in. equals 10 ft, and 1 in. equals 20 ft, which happens to be the scale of the original drawing. Thus, the Northwestern template could be (and was) used to quickly and easily add to the drawing some representations of the vehicles in intermediate positions. Again, refer to Figure 4.6.

It has been this author's practice for many years to create scale drawings of the damaged vehicles by reducing full-size tracings of the actual accident vehicles to scale, showing the damaged shape and the positions of the wheels. (Also, a 1-in.-to-5-ft scale is used so as to more easily see the detail.) This was not done here because no such full-size vehicle maps were available. However, vehicle drawings made using the traffic template were used so as to illustrate the procedure.

Aside from representing the path curvature, another purpose of intermediate vehicle positions is to get a good representation of the path direction taken as the vehicles separated (since the rest position is not a good indicator of that initial path direction if the path was curved). After all, the separation velocity vector is tangent to the path taken by the CG as it emerges from the crash, and is thus approximately parallel to the first trajectory segment.

Another use of the intermediate positions is to obtain the progression of vehicle heading (yaw), in cases where there is significant yaw change between separation and rest. That was the case with both vehicles in RICSAC Test No. 4.

In Figure 2.11 of the published report, the drawing scale is not in exact inches, presumably due to distortions during the reproduction process. Therefore, distance measurements were made by using a Gerber Variable

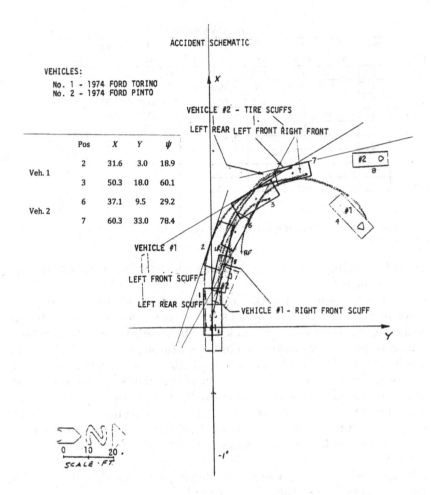

ACCIDENT SCHEMATIC

VEHICLES:
No. 1 - 1974 FORD TORINO
No. 2 - 1974 FORD PINTO

VEHICLE #2 - TIRE SCUFFS
LEFT REAR LEFT FRONT RIGHT FRONT

	Pos	X	Y	ψ
Veh. 1	2	31.6	3.0	18.9
	3	50.3	18.0	60.1
Veh. 2	6	37.1	9.5	29.2
	7	60.3	33.0	78.4

VEHICLE #1

LEFT FRONT SCUFF

LEFT REAR SCUFF

VEHICLE #1 - RIGHT FRONT SCUFF

0 10 20
SCALE · FT.

FIGURE 4.6
Annotated diagram of RICSAC test no. 4.

Scale, Model TP007100B, manufactured by the Gerber Scientific Instrument Company, Hartford, Connecticut. In this device, the scale graduations are attached to a spring, which can be adjusted so that the graduations fit a known distance represented on the drawing. The use of a variable scale is highly recommended to facilitate taking measurements on hardcopy prints of graphs and scale drawings, since it accounts for any scale distortions created when the drawing was reproduced.

Angle measurements were made with the protractor that is part of the Northwestern University traffic template.

The intermediate positions are not intended to represent the vehicles at any particular point in time, except for the separation positions. Note that at separation, the drawings, which represent the undamaged vehicles, overlap (because of the vehicle damage that has taken place). Otherwise, overlap of

Key Trajectory Points								
Key Pt #	X feet	Y feet	Z feet	ψ deg	Frict. Coeff.	Lock Frac	Δ Key S feet	Key S feet
1	6.00	−0.20	0.00	1.2				0.000
2	31.60	3.00	0.00	18.9	0.700	0.330	25.799	25.799
3	50.30	18.00	0.00	60.1	0.700	0.330	23.973	49.772
4	42.80	54.50	0.00	137.5	0.700	0.335	37.263	87.034

FIGURE 4.7
Key point and segment information for RICSAC 4, vehicle 1.

the vehicle drawings is of no significance, because the positions were reached at various time values.

It is this author's usual practice to number a vehicle's impact position as 0 (zero), and its separation position as 1, with other points being numbered sequentially. An exception was made here because two vehicle trajectories are shown; vehicle 1 has trajectory positions 1 through 4, and vehicle 2 has 5 through 8. Positions 2 and 5 were selected because they were the first places that vehicle drawings could be set on their tire marks so that yaw angles could be measured. Positions 3 and 6 were the last places that vehicles could be placed on their right front tire marks, again allowing the determination of yaw. Notice that the left front tire mark of vehicle 1 is outside the car drawing at rest. This indicates some noticeable crab angle toward the end of its trajectory.

Vehicle coordinates for the intermediate points are noted on the drawing. Figure 4.7 shows the key values for Vehicle 1.

The alert reader will notice that Figure 4.7 is identical to Figure 4.1, except that between key points 3 and 4, the friction coefficient is 0.700 instead of 0.650, and the lock fraction is changed accordingly. Refer to Chapter 2, and the discussion surrounding Figure 2.2.

As an exercise, the reader is encouraged to perform the calculations with the last friction coefficient set to 0.700 instead of 0.650, and verify that the separation velocity changes from 25.04 to 25.39 mph.

Application of the reverse trajectory analysis to the two vehicles in RICSAC Test 4 leads to separation velocity reconstructions that are compared in Table 4.1 to the interpretations that have been provided by McHenry and McHenry, and Brach and Smith. Note that the word "interpretations" has been used instead of "observations," because the speeds reported in the above papers were not directly observed, but calculated using two different procedures.

For the present reconstruction, the separation velocities are taken to be parallel to the first trajectory segment in the run-outs. Therefore, the relative

X and Y components are purely a function of the initial path taken by the vehicle after separation. It is not known why the McHenry Y component for vehicle 2 is in the opposite direction, as that would imply noticeable differences in the run-out trajectory being analyzed. The present reconstruction agrees with the Brach interpretation on that score. The X component and the resultant velocity in the present reconstruction for vehicle 2 are slightly lower than those of McHenry and Brach.

For vehicle 1, the separation velocities are almost parallel to the X-axis, and thus the Y components are within 0.2 mph. On the other hand, the X component of the reconstructed separation velocity is higher than the two interpretations by more than 5 mph. So while the initial trajectory angles agree closely, perhaps the subsequent yaw angles lead to larger calculated crab angles (and higher drag factors). Perhaps uncertainties in the test interpretations by McHenry and Brach are at work. It is probably safe to say that no one will ever know for sure.

Results of Reverse Trajectory Validation

The lock fraction inputs used for the 23 vehicles analyzed in the 12 tests are shown in Table 4.2.

A brief summary of the reconstruction results is presented in Table 4.3. A comparison with the separation speeds reported in McHenry and McHenry is shown in Figure 4.8, in which a point on the diagonal line means perfect agreement between test and calculation. Obviously, there is scatter in the results. Remember also that there is also scatter in the comparison of observed separation speeds by Brach and Smith vs. those of McHenry and McHenry. One will notice, however, that in both comparisons, car 1 in test 9 appears to be an outlier. This is not surprising, given the considerations discussed above. This car's trajectory produced some disagreement between the McHenry and the Brach interpretations (Figure 4.5), so it is seen that there is uncertainty still present in the test results, despite major efforts to clean up the data.

It is important to remember that the validation discussed here involves the separation speeds only—not impact speeds or ΔVs. It is not known how other trajectory reconstructions would compare, because validations of separation speeds reconstructed using other methods have not been published. All things considered, the reverse trajectory reconstruction technique has been validated at least as well as any other method, and has the great advantage of working in the reverse direction—backwards in time from the knowns (rest positions) to the unknowns (separation speeds)—without the necessity of iterating through successive guesses until a satisfactory answer is found.

TABLE 4.1

Test 4 Separation Velocities in mph: Reconstruction Results vs. Test Interpretations

	Present Reconstruction			McHenry and McHenry Interpretation			Brach and Smith Interpretation		
	X-Direction	Y-Direction	Resultant	X-Direction	Y-Direction	Resultant	X-Direction	Y-Direction	Resultant
Vehicle 1	25.39	0.26	25.39	19.80	0.20	19.80	20.10	0.30	20.10
Vehicle 2	20.41	4.79	20.97	21.90	-3.20	22.13	20.93	5.37	21.61

TABLE 4.2

Lock Fractions Used in the RICSAC Reconstructions

Car No.	Test 1	Test 2	Test 3	Test 4	Test 5	Test 6	Test 7	Test 8	Test 9	Test 10	Test 11	Test 12
1	0.072	0.344	0.075	0.330	0.075	0.074	0.074	0.059	0.064	0.064	0.329	0.332
2	0.077	0.291	0.068	0.226	0.475	0.291	0.241	0.281	0.074	0.075	0.346	0.342

TABLE 4.3

Reconstructed RICSAC Separation Speeds, mph

Car No.	Test 1	Test 2	Test 3	Test 4	Test 5	Test 6	Test 7	Test 8	Test 9	Test 10	Test 11	Test 12
1	4.2	13.4	14.5	25.4	20.7	9.1	11.5	10.0	10.1	8.3	9.0	10.4
2	12.2	21.1	16.3	21.0	27.5	19.5	21.1	16.9	15.5	–	6.7	7.9

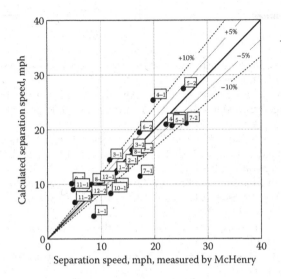

FIGURE 4.8
Comparison of separation speeds: reconstructed vs. interpreted.

In addition, the reverse trajectory technique allows the reconstruction to focus on the crash itself, where the injuries are usually produced. By contrast, time-forward simulations can cause the reconstructionist to spend much effort on such issues as steer angles and cornering coefficients in an attempt to get a time-forward trajectory simulation to produce outputs that comprise a reasonable match with the known facts. For example, consider that Jones and Baum were speaking of SMAC simulations when they said: "This means that even if the cornering stiffnesses and torques on one side of the vehicle are set to zero to simulate the airborne wheels, the forces on the two remaining wheels will not necessarily be realistic." Setting the cornering coefficient to zero is a mis-use of that parameter, which was defined for small slip angles (see Chapter 2); a tire experiencing large slip angles is outside the scope of such a tire model. Therefore, cornering stiffness is clearly not a knob designed for tweaking results where significant crab angles are involved. In this author's experience, such tweaks often occur. Cornering stiffness is an example of an interesting topic that usually has little to do with the issue that brought the accident to the attention of the reconstructionist in the first place.

One exception to the above has to do with driver steering inputs. The reverse trajectory procedure was not designed to reveal what those inputs were. Thus, while brake demand can be entered, there is no provision for steering inputs in the analysis. If there is a question of how the driver steered (usually pre-crash), a different procedure (probably a time-forward simulation) will have to be employed if an insight is to be gained.

References

1. Marquard, E., Progress in the calculations of vehicle collisions, *Automobiltechnische Zeitschrift*, 68(3), 1966.
2. Shoemaker, N.E., *Research Input for Computer Simulation of Automobile Collisions*, Vol. II, Staged Collisions, Tests No. 1 through No. 5, Report DOT-HS-805 038, Calspan Corporation, 1978.
3. Shoemaker, N.E., *Research Input for Computer Simulation of Automobile Collisions*, Vol. III, Staged Collisions, Tests No. 6 through No. 12, Report DOT-HS-805 039, Calspan Corporation, 1978.
4. Jones, I.S. and Baum, A.S., *Research Input for Computer Simulation of Automobile Collisions*, Vol. IV, Staged Collision Reconstructions, Report DOT-HS-805 040, Calspan Corporation, 1978.
5. Brach, R.M., Energy loss in vehicle collisions, SAE Paper 871993, SAE International, 1987.
6. Woolley, R.L., The 'IMPAC' program for collision analysis, SAE Paper 870046, SAE International, 1987.
7. Day, T.D. and Hargens, R.L., Further validation of EDSMAC using the RICSAC staged collisions, SAE Paper 900102, SAE International, 1990.
8. Cliff, W.E. and Montgomery, D.T., Validation of PC-crash—A momentum-based accident reconstruction program, SAE Paper 960885, SAE International, 1996.
9. Bundorf, R.T., Analysis and calculation of delta-V from crash test data, SAE Paper 960899, SAE International, 1996.
10. McHenry, B.G. and McHenry, R.R., RICSAC-97—A reevaluation of the reference set of full scale crash tests, SAE Paper 970961, SAE International, 1997.
11. Brach, R.M. and Smith, R.A., Re-analysis of the RICSAC car crash accelerometer data, SAE Paper 2001-01-1305, SAE International, 2001.
12. Baker, J.S., *Traffic Accident Investigation Manual*, Northwestern University, Evanston, IL, 1975.
13. Wambold, J.C. and Kulakowski, B.T., Limitations of using skid number in accident analysis and pavement management, Transportation Research Board, 1991.
14. Cliff, W.E. and Bowler, J.J., The measured rolling resistance of vehicles for accident reconstruction, SAE Paper 980368, SAE International, 1998.

5

Time–Distance Studies

Purpose

A time–distance study is the determination of the times at which one or more objects are positioned at one or more locations or times at which various events occur. A time–distance study is often not a part of the reconstructionist's assignment, but when such a study is requested, there is usually a question to be answered. Examples include:

The driver of which vehicle had the last opportunity to avoid the crash?

What are the effects of earlier action (often accident avoidance) by the driver?

What if the driver had been going the speed limit instead of the reconstructed speed?

Once the hazard came into view, did the driver have time to avoid the crash?

Which of the conflicting testimonies is in accordance with the reconstruction?

Could the vehicle actually accelerate fast enough to reach the reconstructed speed from a standing stop?

At what time did a line of sight exist between the vehicles? (Note that the question was NOT when the drivers could see each other. The reconstructionist is well advised to stay away from such questions as what a person saw, what they thought, etc.)

Each object in a time–distance study has its own time line, and the study is often based on event, such as a collision, that ties the time lines together. The time–distance study may examine the reasonableness of various assumptions that are being made. (E.g., if the vehicles are going at speed 1 and speed 2, they must be a certain distance apart at T seconds before impact. Would there be a line of sight available at that time, consistent with what one witness said?)

Or a more complicated question could be asked. Given that there was a collision, the cars could be backed up along their trajectories to a certain time *T* before impact. Then the parameters of the motion could be changed, and the clock started in the forward direction again. Would the collision still happen? Would it be at a lower speed? Would the accident mode change? Would the bullet car become the target car, and vice versa? Would the vehicles "whiff" altogether?

As can be seen from the above examples, the questions often have a "what if" nature. There may be several postulates that a client wants explored. Often the questions are asked late in the reconstructionist's work up—just before deposition or trial, say, in response to what the opposition is saying. Many times they are not the central questions (e.g., in a products liability case), so the client may not want to expend much time and effort on them. It behooves the reconstructionist to work them out on a computer spreadsheet, in the hopes that simply tweaking one or a few parameters will provide a quick answer to "one last question."

Perception and Reaction

Many times the question involves driver actions: how quickly they can perceive and react, how quickly they can apply the brakes, how quickly they can turn the steering wheel (steering degrees per second), how many degrees they can turn the steering wheel through in a panic situation, how far down the road they can see at night, and so on. These are all topics that come under the rubric of "human factors." Often a human factors expert is called in to address such issues. The reconstructionist may need to rely on the opinions of such a person, or at least rely on research findings that have been cited in the open literature. As with all topics, the reconstructionist is well advised to be ready for questions as to where the numbers came from.

Consider what happens when a driver operates a vehicle. The driver is a control system that accepts certain inputs, called stimuli. Each stimulus is entered into the system through the senses. Some stimuli are normal and non-threatening, such as the passing scenery, or a girl in a bright-red dress. Others may require further evaluation to see if a hazard is developing; still others may require immediate action. All of this has to be sorted out and prioritized. Either way, the driver may or may not notice the stimulus until some time after it occurs, if at all. The time it takes to notice the stimulus is called "perception." Perception time can depend on a multitude of factors, such as the strength of the stimulus, or whether it is in the driver's field of view, for example. Other factors involve the particulars of the driver: drowsiness, distraction, sleepiness, age, health, mental stress, driving experience, alcohol, drugs, and so forth. The list goes on and on, and makes for a wide variation among drivers.

If emergency action is required, then the brain has to decide which action to take. Brake? Accelerate? How hard? Steer left? Steer right? Honk the horn? Grab at a child in the right front seat? Drop the cell phone? Do nothing? In addition to the factors mentioned above, the decision can depend on how unusual or unexpected the stimulus is, and how many responses are under consideration. This all takes time—decision-making time.

Once the decision is made, further time is required to take action, such as moving the foot from the accelerator pedal to the brake pedal and applying brake force, or turning the steering wheel. Then the vehicle has some response time of its own, depending on what kind of vehicle it is and how well it has been maintained. These human and vehicle response times are often lumped together and combined with the decision-making time, under the heading of "reaction."

Obtaining a representative measure of perception–reaction time is fraught with difficulty. Can a person in a simulator, knowing that he or she is doing nothing other than participating in some kind of experiment, be expected to behave the same when doing routine driving in a car? Can the driver of a car on a test track be expected to respond the same to an artificial hazard (such as a cardboard box being thrown into his path) as he would to a child darting into the street? The interested reader is referred to Sens et al.[1] and Olson and Farber.[2]

Since it is generally not known what the driver's perception and reaction times were in a particular situation, it is common practice to assume 3/4 sec for perception and another 3/4 sec for reaction, for a combined total of 1½ sec, while recognizing that variations are not only possible, but likely. Therefore, what–if questions may focus on the effects of varying the perception–reaction time from 1/2 to 2½ sec, say.

Constant Acceleration

The simplest kind of motion is at constant velocity, for which "rest" is simply a special case. Most often, though, time–distance studies involve vehicles that are being accelerated. The type of acceleration that is simplest to analyze is a constant, for which constant velocity (zero acceleration) is a special case. If a trajectory can be divided into segments with constant acceleration (as has been done in Chapters 3 and 4), the resulting analysis will cover the vast majority of the questions asked of the reconstructionist. Generally, it is not necessary to make the segments small, but of course the velocity and the travel distance need to be continuous functions over the entire path.

It is possible to write a general-purpose computer program in which different formulas are used in various segments, depending on what is known and what is unknown in any given segment. One can even do a forward-directed calculation or a backward-directed calculation, depending on whether the velocity at the end is known (a final-value problem), or the velocity at the beginning is known (an initial-value problem). Obviously, a lot of IF–THEN–ELSE constructs must be employed. In all cases, the object is to wind up knowing the velocity at each of the segment boundaries, otherwise known as the key points. In addition, the user should have both the elapsed time (which is a forward measure), and the time remaining (which is a backward measure). Often the final event is a collision, though it could also be a vehicle coming to rest or even something else. Similarly, the user should know or be able to calculate the elapsed distance and the distance remaining. In each segment, the user should know or be able to calculate ΔS, the segment length, and Δt, the segment time.

The time–distance study can (and usually does) involve more than one vehicle. The final event may be a collision between them, which is the stake in the ground, so to speak, that ties the two time lines together. Obviously, each vehicle has its own trajectory, with its own key points and segments.

Each trajectory can be plotted on a scale drawing of a scene. Key points then become associated with X and Y values, and each segment is represented as a smooth curve between (X,Y) coordinate pairs that represent the end points. Of course, the coordinates are not always known at the outset. Sometimes, it is the time that is known; then the spatial coordinates must be calculated.

The vehicle does not necessarily have to be moving in a straight line. Very often, at least one of the vehicles is in a turning maneuver. In that case, it is advantageous to use an arc-length coordinate to represent the position of the vehicle, which then allows the trajectory to be "flattened out" to a straight line for analysis purposes. If lateral forces are well within adhesion limits, the motion along that line is computed using the same formulas as for a straight line.

If the path has complicated geometry, the methods discussed in Chapters 3 and 4 may be used to analyze its motion (although the specification of time values will cause some complications). Usually, however, the trajectory is not known with that kind of exactitude. The vehicle is often not in an emergency maneuver (so as to leave tire marks), and a loss of vehicle control, by definition, leaves the driver unable to alter the outcome. It is usually sufficient to use a circular arc to represent the path traveled by a turning vehicle. It is then convenient to use an azimuth angle, in combination with a turn radius, to describe the vehicle's position. The calculation can be folded in directly to the time–distance study.

Unfortunately, no two studies are alike. The questions are unique, and so are the accidents being analyzed. It is not possible to build a template for the computations, but on the other hand a spreadsheet can easily be manipulated by hand. For example, it takes about the same amount of effort to set up a set of forward-directed calculations as it does a reverse-directed one. Instead of employing computer programming logic to account for a variety of parameters being unknown, the user can perform the logic himself or herself, and choose the appropriate formulas to be put in any given spreadsheet cell.

In the following, two adjacent key points are denoted by i and $i–1$. The segment between them has an arc length of ΔS, and a drag factor f. The time required to traverse the segment is Δt. At point i, the elapsed distance is S_i, the elapsed time is t_i, and the velocity is V_i. The deceleration (negative acceleration) in the segment is fg, where g is the acceleration of gravity. Quantities applying to the segment as a whole will not be subscripted to avoid confusion, but they will generally vary from one segment to another. We shall assume the calculations will be done in reverse, meaning that by the time we get to a particular segment, the quantities subscripted with $i + 1$ are already calculated, if they were not known in the first place.

Suppose the segment has a known time duration Δt (as would be the case for an assumed perception–reaction time). Then, we have

$$t_i = t_{i+1} - \Delta t$$

$$\Delta V = -fg\Delta t$$

$$V_i = V_{i+1} - \Delta V$$

$$V_{avg} = (V_i + V_{i+1})/2 \tag{5.1}$$

$$\Delta S = V_{avg}\Delta t$$

$$S_i = S_{i+1} - \Delta S$$

Of course, some of these intermediate variables can be eliminated in certain cells.

On the other hand, suppose that S_i is known, as it would be if it pertains to a certain scene feature. Since S_{i+1} is already known from the previous segment, we can write

$$\Delta S = S_{i+1} - S_i$$

$$V_i = \sqrt{V_{i+1}^2 + 2fg\Delta S}$$

$$\Delta V = V_{i+1} - V_i$$

$$V_{avg} = (V_i + V_{i+1})/2 \tag{5.2}$$

$$\Delta t = \Delta S/V_{avg}$$

$$t_i = t_{i+1} - \Delta t$$

Again, some of these intermediate variables can be eliminated in certain cells: ΔV and V_{avg}, in particular.

If forward-directed calculations were being done, one would develop equations in an analogous manner, except that in a given segment the quantities subscripted with i would already be known, and one would solve the equations for those values with subscripts $i + 1$.

Suppose each key point is set up as a rectangular pattern of spread sheet cells containing the pertinent information: position, elapsed arc length distance, and velocity. Between each pair of key points is another rectangular pattern of cells containing the pertinent information for each segment: segment distance and segment time. The formulas, either Equation 5.1 or Equation 5.2, depending on which the variables are known, are all contained in the appropriate cells.

The layout suggested above would make a spreadsheet much greater in one dimension than another. Therefore, it has been the author's practice to make the key position rectangle in one vertical pattern and the segment rectangles in another, to make the entire time line easier to read on the page. Also included in a separate area are the various input parameters and constants, so that the reader can more readily grasp what assumptions are being made.

Example of Constant Acceleration Time–Distance Study

Shown in Figure 5.1 is the first page of a time–distance study developed for an accident involving a truck that was involved in a collision with a bicyclist. The data on the page pertain to the truck, which was stopped for a stop light, 10° into a circular arc right turn onto a boulevard. The bicyclist approached from behind the truck, maneuvered around some obstacles and onto a sidewalk, and continued without slowing into a crosswalk on a red light. Meanwhile, the truck started his turn, struck the bicyclist, and then proceeded some distance before reacting to bystander alerts and braking to a halt.

In this case, there were issues concerning when a line of sight could have existed from the truck driver's seat, and the times and distances involved after the impact until the truck came to rest. Therefore this study, unlike most others, extended both before and after the impact, starting with truck accelerating from a stop at the red light (key position 1), and ending when it was braked to a stop (key position 7).

The truck's turn was laid in as a circular arc, taking the vehicle from its stopped position testified by the driver, to its position at impact (in the crosswalk) as captured by police photographs and later quantified by photogrammetry, and into the right-hand travel lane. In the square titled "Input Parameters," the first three numbers were thus determined. Key positions 2 and 3 were on the arc, demarcating the impact portion of the event. Key position 4 marks the transition from the circular arc to a straight line. The Y coordinates of positions 5 and 6 were determined by calculating a straight line from position 4 to position 7, the location of which was determined by physical evidence and photogrammetry. All direct inputs from the user are highlighted in gray; the numbers not highlighted are calculated.

Position 1: Truck stopped at light					Input parameters		
Position on arc =	10	deg					
Elapsed time =	0.000	sec		Turn radius =	39.341	ft
Elapsed distance =	0.00	ft		X @ center of turn =	111.876	ft
Velocity =	0.00	mph		Y @ center of turn =	-35.067	ft
Position 2: Truck at impact				Truck acceleration =	0.05	G
Position on arc =	57	deg		Truck deceleration =	0.1	G
Elapsed time =	**6.332**	sec				
Elapsed distance =	32.27	ft				
Velocity =	6.95	mph				
Position 3: Truck at separation						
Position on arc =	60	deg		**Computed constants**		
Elapsed time =	6.531	sec		Mi/hr to ft/sec =	1.4667	
Elapsed distance =	34.33	ft		Coasting decel =	0.0497	G
Velocity =	7.17	mph				
Position 4: Truck at point on tangent						
Position on arc =	90	deg				
X =	111.88	ft				
Y =	4.27	ft				
Elapsed time =	8.261	sec		**Trajectory segment information**		
Elapsed distance =	54.93	ft				
Velocity =	9.07	mph				
Position 5: Begin perception/reaction (LP)				**Position 1 to Position 2**		
X =	131.30	ft		Segment distance =	32.27 ft	
Y =	4.04	ft		Segment time =	6.332 sec	
Elapsed time =	9.611	sec		**Position 2 to Position 3**		
Elapsed distance =	74.36	ft		Segment distance =	2.06 ft	
Velocity =	10.55	mph		Segment time =	0.199 sec	
Position 6: Truck begins braking (LB)				**Position 3 to Position 4**		
X =	153.84	ft		Segment distance =	20.60 ft	
Y =	2.83	ft		Segment time =	1.730 sec	
Elapsed time =	11.200	sec		**Position 4 to Position 5**		
Elapsed distance =	96.93	ft		Segment distance =	19.43 ft	
Velocity =	8.82	mph		Segment time =	1.350 sec	
Position 7: Truck at rest				**Position 5 to Position 6: coasting**		
X =	179.76	ft		Segment distance =	22.58 ft	
Y =	1.44	ft		Segment time =	1.590 sec	
Elapsed time =	15.216	sec		**Position 6 to Position 7: braking**		
Elapsed distance =	122.89	ft		Segment distance =	25.96 ft	
Velocity =	0.00	mph		Segment time =	4.015 sec	

FIGURE 5.1

Truck/bicyclist accident, page 1: Time line for truck.

The truck's acceleration and braking deceleration were based on data published by J. Stannard Baker,[3] in addition to testimony from eye witnesses. The truck acceleration shown, 0.05 G, was taken as a lower bound for the truck velocity–time profile. An additional time–distance study was performed using 0.10 G acceleration for the truck, taken as an upper bound.

The bicycle and photos taken at the scene were examined carefully to determine what gear it was in at impact. Using the gear ratio and information on pedal rates in the open literature, in addition to witness testimony, the bicyclist was determined to be traveling between 12 and 17 mph. Those two numbers were taken as lower and upper bounds, respectively, on the bicycle speed. The upper and lower bounds on the two vehicles made up a matrix of four scenarios, which were explored by time–distance studies. For the purposes of analysis, the bicycle was assumed to be traveling at a constant speed.

Perception–reaction time was not assumed, but calculated from the coasting segment (5–6), and checked for reasonableness. The coasting deceleration was also calculated for that segment, and checked for reasonableness.

The accident sequence was assumed to start when the truck accelerated from a standing stop (position 1). The truck traveled about 32.3 ft to the point of impact (position 2). As a lower bound, the truck was found to be traveling about 7 mph at that point, and required about 6.3 sec to get there (see the segment labeled position 1 to position 2). The truck maximum speed in this scenario was about 10½ mph (position 5), after which it coasted in gear briefly while the driver went through the perception/reaction time (position 5–6—about 1.6 sec in this case). The truck then braked to a stop at 0.10 G (position 6–7).

Building a time line for the bicyclist was quite easy because of the bicyclist's constant speed. Both the upper and lower bounds on bicycle speed are shown on page 2 of the time–distance study, seen in Figure 5.2. Again, the directly input numbers are highlighted in gray. The sequence started when the truck started to move, which was about 6.3 sec before impact in the truck velocity lower bound scenario. Therefore, the bicycle time line started at about $t = -6.3$ sec, and counted down until impact, at which point the time line ended. For the upper bound truck velocity scenario, a separate but similar bicycle time line was created, with a start time at about –4.5 sec.

We see that when the truck started to move, the bicycle was well behind it, being from 111 to 158 ft from the point of impact. Obviously, the truck was a far larger visual target than was the bicyclist. Clearly, no line of sight would pass through the truck's windshield until the very last instant. So, a line of sight was explored through the truck's right side window, at right angles. On a scale drawing, the truck was placed at the positions indicated (except for –1.6 and –0.6 sec) and a line of sight through the right-side window was drawn; its intersection with the bicycle path was noted in the right-most column. The columns labeled "Bicycle visible?" were filled in using a logic formula that compared the bicycle's distance from impact to the right-most column. The results are listed directly under the table: a line of sight to the bicycle through the right-side window came into existence about 0.8–1.3 sec before impact, depending on the speed of the bicycle.

In a further development, the time–distance studies were used to create four animations—one for each of the four limit scenarios. A line of sight to the bicyclist did not exist through the truck windshield until about 0.2–0.5 sec before impact.

Time before impact Sec	Bicycle distance from impact @ 12 mph Feet	Bicycle visible? Y/N	Bicycle distance from impact @ 17 mph Feet	Bicycle visible? Y/N	Truck distance from impact Feet	Truck angular position Degrees	Line of sight crosses bicycle path at Feet
-6.332	**111.44**		**157.87**		32.27	10.00	
-1.7	29.92	N	42.39	N	15.00	35.15	25.22
-1.6	28.16		39.89		14.25	36.25	
-1.5	26.40	N	37.40	N	13.48	37.37	24.40
-1.4	24.64	**N**	34.91	N	12.69	38.51	23.96
-1.3	22.88	**Y**	32.41	N	11.89	39.68	23.50
-1.2	21.12	Y	29.92	N	11.07	40.87	23.02
-1.1	19.36	Y	27.43	N	10.24	42.09	22.50
-1.0	17.60	Y	24.93	N	9.39	43.33	21.96
-0.9	15.84	Y	22.44	**N**	8.52	44.59	21.39
-0.8	14.08	Y	19.95	**Y**	7.64	45.87	20.78
-0.7	12.32	Y	17.45	Y	6.74	47.18	20.12
-0.6	10.56		14.96		5.83	48.51	
-0.5	8.80	Y	12.47	Y	4.90	49.87	18.67
-0.4	7.04		9.97		3.95	51.25	
-0.3	5.28		7.48		2.99	52.65	
-0.2	3.52		4.99		2.01	54.08	
-0.1	1.76		2.49		1.01	55.53	
0.0	0.00		0.00		0.00	57.00	

Approximate time, in seconds, when a line of sight exists to the bicycle:
   -1.348   -0.844

At 12 mph bike speed,	**6.332**	sec	means L2 is at	**111.44** ft
At 17 mph bike speed,	**6.332**	sec	means L1 is at	**157.87** ft

FIGURE 5.2
Truck/bicyclist accident, page 2: Time line for bicyclist.

Variable Acceleration

Physics students are taught in high school and beyond that time–distance problems can be solved using constant acceleration. We have just discussed such a case. However, consider the following: with the accelerator floored, this author drove a truck from a standing start, through a distance of 330 ft (on a different case than discussed above). The velocity at the 330-ft mark was recorded as 32.5 mph. What was the truck's acceleration?

One way to answer the question is to use the time and distance information. If the conditions at the start of the run are subscripted with a 0, then $t_0 = 0.0$, $S_0 = 0.0$, and $V_0 = 0.0$. If the conditions at the end of the run are subscripted with a 1, then $t_1 = 10.2$ sec and $S_1 = 330$ ft. From basic physics, we know that

$$S_1 = S_0 + V_0 t_1 + \tfrac{1}{2} a t_1^2 = \tfrac{1}{2} a t_1^2 \qquad (5.3)$$

from which we obtain

$$a = \frac{2S_1}{t_1^2} = \frac{2(330)}{(10.2)^2} = 6.34\,\text{ft/sec}^2 = 0.197\,G \tag{5.4}$$

On the contrary, we can answer the question using the speed and distance information. From Equation 4.3 of Chapter 4, we have

$$V_1^2 = V_0^2 + 29.938\left(\frac{a}{g}\right)S_1 = 29.938\left(\frac{a}{g}\right)S_1 \tag{5.5}$$

where the Vs are in mph. This leads to

$$\frac{a}{g} = \frac{V_1^2}{29.938\,S_1} = \frac{(32.5)^2}{29.938(330)} = 0.106\,G \tag{5.6}$$

What gives here? (Yes, the formulas are correct.) All the squirming on the witness stand and talking about "average acceleration" will probably not get the expert witness off the hook of two different answers, even though a clever answer would entail a discussion of time-averaging versus distance-averaging (which the jury will surely not understand).

No, the situation here is that there is no single acceleration value. The truck, and all motor vehicles, cannot produce a constant acceleration at full throttle. This can be seen by consulting car enthusiast magazines, which regularly produce acceleration performance test curves that typically look like Figure 5.3.

Now if the acceleration was constant, the curve of velocity vs. time would be linear, but instead we see that the curve is anything but linear. The curve is the steepest at the beginning and steadily diminishes as time goes on and the speed increases. Indeed, the acceleration eventually goes to zero (the curve flattens out completely) as the top speed is reached.

As opposed to the truck/bicyclist case, an important question in this case was the acceleration performance of the truck under full throttle. It would simply not do to use a constant acceleration in a time–distance study, because the use of actual test data would lead to just the sort of inconsistency described above.

What to do? The answer was to use a variable acceleration of the form

$$a = a_0 + a_1t + a_2t^2 + a_3t^3 \tag{5.7}$$

where the coefficients a_0 through a_3 could be determined by fitting the equation to the test data. Integrating this equation, subject to the initial conditions $t_0 = 0.0$, $S_0 = 0.0$, and $V_0 = 0.0$, leads to

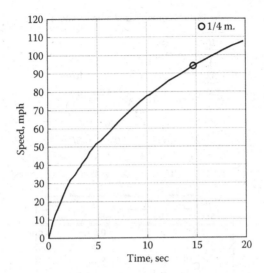

FIGURE 5.3
Automobile acceleration performance curve.

$$V = a_0 t + \tfrac{1}{2} a_1 t^2 + \tfrac{1}{3} a t^3 + \tfrac{1}{4} a t^4$$

$$S = \tfrac{1}{2} a_0 t^2 + \tfrac{1}{6} a_1 t^3 + \tfrac{1}{12} a_2 t^4 + \tfrac{1}{20} a_3 t^5 \tag{5.8}$$

where

$$a_0 = C - a_1 t_1 - a_2 t_1^2 - a_3 t_1^3$$

$$a_1 = -2 a_2 t_1 - 3 a_3 t_1^2$$

$$a_2 = \frac{6\left(7 C t_1^2 + 30 S_1 - 22 V_1 t_1\right)}{t_1^4} \tag{5.9}$$

$$a_3 = \frac{20\left(3 V_1 t_1 - 4 S_1 - C t_1^2\right)}{t_1^5}$$

and where $C = 0.322$ ft/sec^2 (0.010 G). A sample time–distance chart that matches the measurements taken in the test run is shown in Figure 5.4. Again, the numbers directly entered by the user are highlighted with bold type, and the cells are shaded in light gray. V_1 was converted to ft/sec and C was converted to ft/sec^2 to be in consistent units. In the chart, the acceleration was converted back to Gs and the speed to mph in order to be more meaningful to the jury.

Note that both the speed and the distance are matched to the measurements at the end of the run at 10.2 sec. The acceleration starts off vigorously

Time-Distance Study Fitted to Experimental Data

t_1 =	**10.2**	sec		
S_1 =	**330**	ft		
V_1 =	**32.5**	mph =	47.7	ft/sec
C =	**0.010**	G =	0.322	ft/sec^2

a_3 = 0.01904 ft/sec^5 = 0.000591 G/sec^3
a_2 = −0.31146 ft/sec^4 = −0.009673 G/sec^2
a_1 = 0.41159 ft/sec^3 = 0.012782 G/sec
a_0 = 8.32471 ft/sec^2 = 0.258531 G

Time t	Accel	Speed V	Distance
Sec	Gs	mph	ft
0.0	0.2585	0.00	0.0
0.5	0.2626	2.86	1.0
1.0	0.2622	5.75	4.2
2.0	0.2501	11.40	16.8
3.0	0.2258	16.64	37.4
4.0	0.1927	21.25	65.3
5.0	0.1545	25.07	99.4
6.0	0.1147	28.02	138.4
7.0	0.0768	30.12	181.2
8.0	0.0445	31.44	226.4
9.0	0.0211	32.14	273.1
9.5	0.0139	32.33	296.7
10.0	0.0103	32.46	320.5
10.2	0.0100	32.50	330.0

FIGURE 5.4
Time–distance study with variable acceleration.

at slightly more than 1/4 G, but tails off as the vehicle speeds up. The time lines for the other two vehicles in the subject accident were set up in the usual way with constant accelerations, and the times synchronized as illustrated in the previous section.

References

1. Sens, M.J., Cheng, P.H., Weichel, J.F., and Guenther, D.A., Perception/reaction time values for accident reconstruction, SAE Paper 890732, SAE International, 1989.
2. Olson, P.L. and Farber, E., *Forensic Aspects of Driver Perception and Response*, Second Edition, Lawyers and Judges Publishing Company, Inc., Tucson, AZ, 2003.
3. Baker, J.S., *Traffic Accident Investigation Manual*, Northwestern University, Evanston, IL, 1975.

6

Vehicle Data Sources for the Accident Reconstructionist

Introduction

In books and other treatments of accident reconstruction, much emphasis is placed on investigation, measurement, and analysis. This is rightly so, since there are many challenges and much understanding to be gained in these areas, but one quickly encounters the need for more information still. Not all of the information that the reconstructionist needs for analysis is obtainable from investigation and measurement. One needs an arsenal of data that can be tapped quickly when the time comes to apply physics principles, study test data, analyze statistical trends, and so on. This chapter illustrates some data sources that will be useful and/or necessary to the reconstructionist.

Nomenclature and Terminology

When asking a question, looking for information, or discussing one's work with fellow professionals, it is important to be able to speak the language—calling things by their correct names. This is not for the purpose of wrapping ourselves in arcane acronyms, obfuscating, or keeping information from others—quite the opposite. We just want the meaning to be clear and to avoid embarrassing ourselves. After all, accident reconstruction deals with vehicles and roadways, so it is important to learn some of the language used in the industries that specialize in these things. A few of the terms most important to reconstructionists are presented herein, to help the reader to learn how to describe accident scenes and vehicles, and to help lay persons to understand what is being said. The reader is directed to additional sources that are far more definitive.

Body type: The general configuration of a vehicle, such as sedan, convertible, pickup, van, and so forth. Often the description includes the number of doors. Unfortunately, sometimes a rear hatch is included in the number of doors; sometimes it is not. A fastback has a long sloping surface from the roof to the back surface, whereas a notchback has a more vertical surface for the back light. In a hatchback, the upper rear surface is hinged at the rear of the roof, and the luggage area is incorporated into the passenger compartment; the rear seat backs often fold down. A sedan may have two or four side doors.

Make: A distinctive name (e.g., Chevrolet) applied to a group of vehicles from one manufacturer (e.g., GM).

Car line: A name (e.g., Malibu) denoting a family of vehicles having some degree of commonality in construction (e.g., unibody or body-on-frame, front- or rear-wheel drive, etc.)

Platform: A designation used by a manufacturer to indicate a basic configuration upon which variants may be evolved. Usually, platforms are used across a manufacturer's lineup of vehicles. Different body types, makes, and car lines may be built from a single platform, so as to simplify the manufacture of parts and their assembly into vehicles. Thus, one cannot go by styling or even the number of doors in identifying the platform. Platforms are given letter names, such as A, B, and so on, or even names, such as LH, Epsilon, and the like. Since occupant packaging (seating) is so dependent on the wheelbase (see below), the best indicator that two vehicles are built on different platforms is that their wheel base specification is different. The platform may enter into the conversation when an alleged defect in one car line is claimed to exist in another car line built on the same platform.

Series: A group within a car line having distinct marketing characteristics (e.g., F-150).

Model year: A year designation used by the manufacturer for marketing purposes. The model year often starts September 1 of the previous calendar year and ends August 31 of the calendar year. Government regulations often reflect these dates. However, there are numerous exceptions in model year start and stop dates. In some cases, the model year may continue for more than 12 months. To further complicate the analysis of vehicles in use, manufacturing, sales, and registration data are generally reported by calendar year, not model year.

Trim level: A designation (e.g., Eddie Bauer, ES) denoting a package of options offered for sale. Often, the same trim level designations are used for various car lines in a particular make. The easiest way to identify the trim level is to look for an emblem on a rear body surface, or perhaps on the front fenders.

Emblem: A decorative element, often made of plastic but often having a bright metal appearance, that is affixed to the exterior surface to identify certain features of the vehicle, such as engine type, trim level, and so on.

Spin axis: The axis of rotation of a road wheel. This term applies whether or not there is an axle present; thus, the term "axle" may be overly restrictive.

Ackerman steering, or *Ackerman geometry*: For a given constant-radius turn, the idealized geometry that all four wheels would have to obey in order to avoid wheel slip at infinitesimal forward speeds (negligible lateral forces). Thus, the center of curvature would necessarily exist on the straight-line extension of the (unsteered) rear spin axis. For zero wheel slip, the spin axes of the steered wheels would necessarily pass through the center of curvature as well. Thus, drawing a line from the center of curvature to the rear wheel spin axis, then to the front wheel spin axis on the same side of the vehicle, and then back to the center of curvature forms a right triangle. The line from the center of curvature to the rear wheel is the base of this triangle, and the line from the center of curvature to the front wheel is the hypotenuse. Therefore, the rear wheel traverses a smaller circle than does the front (steered) wheel on the same side of the vehicle. The angle of the front wheel is called the Ackerman steer angle, the tangent of which is the wheelbase divided by the radius of the turn. Since the paths of the inside and outside front wheels necessarily have different radii of curvature, their Ackerman angles are different.

Understeer, neutral steer, and oversteer: In crude terms, a vehicle is said to be in understeer if its turning response becomes less sensitive to the steering wheel input as the steering wheel angle increases, and in oversteer if it becomes more sensitive. A truly neutral steer vehicle would always be on the cusp between these two. In the limit, an understeering vehicle could lose all sensitivity to steer inputs, and "plow out"; an oversteering vehicle could become so sensitive to steer inputs that its yaw rate cannot be controlled, and it "spins out." More precisely, "a vehicle is in understeer ...if the ratio of the steering wheel angle gradient to the overall steering ratio is greater than the Ackerman steer angle gradient"[1] (p. 14). Vice versa for oversteer. Since the vehicle is a nonlinear system, a vehicle could be in understeer for some inputs and oversteer for others. For more information, see SAE Recommended Practice J670e.[2]

Off-tracking: The condition by which a wheel on a trailer travels on a path having a smaller radius of curvature than a wheel on the same side of the towing vehicle. The trailer wheel "turns inside" the towing vehicle wheel. See the discussion of Ackerman steering above.

Sprung weight and *unsprung weight*: The sprung weight is all of the weight supported by the suspension, including the weight of a portion of the suspension. The unsprung weight is everything else.

Sprung mass and *unsprung masses*: The sprung mass is a rigid body having the same inertial properties (mass and mass moments of inertia about the same axes) as the total sprung weight. The unsprung masses move with, and are supported by, the tires (which of course have vertical compliance of their own).

Suspension rate or wheel rate: The change of wheel load per unit relative displacement between the sprung and unsprung masses.

Static loaded radius: The loaded radius of a stationary tire inflated to normal recommended pressure. This is different from the rolling radius. The static radius of a tire rolled into position may be different from that of a tire loaded without being rolled.

Wheel hop, brake hop: Wheel hop is the vertical oscillatory motion occurring between the road and the sprung mass. Brake hop occurs when that motion is triggered by the braking action of a moving vehicle. It is most often encountered when the brakes are applied on an empty semitrailer, with the reduced wheel loading making the brakes easier to lock, and the high spring stiffness combining with the reduced sprung mass to produce a higher natural frequency. In such instances, one may see dual tire marks on the road that look like dotted lines.

Jounce, rebound: Jounce is the relative motion in which the sprung and unsprung masses move closer together, which is analogous to the compression of a coil spring (though many suspensions do not have a coil spring). Rebound is the opposite of jounce.

Bump stop, jounce stop, rebound stop: A bump stop or jounce stop is an elastic member which increases the suspension rate near the end of the allowable jounce motion, and may be thought of as a compliant bumper to prevent metal-to-metal contact. A rebound stop acts in the same way for rebound motions. These elements may exhibit permanent deformations or other evidence due to the extreme jounce or rebound motions experienced in a crash, especially rollovers.

Self-aligning torque: The torque on a wheel (usually, a steered wheel) about a vertical axis that acts in the opposite direction of the steer angle, thus tending to return the wheel to zero steer.

Caster: When viewed from the side, the angle of the steering axis from the vertical. Positive caster has the axis more aft at the bottom than at the top, and contributes to self-aligning torque on the steered wheel. The term can be easily remembered by thinking of a castor on a shopping cart (but notice the difference in spelling!). The caster generally changes due to suspension motions.

Camber: The inclination of the wheel plane from the vertical, considered positive when the wheel is more outboard at the top than at the bottom, and negative when the opposite is true. The angle will generally change as the suspension changes its geometry in response to forces in various directions, particularly vertical.

Toe-in and Toe-out: The angle between the vehicle longitudinal axis and the intersection of the wheel plane with the road. As on a person, toe-in means that the front of the wheel is turned toward the vehicle center line; toe-out is the opposite. As with the other angles, toe-in or toe-out generally changes with wheel loading.

Service brakes: The primary brake system used for stopping or slowing the vehicle.

Parking brake system: A system used to keep one or more brakes (the two rear brakes on a passenger car) in an applied position, regardless of whether the ignition is "on." Federal Motor Vehicle Safety Standard 105[3] requires that parking brakes be able to hold vehicles stationary on slopes up to, and including, 20%. The traditional means of actuation has been a cable system running from the parking brake handle or pedal to the rear brakes, but some newer vehicles have electric parking brakes.

Fifth wheel: A calibrated wheel and axle assembly mounted to a vehicle through hinges so that it stays in contact with the road, enabling the accurate determination of true vehicle speed. The wheel itself often has the appearance of a bicycle wheel.

H-point: The hinge between the torso and the thigh on the two- and three-dimensional devices used to establish or determine a vehicle's seating accommodation. It can easily be remembered by thinking of one's hip. For reconstruction purposes, it can be taken as the location of a person's CG in a vertical longitudinal (x–z) plane.

H-point template, H-point machine: The H-point template is used to locate the H-point on a side view drawing. The H-point machine is an apparatus with body segment weights and back and seat pan representations of deflected seat contours of adult male individuals. It can be positioned so that the actual physical H-point location and other dimensions can be measured in the vehicle. See SAE J826 JUN92[4] and J1100 JUN93[5] for more information.

Seating reference point: Abbreviated SgRP, a design reference point that, for each designated seating position, locates the H-point in the vehicle, considering all modes of seat adjustment. It plays a key role in defining and determining the vehicle's seating accommodation.

H-point travel: The range of travel of the H-point that occurs when the various seat adjustments are employed.

H-point couple: The longitudinal distance between first- and second-row H-points.

Water line: A horizontal line at a specified distance above the ground.

Backlight: The primary glazed surface at the rear of the roof panel.

Belt line: A line around the exterior of the vehicle at about the level of the base of the windshield and the base of the backlight. So named because originally, it was at about the level of a man's belt.

Greenhouse: The vehicle structure that exists above the belt line.

Glazing: Transparent material used in window openings. Not all glazing material is glass (as in the backlight of a convertible, for example), so glazing is a more inclusive term.

Daylight opening (DLO): A line on the exterior glazing surface that defines the minimum unobstructed opening through any glazed aperture, including adjoining reveal or garnish moldings.

Laminated glass: Two or more sheets of glass held together by an appropriate number of interlayers of plastic material, such as polyvinyl butyral (PVB).

Tempered glass: Glass possessing great mechanical strength, by virtue of treating the surface of the glass, either by chemical means or by chilling it while the glass is still hot. As the interior of the glass cools and contracts, it places the surfaces in compression, which increases the strength because any surface flaws tend to be pressed closed by the compressive stresses. When broken, the glass breaks up into small granules instead of jagged shards.

Body in white: The assembly stage of a car body in which the welded metal parts are present and welded before painting, but not the nonmetallic, bolted, screwed, or clipped-on parts. It does not include moving parts (doors, hoods, and deck lids) or fenders, the motor, chassis sub-assemblies, or trim (glass, seats, upholstery, electronics, and so on).

Sill, rocker panel: The longitudinal body member, hollow and constructed of stamped sheet metal panels spot welded together, that is positioned at the bottom of the door opening, against which the bottom of the door seals. Note that vehicle windows do not have sills.

A-, B-, and C-pillars: Pillars are members that are more or less vertical below the belt line, hollow and constructed of stamped sheet metal panels spot welded together, that bracket the door openings. They are labeled A, B, C, and so on, from front to back. Thus, the A-pillar is the front door hinge pillar, and the B-pillar is the front door latch pillar (and the rear door hinge pillar as well, if there is a rear door). Some vehicles (such as vans) have D-pillars as well.

Roof rail: The more or less longitudinal member, often curved, hollow and constructed of spot welded sheet metal panels, that is welded to the side of the roof panel, welded to the tops of the pillars, and forms the tops of the door openings.

Windshield header, backlight header: Lateral members, often curved, to which the front and rear of the roof panel is attached, and which are welded to the tops of the appropriate pillars. They form the tops of the front and rear daylight openings, respectively.

Deck lid: The panel, hinged near the base of the backlight that covers the top and rear of the luggage compartment.

Dash panel: Sometimes called the firewall, it is the vertical panel separating the engine and passenger compartments.

Toe board: Located between the dash panel and the floor panel, it is the panel that is slanted to provide a comfortable place for the front seat passengers to put their feet.

Tunnel: The longitudinal "hump" near the middle of the floor panel. For rear wheel drive vehicles, it provides space for the transmission and the drive shaft. Even so, most front wheel drive passenger cars have a tunnel, which accommodates the engine exhaust system.

Shotgun structure, catwalk: Also called the upper longitudinal member, it runs more or less longitudinally along the hood opening. The front fender is attached to it. Its crush characteristics are important in controlling pitch motions during frontal crashes.

Core support: The lateral structure to which the radiator core is attached. There can be an upper and lower core support.

Ladder frame: A vehicle frame that has the general appearance of a ladder. There are two longitudinal members, necessarily inside of the wheel wells, that run the length of the vehicle, with cross-members running laterally between them. This architecture, found mostly in trucks, tends to result in vehicles with higher ground clearances.

Perimeter frame: A vehicle frame in which the longitudinal members run along the sides of the passenger compartment, near the rocker panels. This configuration allows lower ground clearances relative to a ladder frame. However, forward and aft of the passenger compartment, the longitudinal members must be located inboard of the wheel wells. In front, the distance between the longitudinal members is less than that in the rear, to allow room for movement of the wheels due to steering.

Torque box: The portion of the frame positioned just aft of the front wheel, providing a transition from the forward portion inside the front wheel well, to the middle portion near the rocker panel. To the degree that this portion of the frame has a lateral orientation, vertical loads on the longitudinal portions of the frame cause torques to be imposed on the member.

Body mount: A component by which the body is mounted to the frame in a body-on-frame vehicle. It usually has metal attachment surfaces separated by an elastomeric material, so as to allow some isolation from frame vibrations.

Frame versus unitized or unibody construction: Traditionally, the passenger compartment enclosed the passengers, while a separate frame structure running under the passenger compartment was relied on to carry the road loads arising from the engine, brakes, and road disturbances. In unitized or unitbody construction, these functions are combined; the passenger compartment also carries road loads. This is a more efficient use of materials, but makes the job of isolating the passengers from exterior noise, vibration, and harshness (NVH) more difficult.

Sub-frame: Even though a unitized vehicle has no frame (by definition), there must still be hard points to carry road loads. Thus, front and rear subframes are used for mounting the engine, transmission, suspension, and bumpers, among other components. They have longitudinal members that are welded to the floor panel. The front sub-frames may extend under the floor panel aft to the rear passenger foot well, so as to distribute the loads over a large area.

Kick-up: Straight frames and sub-frames would be the easiest to build and the most efficient for carrying loads. Look at the frame on a Model T Ford, for example, or a modern big-rig truck. However, these load-carrying elements must provide clearance above the suspension, axles, half-shafts, and so on at their uppermost excursion. This would make the vehicle sit high off the ground, were it not for kick-ups. The kick-ups allow the frame to pass under the compartment, over these elements, and back down again to the bumpers.

Transaxle: The combined transmission and front axle, for front wheel drive vehicles. The axle shaft is in two sections—a half-shaft extending outward from each side of the transmission, through CV joints, to the front wheels.

Engine cradle: A frame structure supporting the engine and transaxle in a front wheel drive vehicle. The use of an engine cradle allows the engine and transaxle to be pre-mounted, and then installed as a unit into the vehicle during the assembly process.

CV joint: A constant-velocity universal joint that maintains the same angular velocity in the output shaft as in the input shaft, even when the shafts are not parallel. Thus, a constant velocity in the input shaft does not get transformed into vibratory motion when the output shaft is at an angle.

Curb weight: The vehicle weight in the drive-away condition (i.e., idling at the curb), filled to at least 90% capacity with fuel, lubricants, and coolants, and with all standard equipment, but without passengers or cargo.

Gross vehicle weight rating (GVWR): The value specified by the manufacturer as the maximum loaded weight, measured at the tire–ground interfaces.

Gross axle weight rating (GAWR): The value specified by the manufacturer as the maximum loaded weight on a single axle, measured at the tire–ground interfaces. The sum of the GAWRs does not necessarily equal the GVWR.

Drip rail: A protruding flange on the vehicle exterior surface, near the top of a door or window opening, designed to guide water away from the opening. Particularly subject to damage during a rollover, the drip rail damage can provide important evidence.

Cowl: The sheet metal panel between the hood and the windshield.

Plenum, or plenum chamber: A recessed area between the hood and the windshield, designed to capture water and direct it to the sides of the vehicle. It is commonly covered by slots formed in the cowl.

Instrument panel: The portion of the passenger compartment interior that is forward of the front-seat occupants, above the front foot wells, and below the windshield. It is commonly referred to by lay persons as the "dashboard."

Instrument cluster: The portion of the instrument panel, immediately forward of the steering wheel that contains the instruments essential to operating the vehicle.

Latch: A mechanical device used to secure a hinged body panel in a closed position relative to the fixed portion of the body, with provisions for intentional operation.

Striker: A mechanical device that the latch engages on the fixed portion of the body. It is often in the form of a U-shaped bar, or a cantilevered pin.

Secondary latched position: The attitude that exists between the latch and striker when the latch holds the door or hood in a position less than fully closed.

Primary latched position, fully latched position: Fully closed. These terms are used to distinguish from the secondary latched position. Close inspection of the latch mechanism may reveal which position the latch was in at the time of a crash.

Tractor: Literally, a vehicle that provides traction, for the purpose of pulling or towing another vehicle or trailer.

Semitrailer: A truck trailer equipped with one or more axles, and so constructed that the front end and a substantial part of its own weight and that of its load rests on a tractor. See SAE J687 JUN88.[6]

Slack adjustor: An adjustable member that transmits brake application force and permits compensation for brake lining wear. See SAE J656 APR 88.[7] Because it is a maintenance point on trucks, the slack adjustor is often a primary subject of safety inspections and accident investigations.

Section: A vertical cross-section taken perpendicular to the center line of the road.

Profile: A vertical section taken along the center line of the road, and presented in true view (i.e., as if the line around any horizontal curve in the road were stretched out flat).

Horizontal curvature: The curvature of a line down the center of the road, as seen from overhead. Its reciprocal is the radius of curvature, which may often be ascertained from an as-built drawing. One will find that any given curve has a constant radius of curvature, which makes it much less difficult to survey during construction.

Vertical curvature: The curvature of a line down the center of the road, as seen in profile. The radius of vertical curvature, its reciprocal, may often be ascertained from an as-built drawing. Again, constant radius of curvature is the norm.

Point of tangency, point on tangent: The point at which the turn ceases and the straight portion begins (on tangent, of course).

As-built: Literally, a drawing or drawings kept by the highway department for the purpose of representing the road as it was built (as opposed to the way it was designed). There may be notations on the drawing(s) indicating deviations, approved by the responsible highway engineer, from the design, but these are rare. There is generally a new set of as-builts generated every time there is a construction project on the road. Old as-builts are generally retained, so it is incumbent on the investigator to avoid obtaining the wrong set of as-builts.

As-builts show all aspects of the construction job, including lighting, drainage, signage, and so on. Often the roadway is represented only by its center line and typical sections. Generally, as-builts do not contain enough information to assist in a reconstruction—certainly not enough to alleviate the need for a site inspection. But they do provide an accurate assessment of any radii of curvature. For further information, see SAE Paper 940569.[8]

Station: In as-builts, the distance along the center of the road is indicated by station numbers. Each station is 100 ft. Station numbers have a unique form that contains a plus sign. For example, Station 0+00 is the reference point, Station 1+79.66 is 179.66 ft away. This form is occasionally used by law enforcement personnel in accident investigations, except that stations may be marked along the edge of a travel lane instead of the center line.

Right-of-way: The property owned by the entity that operates the road. It is often marked by right-of-way fences.

Fog line: A solid white line painted near the edge of the pavement on the right side of the road, so as to demarcate the rightward extent of the travel lane or lanes. Not all roads have fog lines. In accident investigations, the fog line is usually a better reference line than the edge of pavement, because it is more regular. It is subject to being painted or paved over, but then the edge of the pavement is not immutable either.

Slope: The inclination of a surface, paved or otherwise, calculated as the ratio of rise (vertical distance) divided by run (horizontal distance), often expressed as a percentage. The slope is sometimes expressed by lay persons in terms of an angle in degrees.

Crown: The feature of a paved surface that makes the center of the road higher than the center of a plane passing through its edges. It is the deviation from a plane, level or otherwise. Designed to promote the drainage of water off the road, crown is a factor to be aware of in analyzing fluid spills, and applying photogrammetry to accident scene photographs.

Superelevation: The lateral slope of a roadway, measured from one edge to the other. It is usually designed into the roadway to compensate for centrifugal forces that exist for a vehicle moving through the turn at a specified speed.

Total station: A surveying device that measures a point's location by measuring a distance, a horizontal angle, and a vertical angle. The angles are measured by the theodolite portion, in which the operator adjusts the angles until the point appears in the cross hairs of a sighting telescope, and signals for the data to be recorded. The distance to the point is measured with the electronic distance meter, which measures the time required by an infrared light beam to travel to the point and return. The device may be used with a survey prism, some other type of reflector, or no reflector at all. The ability to take points without a prism or reflector depends on the point not being on too distant a surface, the nature and orientation of the surface, and the atmospheric conditions.

SAE Standard Dimensions

For a system as complicated as a motor vehicle, a large number of dimensions are required to describe it adequately. It is not hard to imagine the confusion that would result with manufacturers reporting vehicle characteristics in such a way as to gain an advantage in a very competitive marketplace. Fortunately, the SAE has evolved a standardized set of dimensions, definitions, and procedures for making measurements, which are spelled out in Recommended Practice SAE J1100. Linear dimensions have a prefix letter and a number. The prefix letter denotes the direction of the measurement, as follows:

L: Length dimension, in the x-direction

W: Width dimension, in the y-direction

H: Height dimension, in the z-direction

There are other prefixes, but these are the ones of most interest to the reconstructionist. The numbers are organized as follows:

1–99 Interior dimensions

100–199 Exterior dimensions

200–299 Cargo or luggage dimensions

300–399 Interior dimensions (trucks and multi-purpose vehicles only)

400–499 Exterior dimensions (trucks and MPVs only)

500–599 Cargo dimensions (trucks and MPVs only)

Here are a few dimensions of most interest to reconstructionists:

Overall length (L103): Abbreviated herein as OAL, it is the maximum dimension measured longitudinally between the foremost point and the rearmost point on the vehicle, including bumper, bumper guards, tow hooks, and/or rub strips, if standard equipment.

Wheelbase (L101): Abbreviated herein as WB, it is the dimension measured longitudinally between the front and rear wheel centerlines. In the case of dual rear axles, the measurement is to the midpoint of the centerlines of the rear wheels.

Front overhang (L104): Abbreviated herein as FOH, it is the dimension measured longitudinally from the centerline of the front wheels to the foremost point on the vehicle, again including bumper, bumper guards, and so on.

Rear overhang (L105): Abbreviated herein as ROH, it is the dimension measured longitudinally from the centerline of the rear wheels, or in the case of dual rear axles, the dimension is from the midpoint of the centerlines of the rear wheels, to the rearmost point on the vehicle, again including bumper, bumper guards, and so on.

Front bumper height (H102): The minimum dimension measured vertically from the lowest point on the front bumper to the ground, including bumper guards, if standard equipment. A somewhat different measurement, emphasizing frontal crash load paths, should be used for frontal underride/override crashes, as discussed in Chapter 26.

Rear bumper height (H104): The minimum dimension measured vertically from the lowest point on the rear bumper to the ground, including bumper guards, if standard equipment. A somewhat different measurement, emphasizing rear crash load paths, should be used for rear underride/override crashes, as discussed in Chapter 26.

Angle of approach (H106): The angle measured between a line tangent to the front tire static loaded radius arc and the initial point of structural interference forward of the front tire, to the ground.

Angle of departure (H107): The angle measured between a line tangent to the rear tire static loaded radius arc and the initial point of structural interference rearward of the rear tire, to the ground.

Ramp breakover angle (H147): The angle measured between two lines tangent to the front and rear tire static loaded radius arc and intersecting at a point on the underside of the vehicle, which defines the largest ramp over which the vehicle can roll.

Overall width (W103): Abbreviated herein as OAW, it is the maximum dimension measured between the widest point on the vehicle, excluding exterior mirrors, flexible mud flaps, and marker lamps, but including bumpers, moldings, sheet metal protrusions, or dual wheels, if standard equipment.

Front tread (W101): Sometimes called front track width, it is the dimension measured between the tire centerlines at the ground. For ease of measurement in accident reconstructions, it suffices to measure from the inside of one front tire to the outside of the other front tire.

Rear tread (W102): Sometimes called rear track width, it is the dimension measured between the tire centerlines at the ground. For ease of measurement in accident reconstructions, it suffices to measure from the inside of one rear tire to the outside of the other rear tire.

Front head room (H61): The dimension measured along a line 8° rear of vertical from the front SgRP to the headlining, plus 4 in (102 mm).

H-point couple (L50): Also designated as SgRP couple distance, it is the horizontal distance between the driver side front SgRP to the second row SgRP.

Vehicle Identification Numbers

The Vehicle Identification Number, or VIN, is a unique number assigned to each vehicle for registration and identification purposes. VINs have been used since 1954, but different manufacturers used different formats. In the mid-1960s, model year began to be included with a production serial number, and in the early 1970s, the number of digits was standardized at 10. In 1981, the National Highway Traffic Safety Administration (NHTSA) standardized the format, and required that all over-the-road vehicles sold in the United States have a 17-character VIN. Any letters had to be uppercase, but the VIN could not include the letters I, O, or Q, to avoid confusion with the numbers 1 and 0. A one-character code for model year was included, which allowed for a 30-year cycle.

Modern-day VIN formats are based on two related standards, originally issued by the International Organization for Standardization (ISO) in 1979 and 1980: ISO 3779 and ISO 3780, respectively. Compatible but different implementations of these standards have been adopted by the European Union and the United States. Now the applicable regulation in the United States and Canada is Title 49, Part 565, of the Code of Federal Regulations (CFR),[9] while in Europe and many other parts of the world, it is ISO 3779. In Australia, ADR (Australian Design Rule) 61/02 is used, which refers to ISO 3779 and 3780.

For both the ISO and CFR formats, the VIN is comprised of three sections: the World Manufacturer Identifier (WMI), the Vehicle Description Section (VDS), and the Vehicle Identification Section (VIS). The WMI occupies the first three digits, and assigned by SAE International under contract to the NHTSA. The VDS occupies digits 4 through 8, and is assigned by the manufacturer. The VIS occupies digits 9 through 17.

For the VIS, the CFR format uses 10th digit for the model year, and the 11th digit for the manufacturing plant, assigned by the manufacturer. Digits 12 through 17 are the sequential serial number, applicable to the manufacturers producing over 500 vehicles per year, and assigned by the manufacturer. The 9th digit is a check digit that is calculated from the other digits through an algorithm, and is either an X or a number from 0 through 9. As its name implies, it serves as a "check" on whether the digits have been assigned or transcribed correctly.

The model year is of particular interest to the reconstructionist, and is encoded in the 10th digit worldwide. Besides the three letters that are not allowed anywhere in the VIN, the model year encoding also prohibits the use of the U and Z, and the number 0. This leaves the allowable codes at 30 (a 26-letter alphabet plus 10 numbers, minus the 6 disallowed codes), for a 30-year cycle of values for the model year. Since the cycle began in 1980 with the letter A, the cycle was exhausted in 2009. In 2008, to extend the 17-digit VIN format for another 30 years, the 7th digit was appropriated, at least in some sense: If the 7th digit is numeric, the 10th digit refers to a model year in the range 1980 through 2009. If the 7th digit is alphabetic, the 10th digit refers to a model year in the range 2010 through 2039. What will they do in 2040? Tenth digit model year codes are shown in Table 6.1.

For the reconstructionist, the VIN is important for three primary reasons:

1. Theoretically, checking the VIN of the accident vehicle being inspected and comparing it with the VINs in the police report and/or in legal documents would ensure that the investigator is looking at the actual vehicle that was involved in the subject accident. In practice, however, it is not uncommon to find that the VIN has been transcribed incorrectly, even in official legal documents, especially when extensive damage or fire has rendered the VIN hard to read. The letter S and the number 5 are sometimes

TABLE 6.1

Model Year VIN Codes: Tenth Digit

Code	Year	Code	Year	Code	Year	Code	Year	Code	Year	Code	Year
A	1980	L	1990	Y	2000	A	2010	L	2020	Y	2030
B	1981	M	1991	1	2001	B	2011	M	2021	1	2031
C	1982	N	1992	2	2002	C	2012	N	2022	2	2032
D	1983	P	1993	3	2003	D	2013	P	2023	3	2033
E	1984	R	1994	4	2004	E	2014	R	2024	4	2034
F	1985	S	1995	5	2005	F	2015	S	2025	5	2035
G	1986	T	1996	6	2006	G	2016	T	2026	6	2036
H	1987	V	1997	7	2007	H	2017	V	2027	7	2037
J	1988	W	1998	8	2008	J	2018	W	2028	8	2038
K	1989	X	1999	9	2009	K	2019	X	2029	9	2039

confused, particularly when reading someone's handwriting. This is why the check digit is important. As it turns out, looking at photographs may be the best way to make sure that the correct vehicle is being inspected, because every real-world crash produces a unique damage pattern.

2. On the other hand, if the vehicle has significantly changed since photographs were taken (by disassembly, for example, but especially if it has been fully repaired), the VIN becomes the foremost means of proving the vehicle being inspected is, in fact, the accident vehicle.

3. Checking the VIN and getting it correct, and then matching certain of the digits against those of another vehicle being sought for testing or inspection purposes, ensures that a truly comparable vehicle will have been obtained.

Of all the parts of the VIN, the WMI and the model year are probably of most importance to the reconstructionist, and the sequential serial number is of no importance, beyond item (1) above.

Not surprisingly, CFR Part 565 specifies where (as a minimum) the VIN must be displayed: "... clearly and indelibly upon either a part of the vehicle, other than the glazing, that is not designed to be removed except for repair or upon a separate plate or label that is permanently affixed to such a part" and "inside the passenger compartment ... readable ... through the vehicle glazing under daylight lighting conditions by an observer ... outside the vehicle adjacent to the left windshield pillar."

The first requirement is satisfied by permanently marking the VIN on the certification plate, or tag, that is usually mounted on the left-side B-pillar outboard of the weather seal, visible by opening the front door. Of course, if the door is jammed shut due to crash damage, or otherwise unopenable, the certification plate will not be accessible. To further complicate the investigator's life, it is common to find that damage to the windshield or an accumulation of dirt and debris, such as broken glass, has rendered the latter-described VIN label to be unreadable.

However, all is not lost, even then. In 1987, the NHTSA published the theft prevention standard,[10] which now requires manufacturers to inscribe or attach the VIN onto major parts of vehicles, such as engine, transmission, front/rear bumper, left/right front fender, hood, left/right front door, sliding or cargo doors, left/right quarter panel (passenger cars), left/right side assembly (MPVs), pickup box and/or cargo box (light-duty trucks), rear doors, deck lid or hatchback and tailgate. Of course, it is possible that one or more of these original parts were damaged in a prior crash and replaced, but it is almost always the case that a VIN can be found somewhere on the vehicle. If a VIN cannot be found on some exterior part, then that fact in and of itself is important, because it may point to the involvement of the vehicle in a prior accident, with subsequent repair activities.

The importance of decoding the VIN has given rise to computer software that performs that function, including the calculation of the check digit. Used car shoppers are often interested in VINs, so one can even do some decoding on the Internet at no cost. However, you generally get what you pay for, and it is not typical to obtain the sort of digit-by-digit decoding that the reconstructionist often needs, to construct VIN "masks" for finding exemplar vehicles, for example. Moreover, the emphasis tends to be on the vehicle accident or theft history, which of course tends to come at a price.

Web-based VIN decoding services that can be subscribed to include VINPower and VINLink.[11] Software that can be licensed includes Expert VIN DeCoder,[12] which produces a digit-by-digit decoding, and it is DOS based and updated annually. VINLink is a pre-pay service for which the per-vehicle cost depends on the type of report that is run. The detail in the report may include standard equipment, optional equipment, and photographs, for example.

Vehicle Specifications and Market Data

As discussed previously, a standardized system of measurements exists for the full description of vehicles, both inside and out. It would seem useful for the reconstructionist to have a set of such measurements for each vehicle of interest. Indeed, a set of specifications was compiled for each car line by the Motor Vehicle Manufacturing Association (MVMA), the forerunner of the present-day Alliance of Automotive Manufacturers, a trade organization of 12 manufacturers in the United States. While the Alliance continues to compile production and sales data, it is not known whether they gather specifications and make them available outside of its membership. In any case, the MVMA specifications, as they were known, contained far more detail than the reconstructionist needed. Rather than this deluge of data, it would be better to have a table or tables of summary MVMA specifications for all vehicles in each model year, particularly in a study of vehicles having similar attributes (such as curb weight, for example).

For many years, this need was nicely met by Crain Communications, the publishers of *Automotive News,** a weekly trade publication. Once a year, the Automotive News Data Center publishes their Market Data Book, which contains sales and production data for all vehicle series sold in the United States, as well as other data pertaining to suppliers and dealers, and sales in Europe and Japan. Until very recent times, the Market Data Book, available to Automotive News subscribers, also contained summary vehicle specification data. In recent years, this information has been provided by JATO Dynamics, an English company of market researchers, but the data are no longer included in the Market Data Book. Perhaps the Data Center could be persuaded to reverse this decision if there is sufficient interest (personal communication with the director, Automotive News Data Center).

Another source of market data is WardsAuto,[13] which is a subscription service of their Data Center that allows the viewing and downloading of monthly sales and production data around the world. Another source is R. L. Polk & Co.† In addition to sales and production data, Polk offers VIN decoding services and registration data (for contacting owners of vehicles being recalled, for example).

Specification data for passenger cars, pickups, vans, and utility vehicles can be obtained from Expert AutoStats.[14] Like Expert VIN DeCoder, this is DOS-based software that creates an output for a specified vehicle, rather than a table of various vehicles. Unlike Automotive News data, front and rear overhang are included. The dimension definitions do not follow SAE convention, but a description of them is provided with the program documentation. Other parameters such as inertial properties are also presented, but one can see from the documentation that these tend to be calculated from empirical data fits, and should not be interpreted as actual measurements.

Specification data that deal specifically with trucks are published by Truck Index, Inc. Published annually on a subscription basis and available in hard copy, and more recently in CD-ROM, there are three manuals: Gasoline Truck Index, Diesel Truck Index, and Import Truck Index.[15] Information includes digit-by-digit tables for decoding VINs, and such summary specifications as wheel base, overall length, overall width, and curb weight.

---

* Crain Communications, Inc., Detroit, MI.
† R. L. Polk & Co., Southfield, MI.

Vehicle Inertial Properties

The most important inertial property of a vehicle is its mass, which is determined by weighing it, of course. Other inertial properties, such as the mass moments of inertia about the roll, pitch, and yaw axes, are substantially more difficult to measure. Fortunately, the results of measurement efforts have been published from time to time.[16–22]

Even so, the published data cover only a small sample of vehicles, while reconstructionists often need to do an analysis using incomplete data. Therefore, a parametric fit to the measured data would be helpful. Garrott et al. reported the following:[23]

$$I_{Pitch} = 0.99W - 1149.0 \qquad R^2 = 0.89$$

$$I_{Roll} = 0.18W - 150.0 \qquad R^2 = 0.80 \qquad \text{Passenger cars} \qquad (6.1)$$

$$I_{Yaw} = 1.03W - 1206.0 \qquad R^2 = 0.88$$

$$I_{Pitch} = 1.12W - 1657.0 \qquad R^2 = 0.70$$

$$I_{Roll} = 0.22W - 235.0 \qquad R^2 = 0.70 \qquad \text{Light trucks} \qquad (6.2)$$

$$I_{Yaw} = 1.03W - 1343.0 \qquad R^2 = 0.73$$

where W is the vehicle weight. By comparison, Neptune[24] fitted a regression line to measured yaw moments of inertia of vehicles at or near curb weight, based on the formula for a homogeneous rectangular prism:

$$I_{ZZ} = \frac{m}{K_G}(OAL^2 + OAW^2) \qquad (6.3)$$

where m is the vehicle mass and K_G is an empirical constant. K_G was 13.1 for all of 144 vehicles studied, and ranged from 12.2 for utility vehicles to 13.8 for cars, and R^2 varied from 0.92 to 0.99. On the other hand, this author analyzed the data by Rasmussen et al.,[16] and found a best fit of

$$I_{Yaw} = 0.92\frac{m}{12}(OAL^2 + OAW^2) \qquad (6.4)$$

which is used in subsequent reconstruction calculations in this book, and which would correspond to a value of K_G of 13.04. Of course, the value of m used in these calculations should reflect the presence of occupants and cargo.

Another analysis of the Rasmussen data by this author also found that the yaw radius of gyration had a best fit of

$$\rho = 0.49WB \qquad (6.5)$$

an easily remembered result that suggests the vehicles behave roughly as if their mass were concentrated at the front and rear spin axes.

Production Change-Overs and Model Runs

Even though most model years are only 1 year long, it is widely known that between most adjacent model years in a given car line, there are few changes in design and features. Every so often, though, there is a major design change, which requires the assembly plant(s) for that car line to be shut down so the tooling can be changed out. This production change-over only occurs every 5 years or so. The period of time between production change-overs, when there is virtually no structural change in a vehicle, is known as the model run. Model runs for some car lines can be considerably longer than 5 years, especially for pickups and vans.

Within a given model run, specifications and test results apply across model years. Thus, manufacturers do not need to perform a whole new battery of tests every model year, and reconstructionists do not have to rely on tests for a given model year (and indeed, there may be none). Instead, it suffices to use test results from any model year in that model run. The only thing the reconstructionist has to do is know what production run a given vehicle is in, and then search throughout that period of time.

Sisters and Clones

It is also well known that a given car platform may show up in different car lines, even from different manufacturers. This is called "badge engineering." But would it be obvious that in Model Year 2009, the Pontiac Vibe and the Toyota Matrix were virtually identical vehicles? That happened to be the case, so at least most of the specification data for one vehicle could be used for the other. To do a thorough search for applicable test data, one would have to look at both vehicles. Not only that, one would have to look at the entire production runs on both vehicles. (It turns out that the production run for these vehicles was only 2 years: 2009 and 2010, because the Pontiac make came to an end in 2010. In 2011, there was neither a Pontiac Vibe nor a Toyota Matrix.)

Where to get such information, of obvious use to reconstructionists? The answer lies in the Vehicle Year & Model Interchange List, otherwise known as the Sisters & Clones List.[25] It is "provided free of charge as a courtesy to the traffic accident reconstruction community by Gregory C. Anderson of Scalia Safety Engineering." It is updated annually and covers model years from 1974 forward. It is obtainable from various sources on the Internet; one source is Neptune Engineering (http://www.neptuneeng.com). In 2012, it was announced that the publication would be searchable on the Scalia Safety Engineering website at http://www.scaliaanderson.com/ with an annual $50 subscription fee.

The publication not only identifies clones, but it also indicates the start and end model years of the production run. The judgement of whether vehicles are "clones" is made primarily on the basis of the frontal structure, and its performance in frontal crash tests. Therefore, in considering side impact performance or calculating yaw moment of inertia, one should check whether the "clones" have the same wheel base (which is listed for each vehicle); if not, they are probably not "clones" for those purposes.

Other Information Sources

There are other sources of information for reconstructionists. Enthusiast magazines, such as *Road & Track* and *Car and Driver*, contain a great deal of subjective opinions, and the focus is on exotic cars and sports cars that may never be the subject of a reconstruction. However, vehicles that appeal to a wider market also make their ways into articles. Test results routinely include acceleration, braking, and handling in a slalom or on the skid pad, although again the emphasis is on vehicles on the high-performance end of the spectrum with regard to those characteristics (and perhaps on the high-price end, as well). There are also cutaway drawings presented, though they appear to be artists' impressions rather than drawings that could be relied upon for dimensions.

Another source is the Internet. One might be surprised at the wealth of detail compiled by people interested in a particular car line. There might be information on platforms, production runs, and model year changes in designs and features that would otherwise be available only from knowledgeable product planning personnel employed by the manufacturer. Manufacturers generally treat such information as proprietary. Of course, the accuracy and reliability of information obtained over the Internet may be questioned, and should be used with caution.

An additional source is crash test reports, especially those published by a governmental agency and therefore in the public domain. Of course, crash test performance is the focus of the reports, and Chapters 11 through 14 discuss the acquiring and analyzing of crash test data in detail. That is not the purpose of this discussion. However, certain vehicle dimensions are often measured and recorded in the report, although official manufacturer specifications might be more reliable.

An exception is curb weight. The curb weight specification is becoming increasingly difficult to come by because of the array of vehicle models and option packages in a given car line. However, the procedure for calculating the test weight in a crash test is generally prescribed in detail, and the test vehicle is carefully weighed to ensure that the procedure has been followed correctly. The "as-received" weight, with the vehicle topped off with fluids, is a good measure of curb weight, and is generally recorded by the testing agency. If the tested vehicle is representative of the vehicle being studied, then the "as-received" weight can be used as the curb weight. Often, weights at the individual wheels are reported, as well as the vehicle total.

Finally, one can investigate whether a local enthusiast club, or car club, exists for the vehicle being studied, especially if it has acquired a following for some reason. The Edsel, Jaguar XKE, and the Camaro come to mind; this author once contacted a Corvair car club in a case involving a Corvair. For such vehicles, an enthusiast club may be the only means of locating an exemplar vehicle (i.e., a vehicle sufficiently free of damage or modification that it is representative of the vehicle being studied, at the time of its manufacture).

People Sizes

It is of interest to know the weights of the occupants of an accident vehicle, if for no other reason than to calculate the total vehicle weight, CG location, and inertial properties. Standing height, and particularly seated height, is also important for biomechanical studies, or crash simulations predicting occupant movements in a crash. Occupant information is sometimes obtained in the course of legal proceedings, but often the information is not available. The occupants themselves may be deceased, or otherwise unable to tell how tall they were, or how much they weighed. Sometimes the information appears in medical records, although the height is usually not measured, and the weight may change in the course of medical treatment and recuperation. An autopsy weight may be unrepresentative for the same reason. Driver's licences indicate the height and weight in some states, but this is generally for identification purposes only, is not measured but self-reported (often on the high side for height and low side for weight), and may reflect conditions when the license was obtained—not when the accident occurred.

People's weights and dimensions also change as they develop and mature. On average, men tend to be taller and heavier than women, as is well known. When only age and gender are known for vehicle occupants, which is often the case, weight and dimensions from anthropometry ("human-measure")

TABLE 6.2

FMVSS 208 Dummy Weights and Dimensions

FMVSS 208 Dummy Weight and Dimensions	50th %-ile 6 Year Child	50th %-ile 10 Year Child	5th %-ile Adult Female	50th %-ile Adult Male	95th %-ile Adult Male
Weight (lb)	47.3	82.1	102	164 ± 3	215
Erect sitting height (in.)	25.4	28.9	30.9	35.7 ± 0.1	38
Hip breadth (sitting) (in.)	8.4	10.1	12.8	14.7 ± 0.7	16.5

tables may have to be used. These are data collected for a large population, preferably one that could reasonably include the persons of interest.

One source of anthropometry data is the Centers for Disease Control and Prevention, the CDC. Advance Data from Vital and Health Statistics[26] are reported from data compiled by the National Center for Health Statistics. Statistical data are reported by age, gender, and ethnicity, for children and adolescents, and for adults. Among the variables presented are weight, standing height, and body mass index.

It may be of interest to use the size and weight of anthropomorphic test devices (i.e., "crash test dummies") in the representation of a vehicle. There are various sizes of dummies being used for different test requirements, and detailed drawings and data have been published by the NHTSA. Information may also be obtainable from the dummy manufacturers. Weights and dimensions of dummies are included in Federal Motor Vehicle Safety Standard (FMVSS) 208. A summary is shown in Table 6.2.

Note that standing height is not included. That is because these dummies were designed to simulate seated occupants, and have pelvises that are molded in the seated position. Some standing dummies have been developed for specific purposes, such as for pedestrian impact studies, but are not covered by the FMVSS requirements.

References

1. SAE Recommended Practice J670e, "Vehicle Dynamics Terminology," 9.4.9, *Understeer*, SAE International, Warrendale, PA, 1978, p. 14.
2. SAE, Vehicle dynamics terminology, *SAE Handbook*, SAE Recommended Practice J670e, SAE International, Warrendale, PA, 1994.
3. U.S. Department of Transportation, Motor vehicle brake performance, *U.S. Code of Federal Regulations*, Title 49, Part 571, Standard 105-75.

4. SAE, *SAE Handbook,* SAE Recommended Practice J826 JUN92, Devices for use in defining and measuring vehicle seating accommodation, SAE International, Warrendale, PA, 1994.

5. SAE, *SAE Handbook,* SAE Recommended Practice J1100 JUN93, Motor vehicle dimensions, SAE International, Warrendale, PA, 1994.

6. SAE, *SAE Handbook,* SAE Recommended Practice J687 JUN88, Nomenclature—Truck, bus, trailer, SAE International, Warrendale, PA, 1994.

7. SAE, *SAE Handbook,* SAE Recommended Practice J656 APR 88, Automotive brake definitions and nomenclature, SAE International, Warrendale, PA, 1994.

8. Daubert, D.B., Understanding the nomenclature of highway drawings, SAE Paper 940569, SAE International, Warrendale, PA, 1994.

9. U.S. Department of Transportation, Vehicle Identification Number (VIN) Requirements, *U.S. Code of Federal Regulations,* Title 49, Part 565, Chapter V (10-0-11 Edition).

10. U.S. Department of Transportation, Federal motor vehicle theft prevention standard, *U.S. Code of Federal Regulations,* Title 49, Part 541.

11. ESP Data Solutions, Inc., http://www.vinpower.com.

12. 4N6XPRT Systems, La Mesa, CA, http://www.4n6xprt.com.

13. WardsAuto, Penton Media, Inc., http://wardsauto.com

14. 4N6XPRT Systems, La Mesa, CA, http://www.4n6xprt.com.

15. Truck Index, Inc., P.O. Box 10291, Santa Ana, CA, http://www.truck-index.com.

16. Rasmussen, R.E., Hill, F.W., and Riede, P.M., *Typical vehicle parameters for dynamics studies,* GM Publication A-2541, General Motors, Milford, MI, 1970.

17. Winkler, C.B., Measurement of inertial properties and suspension parameters of heavy highway vehicles, SAE Paper 730812, 1973.

18. Riede, P.M., Leffert, R.L., and Cobb, W.A., Typical vehicle parameters for dynamics studies revised for the 1980s, SAE Paper 840561, SAE International, 1984.

19. Love, M.L., *Final Report on Task 6 of Contract DOT-FH-11-9158,* Texas A&M Research Foundation, 1985.

20. Garrott, W.R., Inertial Parameters of Selected 1988 Four Wheel Drive Utility Vehicles, project VRTC-88-0087, Vehicle Research Test Center.

21. Garrott, W.R., Chrstos, J., and Monk, M., Vehicle inertial properties, *Accident Reconstruction Journal,* (May/June) 1989, 1(3), 22, (Victor Craig, Editor).

22. Curzon, A.M., Cooperrider, N.K., and Limbert, D.A., Light truck inertial properties, SAE Paper 910122, SAE International, 1991.

23. Garrott, et al., *Accident Reconstruction Journal,* 1989, 1(3), 22.

24. Neptune, J.A., Overview of an HVE vehicle database, SAE Paper 960896, SAE International, 1996.

25. Anderson, G.C., *Vehicle Year & Model Interchange List (Sisters & Clones List),* Scalia Safety Engineering, Madison, WI, 2012.

26. McDowell, M.A., Fryar, C.D., Hirsch, R., and Odgen, C.L., *Anthropometric reference data for children and adults: U.S. population, 1999–2002,* Advance Data from Vital and Health Statistics, U.S. Department of Health and Human Services, 2005.

7

Accident Investigation

Introduction

Accident investigation is something of an art: there are techniques to learn and a style to develop. Over time, the techniques become individualized, influenced by the equipment being used, the types of accidents being investigated, and the set of skills one brings to the endeavor. This unique blend makes up the personal style. Sometimes, one can tell who did the investigation by the telltale footprints left behind in the file materials, at the scene, or on the vehicle.

The best way to embark on this journey is to serve an apprenticeship: that is, to work with a mentor, or better yet, mentors. A police academy is another good place to start. This chapter does not replicate or replace that experience; it simply sets forth a few observations that may be of use.

Information Gathering

Most accident reconstructionists do their investigations well after—often years after—the accident has occurred. However, sometimes one finds oneself in a rapid-response situation, arriving at the scene perhaps before it has been cleared of wreckage. A few remarks will therefore be directed at that circumstance.

The first rule of any scene visit is to obey any instruction from law enforcement personnel, who take control of the scene as soon as they arrive. Regardless of the seriousness of your purpose, you will be officially considered a bystander, and will not be allowed close. Therefore bring the longest camera lens you have, but be prepared to have your view blocked most of the time. Your main objective is to document photographically the most volatile evidence; for example, roadway marks that will soon be obliterated by traffic, and highway damage that will soon be repaired. Your vehicle must not block traffic or contribute to the congestion, and you must not place yourself in danger. Measurements and a detailed investigation will have to come later, particularly if the accident is at night.

In the typical situation, however, you will not have been at the scene. Therefore, use the police photos, if available, to mentally place yourself there. Make sure you ask for them, and try to schedule any scene inspection(s) after you have a chance to study the photographs. It is very helpful to organize the photos spatially—not chronologically—because something unidentifiable in the distance of one photograph may be clearly seen in the foreground of another. It is also useful to letter all the things one wishes to measure, right on the photograph (or a clear plastic overlay that is attached). To be sure that a clean copy is available for eventual use in exhibits, make a copy solely for investigation purposes, if necessary. Rather than verbal descriptions, the identification symbols on the photographs can be used with consistency for the measurements, field notes, survey notes, and so on (Letters are preferred to numbers, so as to avoid confusion with point numbers on a survey or measurement data sheet.) Once having gone through this process, you are far better prepared for the scene inspection. Ideally, when you get to the scene for the first time, it will seem as if you had been there before.

Typically, at-scene photographs taken at night are not terribly useful because the pictures are so dark as to obscure tire marks and other evidence, and dominated by light sources and reflections from reflectors, highway signs, reflective vests, and so on. Some police agencies demarcate with paint the tire marks and the locations of the tires of the vehicles at rest, and if the accident is at night, return to the scene during daytime to take photographs. A great advantage of that procedure is that the scene and its evidence are well lit and far easier to discern in the ensuing photographs, and the view is not blocked by rescue and law enforcement personnel and emergency vehicles. Of course, the accident vehicles are also gone, but if the tire positions have been painted, that is not a problem. In fact, removal of the accident vehicles makes the rest positions easier to measure, and better documented in the photographs. Law enforcement agencies everywhere are strongly encouraged to follow this procedure, which should also speed up the removal of the accident vehicles, since that can be done without having to wait for measurements to be taken. Moreover, the measurements can be taken at another time, when interference with traffic can be minimized.

Quite often, the condition of the accident vehicles changes, sometimes even before they are removed from the scene. Extrication tools may be used to cut off doors or the roof in order to treat the occupants; dangling parts—such as wheels and bumper covers—may be removed in order to tow a vehicle. Therefore, at-scene pictures of vehicles in serious accidents should be taken as soon as possible, although that task must be given a lower priority than rescuing the occupants, providing traffic safety, securing the scene, and so forth.

The police investigation is distilled into a report. The accident investigator should always request a copy of the police report, and review it thoroughly when the investigation begins. Other materials to request for review are:

- *Insurance photos and repair documents.* These have increased importance if police photos are not available, and may show damage not visible in other photographs. If the vehicle is not available for inspection, these are even more important.

- *Witness testimony in either statement or deposition form.* Witness testimony is notoriously unreliable, and therefore should be subordinated to physical evidence and the laws of physics. Nevertheless, it should be considered. Sometimes, puzzling physical evidence or analytical results can be explained by what the witnesses say.

- *Investigator photos.* In almost all cases, they are taken by a professional investigator, long after the accident, after some or all of the scene evidence is gone, and when the scene may have changed. On the other hand, they offer a representation of the scene that will be more up to date, and therefore may be a considerable help in orienting oneself to the scene before an inspection. Similarly, photos of the vehicle(s) are an aid in getting acquainted with the vehicle. They are not a substitute for an inspection in person.

- *Partial medical records.* Descriptions of the course of treatment and rehabilitation are generally not helpful to the reconstructionist. However, medical records may contain information on the occupant's weight. Admittance records generally indicate how the occupant "presents," in medical terms. Injuries are listed, described, and often diagramed. This information provides the reconstructionist with the main areas that were contacted, which helps look for matching areas in the vehicle. It is not the job of the reconstructionist to explain the injuries, and some other expert may be given the task of describing in detail how the various parts of the body moved (occupant kinematics). However, it is helpful to know at least the general direction of occupant motion. Since the occupant(s) are not rigidly fixed to the vehicle, even if they are properly wearing their seat belts, occupant motions are a reflection of the direction of the impulse applied to the vehicle. This should be verified against any calculations that are subsequently made.

- *Weather data.* The timing and amount of rainfall or other precipitation may affect the reconstruction, through estimates of the friction coefficient. Friction is also strongly affected by ice and snow, so the temperature may be a significant factor in some cases. In other cases, the timing of sunrise or sunset, or even wind direction, may be important. All this information may be gleaned, or at least estimated, from weather records, often hourly, maintained by the National Oceanographic and Atmospheric Administration (NOAA).

Scene Inspection

The first task in a scene inspection is to be sure that you are in the right place, by consulting the police report and studying the available photographs, unless prior study enables you to recognize it when you get there. Needless to say, it is embarrassing to do a detailed inspection of the wrong scene, though sometimes a lack of at-scene photos and/or erroneous information in the police report can conspire to just that end.

The second task is to do a visual survey of the entire scene. Often, it is helpful to do a drive-through in each direction the accident vehicles were traveling. In any case, one should walk the scene on foot, covering the area from impact(s) to the rest positions of all vehicles. Often, one will find vehicle debris in off-pavement areas, but it is often not possible to ascertain if it came from one of involved vehicles. The debris may have come from another accident, or no accident at all. If accident-related debris is found, it should be photographed as is, and a marker placed at the location so that it can be found again when a survey or measurements are made. Photographs with placards can be used to keep the identification of the debris straight.

As discussed in Chapter 15, some debris, such as a pattern of tempered glass particles off the paved surface, can remain relatively undisturbed, and can provide important information, such as where the vehicle was when the glass broke. Other objects, such as a large bumper fascia, are almost sure to be moved. One must be careful, therefore, in analyzing debris seen in an at-scene photograph, especially if the debris is at the side of the road. Debris has been known to "migrate" there, with the assistance of helpful bystanders and scene responders.

A product of almost all scene inspections is a scale drawing, upon which will be shown salient features, evidence, and possibly dimensions. It is helpful to think about how the drawing will be oriented on the scene, and on the paper or computer screen. It is also helpful to have the drawing on one page, which may be facilitated by utilizing a landscape format, especially for long scenes. If a total station is to be used for a survey, the orientation may depend on how the survey is set up, and what reference points are to be used. Any reference points or permanent points that were used in the police investigation, such as utility poles, bridge abutments, and so on, should be included. In fact, anything that is measured in the police investigation, and that is still present, should be included.

Occasionally, official survey benchmarks will be found, or paint marks from a prior inspection. If these are surveyed, there is a good chance that your drawing can be overlaid with drawings from other surveys that have been done, thus allowing the comparison of rest positions, tire marks, and so forth.

When doing the reconstruction calculations, it will generally be necessary to locate all objects and positions in an X–Y grid. Therefore, it has been this author's practice at the time of the scene inspection to decide where the X- and Y-axis will be, and to create permanent reference points for the origin of coordinates, and a point on the X-axis, by driving a "mag nails'"* (which can be obtained from a survey supply store) into the pavement, an expansion joint, or perhaps a wooden utility pole. That way, if an additional scene inspection is required (which has been known to happen), the reference points can be found again and reused, thus making it easy to merge the surveys.

During a survey and subsequent analysis, it is helpful to remember the dog that did not bark. In other words, it is helpful to look for the absence of evidence, as well as existing evidence. If an extremely violent crash occurred at some location, why are there no gouge marks? If a vehicle hit a roadside marker, why are there no marks on the vehicle? Or if there are marks from a sign post on the vehicle, why is the sign still standing? Or was the sign replaced? If so, was the old sign at a slightly different place and cut off, and is there a cut off stub still visible? Is there evidence that something was dug up? If a vehicle supposedly went through some area, why is there no damage to the bark of the tree (not the dog)? Walking the scene on foot is particularly important in answering such questions. Also important is documenting, by photographs and measurements or survey, any objects that could possibly have been involved in the accident, yet show no evidence of it.

If gouges or other pavement damage are a part of the evidence package, that evidence should be permanent. In such cases, it is important to take careful note of the pavement condition. Has the roadway been repaired or resurfaced since the accident? If not, is the pavement damage due to the accident at hand, or something else? Fresh damage is distinguishable from old damage, and police are usually trained to tell the difference. Generally, if they took a good picture of a gouge, it is because they interpreted it to be accident-related, and their opinion on that subject has to be respected. If good at-scene photographs are available, a careful comparison with the actual road will allow you to identify the actual gouge by its details, even though its appearance will have changed. On the other hand, if the pavement has been resurfaced or repaired, the evidence will have been destroyed. Therefore, if you are going to rely on surface damage, you can avoid professional embarrassment by ensuring that there was no re-surfacing since the accident. Sometimes, the comparison of tiny cracks and imperfections with those seen in at-scene photographs can help make that determination.

Of course, any evidence turned up during the scene inspection should be photographed. But tire marks tend to disappear quickly when traffic passes over them, or when it rains, although they can persist where they are out

---

* Manufactured by ChrisNik, Inc., Cincinnati, Ohio.

of the traffic. Gouges change their appearance, and scrapes often disappear altogether. So what is hard to see in real life may not be detectable at all in a photograph. Of course, a good quality camera and lens(es) are a prerequisite, but technique and ambient conditions are also important. Low sun angles (i.e., early in the morning or late in the day) can make surface features much easier to see in one direction than another. If you are taking pictures at mid-day, consider taking another set late in the day. You may be surprised in the difference it makes. Also, rain may turn out to be a blessing in disguise. Water will accumulate in the depressed areas, and dry out more slowly there. As asphaltic concrete (i.e., blacktop) dries out, it goes from black to gray. If your timing is right, all the gouges will be beautifully highlighted by that effect. Paint can also be used for highlighting. However, paint should be kept out of the gouges, in consideration of subsequent investigators. It is better to paint circles around the gouges, where the circles can be worn off by passing traffic.

Roadway striping (i.e., white fog lines, white skip lines, yellow median lines, etc.) often provides important information on the location of evidence. Skip lines depicted on a police diagram usually cannot be reliably scaled off the drawing, unless the ends of the lines have been specifically measured and reflected in the drawing. The lines can be very useful, however, in the interpretation of images and getting measurements from photographs (i.e., photogrammetry, which is discussed in Chapter 8). However, has the road-way been re-striped since the accident? If more than a couple of years have passed, there is a good chance of it. If re-striping has occurred, are the lines in the same place? Generally, the lateral positions do not differ by more than a few inches. The longitudinal positions of skip lines may vary more (to per-haps a foot), but often where there are discrepancies, the old line is still vis-ible. It is recommended to use the old lines for surveying or measuring, as long as they can be seen reliably.

From time to time there is a question of when an accident-involved indi-vidual could see another person, vehicle, or object. Of course, the reconstruc-tionist cannot testify to what anyone saw or when they saw it, any more than he or she could testify to anything else in someone's mind. If it was dark, or there were atmospheric disturbances such as fog, dust, or smoke, those would be beyond the reconstructionist's expertise as well. However, a line of sight is not a question of vision or visibility. It is just a question of whether a straight line can be drawn between two points without passing through an obstacle. That is a question to which the reconstructionist can obtain an answer.

If vertical curvature is involved, the roadway itself may obstruct lines of sight. A survey of the roadway will produce elevation (Z-direction) data, which is necessary to construct a profile, as described in Chapter 6. A locus of eye points above the profile can then be constructed. Often, a long sur-vey will be required—perhaps up to a mile. In such cases, though, 100-foot spacing between survey points will probably suffice. A profile view from an as-built drawing can also be used, again keeping in mind the heights of the eye points above the ground. Generally, the as-built profile will reflect the roadway centerline rather than the vehicle's path of travel, but the differences should be negligible.

Sight lines can also be interrupted by horizontal curvature and intersections. Those situations are a little more complicated, in that the obstructions are almost certainly not represented on as-builts. Moreover, the obstruction may be irregular, like a dirt embankment or foliage. It would take a great deal of surveying to capture enough of the geometry to work out the lines of sight on a drawing. In such situations, two assistants can stand a few hundred feet apart so that they can barely see each other around the obstacle in question. The position of each can be surveyed; then they can move appropriately, and the process can be repeated. On the resulting scene drawing, the set of lines drawn between each pair of points will form an outline of the obstacle. These lines can be extended if necessary, to cross the paths of travel of persons at the time of the accident. The procedure should not require more than a few such lines, and has the advantage of physically evaluating the sight line issue in question, which should appeal to clients and juries. Of course, one has to ensure that any foliage contributing to the line of sight obstruction at the time of the scene inspection is reasonably representative of that at the time of the accident.

Scene topography often comes into the picture when one or more of the vehicles run off the road. Of course, elevation data can be gathered for reconstruction purposes by simply surveying a line along the path of travel of each vehicle. This is the most direct way, but it often happens that the paths of travel are not known at the time of the scene inspection. It may be appropriate to survey sections and profiles at regular intervals, so as to in effect "build" a travel surface for the vehicle or vehicles. Roadway sections that extend to the right-of-way boundaries may be surveyed to show the contour of ditches, embankments, roadway shoulder, and so on. These sections will not tend to look as impressive on a drawing as the terrain does in person. The same can be said for photographs.

The topology of the roadway is important when one attempts to establish vehicle paths or positions by analyzing fluid spills. One needs to remember that water runs downhill, and so does fuel in a fire case. A scene survey will provide the necessary Z-direction data. Also, in fire cases the fire propagation is affected by the wind direction, which can be ascertained from weather data. The survey of a North arrow not only looks good on the drawing; it may also be important to some aspect of the reconstruction.

Vehicle Inspection

One should prepare for a vehicle inspection by reviewing the case materials—most importantly, any photographs taken by the police or perhaps other investigators. The photographs probably do not require the detailed level of study that the scene photos do; nevertheless, it is good to get familiar with the vehicle and plan the inspection, at least mentally. It is not a bad idea to make up a to-do list specifically for the inspection. Sometimes, standardized inspection checklists are used for this purpose, but the reality is that every crashed vehicle is different. It is probably not a good idea to rely too heavily on a standardized checklist, but rather to review it after the inspection is well underway and see if any new thoughts are triggered, or if anything has been forgotten. It is a harsh fact of reality that at least one photograph does not get taken, even with the most thorough preparation.

The first thing to do at a vehicle inspection is check in with the custodian, and get permission to inspect the vehicle. Often, a representative from the opposing counsel's office will need to be present, especially if they have custody of the vehicle. It is not unusual for opposing counsel to make a video recording of the inspection. The next thing to do is take a picture of a placard containing salient information: the name of the case, the date and location of the inspection, and so forth. Some investigators prepare a placard in advance for this purpose.

The next step is to witness the removal of any protective tarps and/ or movement of the vehicle. If one has studied accident vehicle photos in advance, one can nearly always recognize the actual vehicle at first glance. After verifying that the inspection is of the correct vehicle, the investigators usually circle the vehicle, in kind of a choreographed dance, taking overall pictures. Usually they are careful to stay out of each others' pictures. This first series of pictures is often taken from eight angles: front, sides, rear, and quarter views. In that way, the investigators have a visual record of the vehicle condition as they found it, and the assurance that every part of the exterior has been photographed, though probably not in satisfactory detail. There is also visual documentation of any steps that have been observed in preparing the vehicle for inspection.

It is good idea to develop a standard sequence for vehicle inspections, in that it makes the organizing of photographs and notes much easier. It is recommended that blanket photography be used for the vehicle exterior, one panel at a time (e.g., rear bumper and deck lid, backlight, right rear quarter panel, right rear door above the belt line, right rear door below the belt line, and so forth). The purpose is for these photographs to serve as mid-range images: between the overall photos and the close-up ones. That way, the area seen in a close-up photo can be located in a mid-range photo and thus identified. A close-up photo is of no use if you cannot tell what the photo is of, or where the part is located.

Many investigators use portable voice recorders, and dictate their notes. Others rely on pencil and paper. Sometimes Polaroid photographs are taken for the purpose of writing notes, though that practice has become rare. Other investigators use video. Of course, certain areas of the vehicle will be revisited, with closer inspection, depending on the kind of accident, location and extent of damage, and so on. It is helpful to have a command of vehicle terminology (Chapter 6) when writing or dictating notes.

The vehicle identity should be fully documented. VINs can be written down or dictated, but it is helpful to photograph the VIN plate in the windshield, and the certification plate on the driver's B-pillar, to ensure against misreading the information. A close-focusing lens will be required. Emblems that help identify the vehicle—particularly the option package—should be noted and photographed, as necessary.

It is a good idea to bring to the vehicle inspection any at-scene photos of the vehicles. That way, one can compare the photos to the actual vehicle and attempt to determine whether certain damage existed immediately after the accident, or occurred later. It would be embarrassing to base the reconstruction on damage that is clearly not accident related.

Loose parts and other debris may have been cleaned up at the scene, and tossed into the vehicles before they are moved. The loose parts are usually still in the vehicle at the time of inspection (though they may have been removed during a prior inspection, and then replaced). It is usual practice to remove these items from the vehicle, though one must be sure there is permission to do this. If the parts are arrayed in an orderly fashion on the floor, photographs of them may show evidence that otherwise would have been missed. There is always the possibility that some of the parts came from one of the other accident vehicles, or not from the subject crash at all. Some parts can be positioned against the vehicle to determine where they came from (e.g., right or left A-pillar trim). Any part that is large enough should be inspected with its provenance in mind; often, plastic parts have identifying information molded into their surfaces that are not normally visible after the vehicle is assembled. Parts known to be not from the vehicle in question should be kept separate. Since damage may show on both sides of a plastic part, each such part should be photographed from both sides. Obviously, care should be used in taking loose parts out of the vehicle and replacing them, so as not to create additional, nonaccident related, damage.

Missing parts that are normally attached with screws or bolts may have been removed post-accident, or ripped off during the crash. At-scene photos will provide important clues. Also, parts ripped off during the crash will leave behind broken or bent fasteners, torn sheet metal, and so forth. Lack of such evidence will indicate the deliberate removal of a part. Any such evidence, or lack of it, should be noted and photographed.

It is not unusual to find that the vehicle has been scavenged prior to the inspection (i.e., parts have been removed). Parts that are not available for inspection should be noted. Other parts may be present, but partly or wholly detached from the vehicle. Sometimes these can be reattached by applying the fastener nuts finger tight. For example, doing this to a suspension crossmember (with permission, of course) may help in ascertaining the wheel location in the damaged vehicle, and whether it was displaced as a result of the crash.

Tire conditions should be noted and photographed. Tire brand and size should be noted routinely. If pre-crash hydroplaning is an issue, tread depth will be important. In that case, all tread depths should be measured. Extreme tire wear, such as bald spots and exposed cords, should be looked for and photographed if found. If vehicle dynamics are at issue, tire pressure may be important, and some inspection procedures have called for the measurement of all tire pressures. However, how does one know that the observed pressure is representative of the pressure just before the crash? In this author's experience, it suffices to observe visually whether the tire appears fully inflated, partly inflated, or deflated.

As mentioned in Chapter 2, one or more wheels being bound up in the wreckage is an important factor in analyzing the run-out trajectory. To the extent possible, a determination should be made for each of the wheels, and duly recorded, at the time of inspection. The limit conditions—either no damage in the area, or a wheel and de-beaded tire shoved into crumpled up sheet metal—are obvious. Between those limits, a tire may be rubbing on the wheel well when the vehicle is on the ground, but not so when the suspension is in a rebound position. The wheel well and tire tread should be inspected for evidence of rubbing. Sometimes, the tire has been cut by jagged sheet metal.

Nighttime accidents may involve questions about whether the headlights were on at the time of the crash. The position of the headlight switch is an important clue, but not necessarily definitive because of possible post-crash manipulation. To obtain more reliable information, the lamps themselves can be examined. Brake light illumination can be another important issue, whether the crash was at night or not. Again, the lamps—and more importantly, the filaments—may tell the story.

Not uncommonly, lamps are removed by the police, and are unavailable for inspection by the investigator. Close-up photography of accessible bulbs can often show the filament(s) if the lens is focused carefully. Sometimes bulbs can be removed for examination and photography (with permission, of course), but one must be sure to return the bulbs exactly to where they were found. Fractured bulbs should not be touched or disturbed. Before considering attempting a forensic analysis of the lamps, keep in mind that such work is a specialty all of its own, which can be appreciated by reading the book by Rivers and Hochgraf.[1] Accordingly, the subject will not be discussed further here, other than to say that filament examination may not be conclusive.

It is usually the case that a severe crash will result not only in crumpling and tearing the sheet metal, but also in rupturing some of the welded joints that hold the sheet metal panels together. Most of these joints consist of spot welds. It does not hurt to document the ruptured welds photographically, although it can be a tedious process. Very occasionally, there will be an allegation of defective spot welds, either in the design of their spacing, or of the location, size, or quality of welds achieved during manufacture. A metallurgist or weld specialist may be called in to evaluate the situation. Perhaps as a prelude to that, the reconstructionist may be asked to provide a closer weld inspection than usual. In that case, it is a good idea to give the welds in question identifying numbers, using removable labels, and take photographs of the entire joint, or as much of it as can be seen. Close-up photographs should also be taken of the individual welds, which again requires a macro lens. It should be noted whether the "button" or "nugget" of fused material that makes up the weld has stayed intact, ripping the parent metal of one of the panels, whether the nugget has fractured, or whether the nugget was of unusual size, or did not exist in the first place.

The vehicle interior is also of interest to the reconstructionist. As mentioned previously, evidence of occupant contacts may provide an indication of the general direction of occupant movement, which should be checked for consistency with the calculated direction of the impulse applied to the vehicle during the crash. If the crash is severe enough, evidence is often seen from knee contacts to the lower instrument panel, steering column trim, outboard surface of the center console, or glove box door; torso contacts to the steering wheel or inner door panel and trim; and/or head contacts to the windshield header trim, head liner, visors, A-pillar trim roof rail trim, or head restraints. Of course, there are other areas in the interior that can show contact evidence as well. A front seat back may be deformed rearward if the seat was occupied during a rear crash, for example. A windshield may be cracked due to a contact by the head or a hand.

It is usually the case that one of the experts will be asked to develop opinions regarding detailed occupant kinematics (how the various parts of the body moved, what surfaces they contacted, and when contacts were made). A biomechanics expert will often be assigned this task, but very occasionally the reconstructionist may be asked for an opinion. In either case, the kinematics opinions depend on the reconstruction, although the reconstructionist should not be tempted to venture into areas that are outside his or her expertise. For example, the question: Is a contact to the vehicle matched by an injury to the body? gets very quickly into medical issues that the reconstructionist may not be qualified to deal with.

Other related areas of expertise are those of restraint systems and occupant protection. If there is a claim of defect in crashworthiness (the ability to protect an occupant in a crash), an expert in restraints will probably be called in for an evaluation. That individual will also interact with the biomechanics expert with regard to contact evidence in the interior, with an emphasis on the seat belts and air bags, of course. If there is not such a defect claim, the reconstructionist may be asked to examine the belt systems to determine whether they were in use at the time of the crash. That is another specialty all of its own, with a number of references,[2–4] that may be consulted, and will not be discussed further here, except to say a little bit about the inspection process. First of all, the investigator must take care to ensure that no evidence is added to or subtracted from the webbing or the belt system hardware. If more than one belt system is inspected (which is often the case), it is a good idea to make a tiny label which can be placed in the field of view of the camera, indicating which belt system (LF, RF, etc.) is being depicted. The webbing can be photographed in sections, showing perhaps 6 in. to a foot of the webbing each. A cloth dress maker's tape can be used for identifying the locations shown in the pictures. Metal end trim on the tape can be removed (usually by just prying it off) without disturbing the tape, which allows the tape to be threaded through the belt hardware, along with the belt itself (thus providing a truer representation of the belt length compared to trying to use a steel tape measure). That way, locations can be described in terms of their distance from some known point on the belt (e.g., the end of the loop through an end attachment bracket). The numbers on the tape can then appear in the picture with the webbing.

Crush Measurement

For all except minor crashes, the vehicle inspection includes a sufficient measurement of the damage geometry to permit the measurement of crush. Generally, crush measurement cannot be completed at the inspection because of the need to compare to an exemplar vehicle, which is generally not present. Techniques for the measurement of damage geometry and crush are discussed in some detail in Chapter 22.

References

1. Rivers, R.W. and Hochgraf, F.G., *Traffic Accident Investigators' Lamp Analysis Manual*, Charles C. Thomas, Springfield, IL, 2000.
2. Moffatt, C.A., Moffatt, E.A., and Weiman, T.R., Diagnosis of seat belt usage in accidents, SAE Paper 840396, SAE International, 1984.
3. Bready, J.E., Nordhagen, R.P., Kent, R.W., and Jakstis, M.W., Characteristics of seat belt restraint system markings, SAE Paper 2000-01-1317, SAE International, March 2000.
4. Tanner, C.B., Durisek, N.J., Hoover, T.D., and Guenther, D.A., Automotive restraint loading evidence for moderate speed impacts and a variety of restraint conditions, SAE Paper 2006-01-0900, SAE SP-1999, SAE International, 2006.

8

Getting Information from Photographs

Introduction

Very often the reconstructionist will not be called upon until the passage of time from the occurrence of the accident that is measured not in days, weeks, or even months, but years. By that time, all of the volatile evidence will have disappeared. Cars may have been repaired or sent to the crusher, the road may have been repaved, tire marks will have disappeared, fluid spills will have washed or worn away, debris will have been removed, signs may have been moved, foliage will have grown or been cut down, and so forth. But perhaps someone took pictures. Maybe bystanders or newspaper photographers or insurance investigators or parties to a lawsuit took pictures that are unearthed during the discovery process. Perhaps police agencies took photographs, in which case there is often an indication of such in the written police report. (Although police photographs have been known to "not turn out," disappear, or even mysteriously re-appear at the time of trial.)

A word about newspaper photographs: Generally, news media will not provide copies of photographs or videos without a subpoena, and even then, they will only produce materials that have already been published or broadcast. Therefore, it is a job for the attorneys to see whether such materials exist, and attempt to acquire copies. Reconstructionists should be aware that cameras and lenses used by news media can be expected to go far beyond what is available to police agencies or other nonprofessional photographers. Since bystanders are usually kept at a distance at the accident scene, it should not be surprising to find the media using very long lenses.

Why all the interest in photographs? Because there is objective information stored therein, waiting to be mined. Usually the breadth and depth of detailed information in good photographs extend far beyond any measurements or notes that could have been taken contemporaneously, and the information stays permanently recorded, as opposed to being ephemeral or changeable or transitory, like memories. The subject of this chapter is how to extract photographic information that can be used to re-establish the pertinent evidence, so that it can be used in the reconstruction.

Photographic Analysis

Every reconstructionist needs to have a good understanding of how images are formed in a camera. The simplest construct, both physically and mentally, is a pin-hole camera, in which light rays pass through an infinitesimally small pin hole (or lens center) and impinge upon a plane formed by some light-sensitive film (or electronic receptors, these days). This plane is called the film plane or focal plane. For all intents and purposes, a good quality camera lets the light rays through as if they had passed through a pin hole, unaltered, meaning that their paths through the pin hole are straight. (Though with wide-angle lenses, some distortion can occur near the image edges.)

The perpendicular distance from the lens center to the focal plane is the focal length. A light ray passing through the lens center that is perpendicular to the picture plane (the axis of vision) will intersect the focal plane at a point called the center of vision, or principal point. A light ray that passes through the lens center at an angle from the axis of vision will also intersect the focal plane at that angle, at a location away from the center of vision, helping to create the rest of the image.

If an oncoming locomotive is far away, the portion of light rays coming from its front that pass through the pin hole will have to converge at only a slight angle to do so. The same can be said for light rays coming from the railroad tracks under the train. The locomotive front and the tracks will occupy only a small part of the total image, and will thus seem small. On the other hand, when the locomotive is close, some of the light rays coming from its front that pass through the pin hole must be those that converge at a much greater angle than when the train was distant. Similarly, the light rays that come from the railroad tracks in the foreground and pass through the pin hole do so at wider angles than from the distant track. The locomotive and the tracks occupy a larger portion of the image. "Perspective" is this geometry that makes an onrushing locomotive loom larger as it approaches, and that makes (parallel) railroad tracks appear to come together (converge) at a point in the distance.

This perspective is not altered by the choice of camera, lens, focal length, f-stop, and so on. The focal length will affect the range of angles captured in the image. With a short focal length, the range of angles in the entire image is wide, and the angles of rays coming from a given object will occupy a small portion of the entire range. The object will appear small in the image. Conversely, a long lens captures a small range of angles; the angles of rays coming from the same object will occupy a large portion of the entire range. The object will thus appear large.

Except for lines parallel to the picture plane, all lines that are parallel to each other will appear to converge in the distance, at a "vanishing point." Parallel lines on flat ground, such as a pair of railroad tracks, will appear to come together at their own vanishing point, which is obviously on the ground infinitely far away. Another set of railroad tracks oriented in a different direction will appear to come together at a different vanishing point, which must also be on the ground infinitely far away. The locus of vanishing points, for sets of parallel lines in all possible directions in a plane, forms a line of its own: the "horizon line." Since the horizon line is infinitely far away (the earth must be flat after all), a light ray from the horizon line to the center of vision will be parallel to the plane. Thus, all objects in the picture that are intersected by the horizon line are at the same distance from the plane as the viewer. Specifically, all objects in line with the ground horizon are at the picture taker's eye height.

This is true regardless of how the camera is tipped (or "posed"). If the axis of vision is parallel to the plane, the view axis will pass through the horizon line, which will be in the middle of the image. On the other hand, if the camera is rotated so that the axis of vision is tipped sufficiently far away from the plane, no horizon line will appear in the picture. (It still exists; it just will not be included in the image.)

Analyzing a photograph graphically by drawing lines that obey the laws of perspective, in order to ascertain the locations of objects in three-dimensional space, is called photographic analysis.

Although photographic analysis may appear to be turning into a lost art (as it recedes in the distant past?), it is in fact an essential part of the reconstructionist's skill set. Every person setting out on a career as a reconstructionist can do no better than read, mark, and inwardly digest Kerkhoff's paper on the subject.[1] Doing so will make the subject of photogrammetry much easier to understand.

One of the sections of Kerkhoff's paper is titled "Photograph Correlates." This section deals with what Kerkhoff calls "lines of alignment," in which two points in a picture lie on a vertical line. The vertical line represents a plane, seen edge-on, in which the two objects and the camera's lens center lie. If two such lines can be drawn, the X–Y location of the camera is uniquely defined. In other words, if a person can go to a scene and adjust his or her position so that those lines of alignment with still-existing objects exist as they do in a photograph, he or she will be located at the intersection of the planes represented by those lines.

While such an exercise would enable one to duplicate the appearance of all the permanent objects in the photograph, that is somewhat beside the point. However, if some evidence, no longer present, is shown in a photograph to lie on a line of alignment, then the evidence itself must have been positioned in the plane represented by that line. By measuring the position from which the picture was taken, and the position of the permanent feature through which line of alignment is drawn, the line can now be constructed on the scene drawing. If two photographs, taken from different points, each has a line of alignment for the evidence, its location on the drawing can be established at the intersection of the two lines; that is, by triangulation.

It is easily appreciated that with the focal length and the size of the image all defined, a defined geometry exists between the points on the image, the light rays, and the points from which the light rays came. If somehow that geometry can be understood and duplicated or analyzed, the rays that made up the original photograph can be reconstructed, and their intersections with the original objects re-established.

If perspective lines are being drawn on a picture of a physical surface (like a roadway) in order to do photographic analysis, it is important that the surface coincides with the lines (i.e., that the surface be flat). Otherwise, errors will occur. The surface does not have to be level, but it has to be flat (since straight lines are being drawn on it). This is also a central feature of both photographic analysis and two-dimensional photogrammetry. It is, of course, perfectly acceptable to have more than one flat surface in the picture (such as the two sides of a crowned road), as long as they are analyzed as two separate planes.

Mathematical Basis of Photogrammetry

As its name implies, "photogrammetry" is the science of measuring things by using photographs. The mathematics behind the science can be understood from the pin-hole camera analogy, wherein the light rays that pass through the pin hole do so without having their straight-line paths lens changed (bent) as they pass through the lens. Some lens distortion can occur at large ray angles (i.e., near the edges in a wide-angle lens), but for the most part, the assumption that the point on the object in three-dimensional space (object space) is colinear with the point on the image in two-dimensional space (image space) is valid for good quality lenses. Of course, the image is created by sensing and storing the light pattern received on the focal plane of the camera, and projecting it through the optics of a printer, slide projector, or enlarger. In our discussion, all the lens systems are combined, and behave like an ideal pin hole.

The assumption of colinearity gives rise to the so-called "direct linear transformation," or DLT, method, a term first used in 1971.[2] Under this assumption, a point (x,y) in image space has coordinates given by

$$x = \frac{a_1 X + a_2 Y + a_3 Z + \alpha}{c_1 X + c_2 Y + c_3 Z + 1} \tag{8.1}$$

and

$$y = \frac{b_1 X + b_2 Y + b_3 Z + \beta}{c_1 X + c_2 Y + c_3 Z + 1} \tag{8.2}$$

where (X,Y,Z) is the point location in object space, and where the a's, b's, c's, α, and β are the 11 DLT parameters for the particular photograph.[3]

If the object space is a plane (i.e., all the physical points lie in a plane), then the X, Y, and Z coordinates are not independent. Since three points define a plane, we can see that enforcing the condition that all the object-space points lie in a plane is tantamount to imposing three constraint equations on the points (X,Y,Z). Relative to the image plane, the object plane can be in any position and orientation (except edge-on, obviously). However, the planar condition for object space imposes three constraint conditions on the DLT parameters and reduces the number of independent constants by three, to eight in all.

The process of imposing planarity is discussed in the 1994 paper by Tumbas et al.[4] The result is a simplification of transformation Equations 8.1 and 8.2 to the form

$$x = \frac{D_1 X + D_2 Y + D_3}{D_7 X + D_8 Y + 1} \tag{8.3}$$

and

$$y = \frac{D_4 X + D_5 Y + D_6}{D_7 X + D_8 Y + 1} \tag{8.4}$$

where the D's are the eight constants that are derived from the original 11.

Of course, photogrammetry consists in measuring the (x,y) coordinate pairs on the image, and determining the (X,Y) coordinates in object space. To make that determination, Equations 8.3 and 8.4 must be solved for X and Y. As shown by Tumbas et al., the inverted transform takes the form

$$X = \frac{C_1 x + C_2 y + C_3}{C_7 x + C_8 y + 1} \tag{8.5}$$

$$Y = \frac{C_4 x + C_5 y + C_6}{C_7 x + C_8 y + 1} \tag{8.6}$$

These are the same equations cited by Kerkhoff in 1985, although the numbering of the calibration coefficients C_i, $i = 1,2,\ldots,8$, is different.

Two-Dimensional Photogrammetry

The term "two-dimensional" means that the object space is a plane. Thus, Equations 8.3 through 8.6 mathematically describe a plane-to-plane transformation. The geometry of three-dimensional objects can be analyzed only insofar as they have points positioned in a plane (such as the locations of the bottoms of the tire sidewalls on a flat road).

If we can find the calibration coefficients for the particular picture, we can measure points on the photograph, work through the inverted transform, and figure out where the physical points were in the plane.

So how do we find the calibration coefficients? We start by rearranging Equations 8.5 and 8.6 as follows:

$$C_1x + C_2y + C_3 - C_7Xx - C_8Xy - X = 0 \tag{8.7}$$

$$C_4x + C_5y + C_6 - C_7Yy - C_8Yy - Y = 0 \tag{8.8}$$

noting that the transformation is linear (i.e., the transformation is the same everywhere in the picture; the C's are independent of x, y, X, and Y). This means that Equations 8.7 and 8.8 are linear in the C's. The C's could be treated as the unknowns and solved for, if only there were eight equations instead of two.

To find more equations, we recognize that for every picture point (x,y), there existed an object point (X,Y). If the picture point can be measured (using a digitizing tablet, for example), and if the corresponding object point can be measured (using a total station, for example), then Equations 8.7 and 8.8 can be written for that point, with x, y, X, and Y as knowns, leaving the coefficients C1–C8 as the unknowns to be found. Such a point is called a "control" or "reference" or "calibration" point. Of course, each control point results in two equations. Therefore, the selection and processing of four control points results in a system of eight linear equations for the eight unknown calibration coefficients. If the equations are independent (i.e., if no set of three of the control points lie on a straight line), they can be solved for the C's.

After solving for the projection coefficients, the inverted transform can be used to find the (X,Y) locations of any other object point, called a "mapped" point, for which the corresponding picture plane locations (x,y) have been measured. Kerkhoff's paper presents a solution using four sample control points.

Computer programs that have been written to effect the utilization of the plane-to-plane transform for photogrammetry include FotoGram,[5,6] by Brelin et al., TRANS4[7] by Kinney and Magedanz, and PLANTRAN[8] by Smith. The FotoGram program was developed in the 1960s for use on a main frame computer at General Motors, and was later ported to DOS and made available for sale to the general public on a single 5¼ in. floppy disk (which tells something about the evolution of computer technology). Lessons learned from experience with FotoGram will be discussed briefly, as those lessons are applicable to all planar photogrammetry programs.

First of all, it should be recognized that four control points are required for the procedure, but in any given scene, more than four such points may be available. So which points to choose?

FotoGram seems to be somewhat unique in that it considers all possible combinations of four control points chosen from a set of up to 10 points. Only one combination is selected for the final matrix analysis: that which "best represents the projection plane."[6] No explanation is offered as to how the representation is numerically evaluated, or what the selection criteria are. In TRANS4, multiple photographs of a given control point may be used, "judiciously selecting for each point's X- or Y-coordinate the view that provides the most accurate transformation"[4] (p. 468), which photograph is then used in the transformation. In PLANTRAN, the extra control points are used to calculate a least-squares solution for the calibration coefficients.

The selection of control points can also be made satisfactorily by the user, as long as certain guidelines are followed:

1. Pictures taken by a person on the ground, whether standing on the ground or on a vehicle bumper or kneeling, will entail light rays that are nearly parallel to the ground. In that situation, any point that is out of the plane can result in significant errors in the mapped (X,Y) position because of the small angle of the light ray relative to the surface. Control points that are out of the plane will render the entire exercise invalid.

2. The possible exceptions to the above are aerial photos, or photographs taken from a fire department ladder, for example. Here, light rays that are nearly perpendicular to the surface can result in out-of-plane errors that are insignificant.

3. Be aware of roadway crown. If points lie on either side of the crown, it will probably be necessary to use two planes—one for each side. Roadway crown may be hard to see, but total station survey data can be consulted to see if it exists.

4. The greatest accuracy exists for points in the foreground. All else being equal (which they seldom are) points far away from the camera will have larger uncertainties. It is also the case that objects in the distance will be smaller, probably not in as sharp a focus, and may be less well lit (in the shadows, or farther away from the flash, for instance).

5. If there are some points in the distance to be analyzed, try to find a closer photograph, possibly for separate photogrammetry. Sequential photographs can produce results in which overlapping areas are in close agreement.

6. Be sure the photogrammetry process is planned before going to the scene to take measurements. To assist in this process, identify all control points in advance: label them on copies of the photographs (or on overlays) and physically locate (and perhaps label) them at the scene. While at the scene, it may be apparent that alternate control points are required. Be sure that at least four control points are available for each photogrammetry procedure.

7. To the extent possible, select at least four control points that surround or encompass the area of interest. If points to be analyzed are outside of the control point boundary, the point coordinates will be based on an extrapolation from the mathematical plane and may be in error. For points in the foreground, the extrapolation error may well be insignificant, in large part due to the steeper light rays involved. Mapped points in the background, outside of the control point boundary, are almost certain to have noticeable errors.

8. No three of the four control points can be in a straight line.

9. Nighttime outdoor conditions dim the prospects of performing photogrammetry.

One is well advised to keep these guidelines in mind when taking photographs in the first place. A very useful set of recommendations for the photographer appears in the Brelin paper.

By definition, photogrammetry involves the measurement of point locations (x,y) on the photograph. Theoretically, this can be done manually, and the FotoGram program offers a means of entering the measurements from the keyboard, in response to prompts on the screen. As a practical matter, it behooves one to use a digitizing procedure and build an input file in advance of using FotoGram. Digitizing can be done by importing the photographs into AutoCAD®, which has the advantage of being able to zoom in while digitizing points of interest, and also creating a pictorial record (albeit of somewhat reduced quality compared to the original image) of both the control and the mapped points. Otherwise, the picture measurement can be done on a digitizing tablet. In fact, the Fotogram program package comes with a software driver for a Summagraphics MM Series digitizer.[6] This author developed a program called FotoIn3.bas for using a Kurta XGT tablet to digitize a photograph and create an input file that could be read by FotoGram. Grimes et al.[9] included another such program in their 1986 paper, and presented photos of some actual tire marks that were mapped using FotoGram.

Camera Reverse Projection Methods

Evidence captured in photographs, that can be evaluated only through photographs, is encountered often enough that the reconstructionist needs to have access to some means of three-dimensional analysis of the pictures. Perhaps, the most common instance in an accident scene investigation is where a vehicle has left the roadway and traveled over uneven or sloped terrain to its rest position. There may be photographs of the tire tracks, which have long since disappeared, and/or the rest position of the vehicle, which has been removed. It is up to the reconstructionist to figure out what happened, based on this evidence that once existed and was photographed, but is no longer.

A very clever method of solving such problems was presented in 1988 by Whitnall et al.[10] Called "camera reverse projection," this method permits the investigator to peer through the viewfinder of a camera at the scene so that the original image is superposed on the present view. If the camera position, pose, and focal length are not replicated, the lack of alignment between the original image and the present view will be apparent. On the other hand, if the position, pose, and focal length of the camera are replicated, the original image of the permanent features and the present view of them will line up in the viewfinder. If a marker is placed in the view so that it lines up with some evidence seen in the original image, the position of the marker (and that of the missing evidence) can then be measured.

In concept, the means of superposing the original image is to place a film negative of that image under the focusing screen. This requires the use of a single lens reflex camera with a removable pentaprism and viewfinder assembly, so that the focusing screen can be accessed and removed to allow placement of the negative. The Nikon F3 and F4 are two such cameras. In practice, a film negative is difficult to use because the negative blocks too much light. A more satisfactory practice is to make a tracing by computer means (using AutoCAD, for example) that can be printed on clear stock at a 24 × 36 mm size and with an appropriate line width, or by hand on clear material that can be photographed with a PMT camera and reduced to negative size.

Needless to say, the tracing needs to be prepared before the scene inspection if camera reverse projection is to be used. Like most things in this world, the method has its advantages, its limitations, and its drawbacks:

1. The method does not require that still-visible features of the original image be surveyed. Distant mountain tops can be used for alignment purposes (and in fact are very useful).

2. The method requires patience, skill, and time. This is another of those instances where "art" comes into the field of accident reconstruction.

3. Because of the time factor and because the pose and position of the camera must be maintained, a tripod is necessary. The setup cannot be bumped by stray animals (this has been known to occur) or by the analyst. The camera and tripod (and more importantly, the operator) cannot be exposed to passing traffic.

4. It is often the case that police photos are taken while a lane or the entire road has been closed to traffic, and the photographer is positioned close to or in the normal path of traffic. It is very rare that police will agree to repeat such a closure for the reconstructionist, particularly in view of the time required for set-up. Therefore, such

photos that were taken too close to, or in the way of, passing traffic cannot normally be used for camera reverse projection.

5. A photograph taken through the viewfinder of the reverse projection camera, using another camera, can show the superposition of the original image with the current scene, and show the markers that were surveyed. For example, an at-scene photograph is shown in Figure 8.1, and the camera reverse projection setup is shown in Figure 8.2. In Figure 8.2, one can see the outline tracing made of the car and some of the still-present scene features utilized for the alignment: the roadway edge, buildings, signs, background foliage, and utility poles. A marker has been placed where image showed the bottom of the driver-side tire sidewall to be, as seen through the lens. This marker was surveyed, in order to help establish the position of the accident vehicle at its point of rest. The vehicle image has been traced only in outline form, to help the reconstructionist to make comparisons with the photograph at the scene. Through-the-lens photographs such as this show the quality of the alignment, which can be evaluated by independent observers, and they encourage an understanding of the process.

6. The procedure does not rely on mathematical analysis or computer processing, and is thus easily understood by lay persons. It can be demonstrated to juries, as long as the judge permits a camera with a removable pentaprism and viewfinder to be brought into the courtroom.

FIGURE 8.1
Police photo of vehicle at rest.

FIGURE 8.2
Tracing of police photo visible in viewfinder of reverse projection camera.

7. The camera position does not have to be on level ground. However, ground slope complicates the setup because every time the camera (X,Y) position is adjusted by moving the tripod to improve the alignment, the Z coordinate changes as well because of the slope.

The process of using the camera reverse trajectory procedure is best mastered through experience. However, the salient points can be summarized as follows:

1. Prepare the overlay tracing, reduced to a negative size. Thin lines are helpful in seeing alignments, but will be hard to see if they are too thin. Conversely, thick lines stand out, but may obscure details in the scene that need to be seen.

2. Insert the tracing below the focusing screen in the reverse projection camera. Attach a zoom lens of a range that can bracket that required by the view seen in the original photograph.

3. With camera in hand, position oneself at the scene so that all alignment lines are maintained when looking through the viewfinder. Adjust the focal length to provide a rough match. These can be thought of as coarse adjustments. Roughly mark the position, in preparation for setting up a tripod.

4. Mount the camera on a tripod. The tripod should be sturdy, and tall enough to accommodate the Z-position of the camera, with room for adjustment up and down. The tripod head should be capable of smooth and independent adjustment and locking independently in

height, azimuth, roll, and pitch. Adjust all of the tripod head positions to provide a rough match, and lock in place.

5. Using the most distant objects in the tracing, adjust the focal length (for a final time, one hopes). The reason the distant objects are chosen is that movement in all the other camera degrees of freedom will have little or no effect on the chosen focal length; distant objects should therefore be included in the tracing. This makes a mountain ridge line ideal. Without distant objects to work with, the best procedure is to identify the lines of alignment that have the most lateral separation in the photograph, and adjust the focal length until the lines of alignment are the proper distance apart when seen through the camera. It may be helpful to include such lines in the tracing, as long as it does not get too cluttered, or important features are obscured.

6. Now begin the fine adjustments, which consist of moving the tripod and adjusting the degrees of freedom of the tripod head. Repeat until the alignment is complete. Discipline in checking all areas of the view, and all lines in the tracing, is required. It is possible to conclude that the alignment is good, only to find that it is good in some areas but not others.

7. Changes that have been made to the scene since the original photographs were taken will be immediately apparent, and can be confounding if one is not careful. Signs may have been knocked down, removed, or moved. Foliage may have grown, so be careful about too much reliance on it. Foliage may have shrunk, or leaves disappeared, due to seasonal changes. Tree branches that are clearly visible in winter may be obscured at other times. Buildings may have been torn down, erected, or remodeled, and so forth.

8. Looking through the reverse projection camera, place markers (or use paint) to demarcate the points to be located. This is most effectively done by a second person, under the guidance and direction of the person looking through the viewfinder. Survey the points, and the camera location.

9. Multiple reverse projection cameras, multiple tracings, and multiple setups may be required for some scenes.

10. Using a second camera, take photos through the viewfinder of the reverse projection camera(s), in order to document the setups. A macro lens on the second camera may be best for this application. Be aware, though, that an auto-focus mechanism on the second camera will move the lens in and out, and could bump the reverse projection camera. Care is required not to disturb the setup, which is why marking and surveying should be done before the through-the-lens photographs are taken. One may need to move the second camera around a bit to reduce obscuration around the edges of the view.

11. The presence of grass in the area of interest can pose problems, since it may obscure points visible in the original photos, such as tire positions for vehicles at rest. It may be necessary to do some "yard work" before starting the camera reverse projection process. It is helpful to have some appropriate tools in the inspection gear, should this become necessary.

Two-Photograph Camera Reverse Projection

A variation on the camera reverse projection procedure can sometimes be applied to accident vehicles. The method is explained in detail in a 1991 paper by Woolley et al.[11] If two photographs have lines of alignment for vehicle features that are visible in both views, triangulation may be used to find their coordinates in the ground plane, as discussed previously. The Woolley paper explains how the camera reverse projection method can be applied using two photographs to locate points on the crushed surface of the vehicle. Even the Z coordinates can be determined, if necessary. Here, the method does not attempt to ascertain the distance to the point from the camera, or to locate the point relative to some fixed ground reference. Rather, it is merely to observe the lines of alignment that are implicit in the vehicle shape and details, and use triangulation to locate the points on the crush surface. An example is given for two photographs of a crushed vehicle, for which the crush vectors were measured and compared against those determined for the actual vehicle by use of a total station. The errors for the individual crush vectors were not reported, but the average error magnitude was about 0.9 in.

This author's experience has been that while the single-photograph method for scenes and the two-photograph method for vehicles are based on the same camera reverse projection principles, the procedure for vehicles is often more difficult to apply. For one thing, the vehicle photos that one often has to work with were taken under cramped conditions; for example, in a wrecker yard. In such cases, wider lenses are often used, and the resulting tracings are more difficult to line up with. For another, the "background" geometry that one must use for alignment purposes is limited to the undamaged portion of the vehicle, which is not all that far away, and which is often partly hidden from view. This makes independent setting of the focal length difficult, if not impossible. Two camera setups must be used simultaneously, for which there may not be enough indoor space. But time, weather, and surface constraints (e.g., a flat enough space outdoors) tend to force the operation indoors, where lighting conditions may render the vehicle features difficult to see through the lens of a camera.

Analytical Reverse Projection

If one knows (mathematically) the camera location and pose, and the three-dimensional shape of the surface upon which the mapped points are to be located, then where the light rays for those mapped points intersect the surface can be determined mathematically. This is the basis of an analog to the camera reverse projection method, the "analytical reverse projection procedure," which was introduced in 1989 by Smith and Allsop.[12]

As in two-dimensional photogrammetry, the locations of the control points must be known, but in three dimensions in this case. They need not lie in a plane or the roadway surface, but unlike the single view camera reverse projection method, the control points cannot be on a distant mountain ridge line, either. The locations of six or more physical control points in three dimensions, and the locations of their images in two dimensions on the photograph, determine the camera pose, focal length, and lens center position. A light ray passing from a point on the image, through the lens center, must also pass through the physical point. If two photographs are available showing the point, then the intersection of light rays determines the location of the point.

If only one photograph is available, however, or if a feature cannot be defined by a discrete point, then this procedure utilizes the intersection of the light ray with a mathematical representation of the surface, instead of the intersection of two light rays.

A computer program called CAMPOSE does the heavy lifting as far as the camera projection system position and pose are concerned. It also calculates the rays from any mapped point through the lens center. The intersection, in three dimensions, of such rays with the object space surface is calculated by a program called SURFINT, which fits an nth-order polynomial to a set of points surveyed in the vicinity of the object points to be located.

Smith and Allsop's paper contains a comparison of this method with its camera reverse projection cousin, and with the two-dimensional programs FotoGram and PLANTRAN. Since the comparison involved a street intersection with considerable contour, which was not accounted for by using multiple planes, the two-dimensional programs did not fare so well, as might be expected. However, the two reverse projection methods performed "significantly better."

Three-Dimensional Multiple-Image Photogrammetry

One method of using two images with camera reverse projection photogrammetry for measuring crush on vehicles was discussed previously.

Other multiple-image methods were presented by Tumbas et al., and an extensive and carefully planned comparison was carried out using them.[4] One method utilized an ADAM Technology MPS-2 analytical stereoplotter to analyze stereo pairs of photographs. A stereo pair of photographs is two images made from slightly different positions and with camera angles nearly parallel, much as would be the case with human eyes. When one image is viewed by each eye, the appearance of depth is created, which may be used by the stereoplotter to quantify the third dimension.

Two other approaches utilized multiple photographs that were not stereo pairs. One of them was called 3-D Analytical-A in the paper, and not otherwise identified. No information was provided (or available, apparently) on how it worked. The other approach was the Rollei-metric MR2 Close Range Photogrammetry System. A brief description is provided in the paper.

Another approach is PhotoModeler,* a software package available since about the mid-1990s, and now fairly well known. As its name implies, PhotoModeler was designed for the purpose of building digital models from photographs, through the use of multiple overlapping photographs. For any given point on the object, the ideal situation is that two of the photos have about a 90° separation; other photos may be used. Details of the inner workings of the software are proprietary, but it appears that two (or more) intersecting light rays are used to triangulate the position of the point. If a collection of such points is known, then a digital representation, or a model, of the object can be constructed.

The most straightforward application of PhotoModeler is a sort of a non-contact measuring device. For this concept, a "calibrated" camera (i.e., one for which the focal length, principal point, and format size are known) is used to take a number of overlapping pictures. Since the camera has been "calibrated," for any picture, the position, pose, and focal length are known, and the location of any light ray on a given image can be calculated. If a point is visible in two or more photographs, the intersection of light rays is used to calculate the position of the point. Given enough distinct points, one obtains the shape of the object; that is, one has a model.

All this begs the question: Why not just survey the points? The fact is that with a modern total station, one can minimize the use of a prism or reflector, and thus avoid or eliminate being exposed to traffic. On the other hand, targets that are visible in the photographs are often needed for photogrammetric purposes, and this means going into the scene to place them. Targets are often used for surveying or for general documentation, however. For agencies that respond to crash scenes, perhaps the issue boils down to whether direct measurement or photogrammetry is faster, easier, or less expensive.

There is a second type of application, however, that looms large for reconstructionists. If one has photographs of evidence, long since disappeared, one can build a model of the evidence in PhotoModeler. The procedure would be

---

* Eos Systems, Inc. Vancouver, BC, Canada.

identical to that described previously if the photographs were taken with a "calibrated" camera. That is hardly ever the case, however. Typically, the reconstructionist receives a set of photos taken by a camera with unknown characteristics. In that situation, a section of PhotoModeler called "Inverse Camera"[14] is utilized to establish the camera characteristics. That is done by finding identifiable control points in the photographs that are still present, measuring them, and having Inverse Camera calculate the focal length, principal point, and format size from knowledge of the control points. The "measurement" might take the form of knowing the size of something in the photographs that can be measured on another object (such as an exemplar vehicle) of the same size.

A different calibration is required for each focal length, even for the same camera. Therefore, the use of zoom lenses is an anathema to PhotoModeler.

The use of PhotoModeler for documenting accident scenes is discussed in a 1997 paper by Fenton and Kerr.[14] Here, scene measurements were obtained by using a set of photographs, including one taken at the scene before the vehicles had been removed. The "old" photograph had been taken with an unknown camera. Control points in the scene were the ends of skip lines that remained in place after the accident. At a later time, other photographs were taken, showing the same control points. The paper does not state what type of camera or lens was used for the later photographs, but presumably a "calibrated" camera was employed.

In PhotoModeler parlance, the old and new photographs were entered into a project. Instead of simply surveying the control points, the new photographs were used to determine their positions in object space. This knowledge was used in Inverse Camera to determine the characteristics of the camera that took the "old" picture. The old picture was then "rectified," which is to say that an overhead view of the model was developed, as if the picture had been taken from above. The rectified view could then be traced, including the vanished evidence, creating an accident scene drawing to scale. The tracing showed the skid marks, which had disappeared by the time the new photographs were taken.

The paper also included a completed scene diagram, included a wrecked tractor and semi-trailer, damaged asphalt area, gouges, skip lines, and displaced Jersey barrier sections. It was not stated whether the additional features were placed on the drawing through the use of PhotoModeler, or by other means. Various field measurements were taken: skip line length, gap between skip lines, lane width, and Jersey barrier segment lengths. Discrepancies between the field measurements and those obtained by PhotoModeler were 0.9% or less.

A 2001 paper by Dierckx et al.[15] discusses the use of PhotoModeler to describe the geometry of complex shapes: an exhaust system and a body-in-white. The application is more for model building than accident reconstruction. Nevertheless, many practical tips and guidelines are presented for obtaining accurate results. According to the authors, the main benefits of photogrammetry are "increased accuracy compared to manual measurements, short immobilization time of the object, simplicity, enhanced visualization and low cost."

A 2010 paper by Randles et al.[16] presents an examination of photogrammetry as applied to the measurement of deformed shapes of vehicles. In this study, three vehicles were subjected to four impacts, resulting in damage areas on two fronts, one side, and one rear. The vehicles were marked with 19 mm circular adhesive targets, each with a 4 mm circular dot in the middle, to indicate the measurement points. The targets were spaced at 6 in. intervals at various heights, and some were located outside the damaged area. Prescribed station lines were set out "in accordance with standard accident reconstruction practice." Groups comprising of 13–14 individuals made "hands on" measurements, meaning that the vehicles, station lines, and targets were physically present. Perpendicular distances to the targets from the station lines were measured. The "hands on" technique is not described, but apparently the individuals had the latitude to "make and report the measurements in a manner consistent with their normal forensic practice."

An imaging total station was used without a prism (for increased accuracy) in order to obtain measurements to be used as a standard of comparison.

In a separate exercise, three individuals were provided photos of the damaged vehicles and were asked to make measurements using PhotoModeler. Three cameras were used with known properties, and a fourth was unknown. The "inverse camera feature in PhotoModeler in conjunction with control points" was used. The control points are not described in the paper. A 3-D DXF file was created in PhotoModeler and overlain onto a baseline model created from total station measurements of the undamaged vehicles. This suggests some sort of attempt to measure crush, but in the paper, only the distances from the station line to the targets on the damaged surface are reported.

According to the paper, "Qualitative examination of both participant hand measurements and PhotoModeler measurement data showed a strong agreement with the baseline total station measurements ..."[16] (p. 167). The graphed data suggest systematic errors in the hand measurements: one damage profile on the low side, and the others on the high side. Possibly the positioning or documenting of the station line was at fault. In any case, the PhotoModeler measurements were closer to the total station measurements. Or, as the paper puts it, "... the general trend was that the measurement error expressed as a percentage of the expected measurement and the values for the mean, standard deviation and maximum deviations were all greater in the participant hand measurements (0.6 ± 1.4 cm) compared to the PhotoModeler (0.1 ± 1.0 cm)"[16] (p. 167). The paper concludes, "The photogrammetric measurements were statistically found to be more accurate than those obtained via hands-on measurement by qualified professionals in the field of accident reconstruction, both in terms of the relative differences and percent accuracy"[16] (p. 168).

One must point out, however, that targets were applied to the vehicles before the measurement exercises, and were therefore visible in the post-crash photographs. Field investigators do not have this luxury; instead, they must make do with whatever vehicle features (e.g., emblems) they can find. Therefore, they would not have as many measurements available for their damage documentation, and it is possible that the remaining damage assessments would be degraded as well. Where landmarks exist in the photos, however, good damage measurements can be obtained.

References

1. Kerkhoff, J.F., Photographic techniques for accident reconstruction, SAE Paper 850248, SAE International, Warrendale, PA, 1985.
2. Abdel-Aziz, Y.I. and Karara, H.M., Direct linear transformation from comparator coordinates into object space coordinates in close-range photogrammetry, *Proceedings of the Symposium on Close-Range Photogrammetry*, American Society of Photogrammetry, 1971.
3. Dermanis, A., Free network solutions with the DLT method, *ISPRS Journal of Photogrammetry and Remote Sensing*, 49(2), 1004, 1994.
4. Tumbas, N.S., Kinney, J.R., and Smith, G.C., Photogrammetry and accident reconstruction: Experimental results, SAE Paper 940925, SAE International, Warrendale, PA, 1994.
5. Brelin, J.M., Cichowski, W.G., and Holcomb, M.P., Photogrammetric analysis using the personal computer, SAE Paper 861416, SAE International, Warrendale, PA, 1986.
6. Brelin, J.M., Cichowski, W.G., and Holcomb, M.P., *FotoGram Instruction Manual*, General Motors Corporation, Warren, MI, 1986.
7. Kinney, J.R. and Magedanz, B., TRANS4—A traffic accident photogrammetric system, description of the system and its inherent errors, SAE Paper 861417, SAE International, Warrendale, PA, 1986.
8. Smith, G.C., *PLANTRAN Users' Manual*, Collision Safety Engineering, Orem, UT, 1988.
9. Grimes, W.D., Culley, C.H., and Cromack, J.R., Field application of photogrammetric analysis techniques: Applications of the FOTOGRAM program, SAE Paper 861418, SAE International, Warrendale, PA, 1986.
10. Whitnall, J., Moffitt, F.H., and Millen-Playter, K., The reverse projection technique in forensic photogrammetry, *Functional Photography*, 23(1), January, 1988.
11. Woolley, R.L., White, K.A., Asay, A.F., and Bready, J.E., Determination of vehicle crush from two photographs and the use of 3D displacement vectors in accident reconstruction, SAE Paper 910118, SAE International, Warrendale, PA, 1991.
12. Smith, G.C. and Allsop, D.L., A case comparison of single-image photogrammetry methods, SAE Paper 890737, SAE International, Warrendale, PA, 1989, pp. 121–128.
13. Eos Systems, Inc., *Photomodeler Pro User Manual*, Eos Systems, Inc., 1997.

14. Fenton, S. and Kerr, R., Accident scene diagramming using new photogrammetric technique, SAE Paper 970944, SAE International, Warrendale, PA, 1997.
15. Dierckx, B., De Veuster, C., and Guidault, P.-A., Measuring a geometry by photogrammetry: Evaluation of the approach in view of experimental modal analysis on automotive structures, SAE Paper 2001-01-1473, SAE International, Warrendale, PA, 2001.
16. Randles, B., Jones, B., Welcher, J., Szabo, T., Elliott, D., and MacAdams, C., The accuracy of photogrammetry versus hands-on measurement techniques used in accident reconstruction, SAE Paper 2010-01-0065, SAE International, Warrendale, PA, 2010.

9

Filtering Impulse Data

Background and Theory

Central to the concept of filtering is the notion that a particular signal (i.e., a function of time $f(t)$) has a frequency content. This is a very familiar notion in music. A higher musical pitch means a higher frequency, and vice versa. Standard musical scales have been set up, including American Standard Pitch, adopted in 1936, International Pitch, adopted in 1891, and the Scientific or Just Scale (no flats or sharps).[1] When at the beginning of a concert the oboe plays a note and the orchestra tunes to it, the players are tuning to "Concert A," 440 Hz, if they are playing in American Standard Pitch.

For a given set of conditions, there are only certain frequencies at which a string can vibrate. The lowest of these is its fundamental frequency, and all other possible frequencies, or harmonics, are multiples of it. In the case of a vibrating string, the multiples are integers. The second harmonic occurs at double the frequency, or one octave higher. Thus, the A above Concert A (one octave up) is 880 Hz. The next A occurs at two octaves up, four times the frequency of Concert A, or 1760 Hz. In fact, when an orchestra tunes the various instruments are probably being tuned to various As in various octaves. We see that octaves form a base-two geometric series: 2^{-2}, 2^{-1}, 2^0, 2^1, 2^2, ... 2^N, where N is any integer between $-\infty$ and $+\infty$.

In engineering, frequency is more often described in terms of decades, which are 10 fold increases or decreases. Decades form a base-10 geometric series: 10^{-2}, 10^{-1}, 10^0, 10^1, 10^2, etc.

In reality, it must be pointed out that a musical instrument does not produce just a single tone. It also produces overtones, which occur at (not necessarily integer) multiples of the frequency. The relative amplitudes of these overtones give the instrument its particular sound quality, or timbre. We thus come to the idea that sound quality implies that a combination of frequencies that are present. The sound quality can be changed by emphasizing some frequency ranges and de-emphasizing others. This is the basic idea behind graphic equalizers.

If such amplitude adjustments, or attenuations, are done in a systematic way according to frequency, the result is a filter, which is a device that can attenuate (i.e., filter out, or stop) certain frequency ranges, while leaving unchanged (i.e., passing) certain others. Common filter types are low pass, high pass, band pass, and band stop.

Analog Filters

When an electrical signal is passed through a circuit containing elements such as resistors and capacitors, the amplitude and phase of the output vary according to how these devices are arranged in series and parallel combinations (their topology), and according to their resistances and capacitances, respectively. Generally speaking, the changes are frequency-dependent, because the resistance to current flow across a capacitor (its capacitive reactance) is inversely proportional to the frequency.

We can contemplate the determination of the response with respect to time of any such circuit by using Kirchhoff's voltage and current laws[2] to write the governing equations, with time as an independent variable. These are differential equations because the current through the capacitor depends on the rate of change of its charge. We would then have to solve the resulting system of simultaneous linear differential equations. If we were to seek an analytical solution (as opposed to a brute-force computer simulation), we would probably use the Laplace transform,[3] which is defined as follows. If a function of time $F(t)$ is defined for all positive values of t, a new function $f(s)$ of the parameter s is obtained by evaluating the following integral:

$$L\{F(t)\} = f(s) = \int_{t=0}^{\infty} e^{-st} F(t) dt \qquad (9.1)$$

Similarly, the Laplace transform of a derivative is given by

$$L\{F'(t)\} = \int_{t=0}^{\infty} e^{-st} F'(t) dt \qquad (9.2)$$

where the prime denotes differentiation with respect to time. If we integrate Equation 9.2 by parts, the result is

$$L\{F'(t)\} = sf(s) - F(0) \qquad (9.3)$$

When applied to a differential equation, this important property of Laplace transforms changes the differential equation to an algebraic equation. For an electrical circuit, the system of N simultaneous differential equations is transformed into a system of N simultaneous algebraic equations having s as the independent variable. If the solution has an oscillatory component, then s is related to the frequency. In any case, the output of the circuit is the product of the input, multiplied by a transfer function. The transfer functions, and hence the output, are functions of frequency, and the solution is said to be in the frequency domain.[4]

It is easily seen (by considering the use of Cramer's rule to solve the simultaneous algebraic equations, for example) that the transfer function will be a ratio of two polynomials. Each polynomial has roots, the number of which depend on the degree of the polynomial. One or more of these roots may be real, but typically they are complex. For conditions corresponding to a root in the numerator, the filter response is zero. This is called a zero of the filter. Similarly, as we approach a root in the denominator, the filter response exhibits a singularity and we have what is called a pole.

To get back to the time domain, we would need to take the inverse Laplace transform of the frequency-domain solution. If one is lucky, this can be done using Laplace transform properties (such as convolution) and a table of Laplace transforms; otherwise, the inversion could be a laborious process, to say the least. The process of analog filter design is to find a circuit layout, and the various device parameters, that result in a solution to the governing equations that has the desired frequency response.

Filter Order

Filters are classified by their order. The order of an analog filter is equal to the number of reactive components (capacitors) in the circuit. Typically, higher-order filters are built by cascading (combining) first- and second-order filters. An Nth order filter has N poles.

Bode Plots

A Bode plot, or frequency response curve, is a graph of the filter gain versus frequency. Usually, a logarithmic scale is used for frequency, and the gain is expressed in decibels, or dB, where

$$\text{Gain in dB} = 20 \log\left(\frac{V_{out}}{V_{in}}\right) \tag{9.4}$$

In the band pass, an ideal passive filter (i.e., one without amplifiers) would perfectly pass signals, V_{out} would equal V_{in}, and the gain would be 0 dB. In the stop band, input signals would be eliminated altogether, and the gain in dB would be a very large negative number. The transition point between such bands is called the corner (or cutoff or break-point) frequency.

In practice, nature abhors sharp corners, and the transition is more gradual. In real filters, various trade-offs are made to get optimum performance for a given application. In many such cases, the corner frequency is defined as the frequency at which the capacitive reactance and the resistance are equal. At this point, the output signal is attenuated by the vector sum of the resistance and the capacitive reactance, or a factor of $\sqrt{1/2}$ and the filter response is approximately −3 dB. Beyond that point, the Bode plot is essentially linear because of the logarithmic scales being used. The slope of the line in this region is called the "roll-off characteristic." A first-order filter has a slope of −20 dB per decade, or approximately −6 dB per octave. An Nth order filter has a roll-off characteristic of −20 N dB/decade, or approximately −6 N dB/ octave.

Figure 9.1 shows an example drawn from SAE Recommended Practice J211.[5] The output of a data processing filter is allowed to be in the cross-hatched area. At frequency F_H, the allowable gain is −1 to +0.5 dB, and at frequency F_N, the allowable gain is −4 to +0.5 dB. Thus, F_N is the corner frequency, according to the above definition. The slopes d and e are −9 and −24 dB/octave, respectively. Therefore, either a second-, third-, or fourth-order filter could be used to satisfy the roll-off requirements.

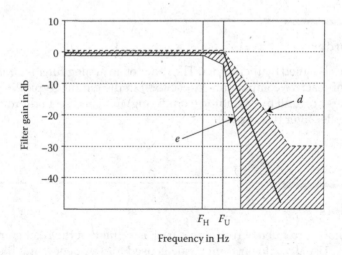

FIGURE 9.1
Filter requirements from SAE J211.

Filter Types

Butterworth filters[6] were first described by Stephen Butterworth in his 1930 paper "On the Theory of Filter Amplifiers." At the time filters generated substantial ripple in the band pass and the choice of component values was highly interactive. Butterworth, who had a reputation for solving "impossible" mathematical problems, solved the equations for two- and four-pole filters, showing how the latter could be cascaded when separated by vacuum tube amplifiers and so enabling the construction of higher-order filters. He also showed that his basic low-pass filter could be modified to give low pass, high pass, band pass, and band-stop functionality.

Butterworth filters are termed maximally flat-magnitude-response filters, optimized for gain flatness (i.e., minimal ripple) in the band pass. The transient response of a Butterworth filter to a pulse input shows moderate overshoot and ringing.

Bessel filters are optimized for maximally flat time delay (or constant-group delay). This means that they have a linear phase response and excellent transient response to a pulse input. This comes at the expense of flatness in the band pass and rate of roll-off.

Chebyshev filters are designed to have ripple in the band pass, but steeper roll-off after the cutoff frequency. Cutoff frequency for such filters is defined as the frequency at which the response falls below the ripple band. For a given filter order, a steeper cutoff can be achieved by allowing more band-pass ripple. The transient response of a Chebyshev filter to a pulse input shows more overshoot and ringing than a Butterworth filter.

Digital Filters

A digital filter is a system that performs mathematical operations on a sampled, discrete-time signal to reduce or enhance certain aspects of that signal.[7] Usually, the digital filter input comes from an analog signal (which may be pre-filtered by an analog filter) that is passed through an analog-to-digital converter. Since digital filters use a sampling process and discrete-time processing, they experience latency (the difference in time between the input and the response). In an analog filter, latency is often negligible; strictly speaking it is the time for an electrical signal to propagate through the filter circuit. In digital filters, latency is a function of the number of delay elements in the system.

FIR Filters

The impulse response is a measurement of how a filter will respond to the Kronecker delta function.[8] A finite impulse response (FIR) filter[9] expresses each output sample as a weighted sum of the last N inputs, where N is the order of the filter. If $x[n]$ is the nth sample of the input signal and $y[n]$ is the nth output of the filter, then

$$y[n] = \sum_{i=0}^{N} b_i x[n-i] = b_0 x[n] + b_1 x[n-1] + \cdots + b_N x[n-N] \qquad (9.5)$$

where b_i are the filter coefficients. The above equation is known as a difference equation, or more precisely, a linear constant-coefficient difference (LCCD) equation.

An Nth order filter has $N+1$ terms on the right-hand side, and so its impulse response will therefore be reduced to zero and remain zero after $N+1$ steps; that is, its response will die out after a finite number of samples. Since FIR filters do not use feedback (i.e., there are no y terms on the right-hand side), they are inherently stable. This is because the output is a sum of a finite number of finite multiples of the input values.

A moving average calculation serves as a very simple FIR filter. In this case, the filter coefficients are

$$b_i = \frac{1}{N+1} \quad \text{for } i = 0, 1, \ldots, N \qquad (9.6)$$

where again N is the order of the filter. All the coefficients are the same; every sample is treated equally. A common example is the 200-day moving average calculation sometimes used to track a stock price, and thereby distinguish trends from the daily stock market ups and downs. In such a calculation, the effects of any particular day's price completely disappear after 200 trading days. It is seen that the area-weighted average method used for data processing in SAE 2009-01-0105[10] is another type of moving-average FIR filter, except in that case b_0 and b_N were half the values of the other b_i, and the filter was forward looking as well as backward looking. In other words, the calculation of $y[n]$ used values of $x[n]$ that came after the nth data point, as well as prior to the nth data point.

IIR Filters

Infinite impulse response, or IIR, filters[11] are the digital counterpart to analog filters. In such a filter, the output is determined by a linear combination of the previous inputs and outputs. They are recursive; in other words, they use feedback, which FIR filters normally do not. In theory, the impulse response of such a filter never dies out completely, hence the name IIR, though this is not true in practice due to the finite resolution of computer arithmetic. Since the phase shift is inherently a nonlinear function of frequency, the time delay through such a filter is frequency dependent.

The difference equation for an IIR filter takes the form

$$\sum_{j=0}^{Q} a_j y[n - j] = \sum_{i=0}^{P} b_i x[n - i] \tag{9.7}$$

where Q is the feedback filter order, a_j are the feedback filter coefficients, P is the feed-forward filter order, b_i are the feed-forward filter coefficients, $x[n]$ is the nth value of the input signal, and $y[n]$ is the nth value of the output signal. The order of the filter is the greater of P and Q. Solving for the nth value of the output signal yields

$$y[n] = \frac{1}{a_0} \left(\sum_{i=0}^{P} b_i x[n - i] - \sum_{j=1}^{Q} a_j y[n - j] \right) \tag{9.8}$$

Use of the Z-transform

To find the transfer function of the filter, we take the Z-transform of each side of Equation 9.7. The bilateral or two-sided Z-transform[12] of a discrete time signal $x[n]$ is the function $X(z)$, defined as

$$Z\{x[n]\} = X(z) = \sum_{n=-\infty}^{\infty} x[n] z^{-n} \tag{9.9}$$

where n is an integer and z is, in general, a complex number:

$$z = Ae^{j\varphi} = A(\cos\varphi + j\sin\varphi) \tag{9.10}$$

where j is the phase in radians. However, it is often the case that $x[n]$ is defined only for $n \geq 0$. Then, we would define a single-sided or unilateral Z-transform as

$$Z\{x[n]\} = X(z) = \sum_{n=0}^{\infty} x[n] z^{-n} \tag{9.11}$$

Taking the Z-transform of each side of Equation 9.7, we obtain

$$\sum_{j=0}^{Q} a_j z^{-j} Y(z) = \sum_{i=0}^{P} b_i z^{-i} X(z) \tag{9.12}$$

The transfer function is defined to be

$$H(z) = \frac{Y(z)}{X(z)} = \frac{\sum_{i=0}^{P} b_i z^{-i}}{\sum_{j=0}^{Q} a_j z^{-j}} \tag{9.13}$$

Most IIR filter designs have the coefficient a_0 set to 1. In that case, the transfer function takes the form

$$H(z) = \frac{\sum_{i=0}^{P} b_i z^{-i}}{1 + \sum_{j=1}^{Q} a_j z^{-j}} \tag{9.14}$$

As with analog filters, we see that the transfer function is a ratio of polynomials. The numerator will have P zeros, corresponding to the zeros of the transfer function. The denominator will have Q zeros, corresponding to the poles of the transfer function. The zeros and poles can be plotted on the complex plane (z-plane).

The unilateral Z-transform is the Laplace transform of the ideally sampled time function

$$x_s(t) = \sum_{n=0}^{\infty} x[n] \delta(t - nT) \tag{9.15}$$

where $x(t)$ is the continuous-time function being sampled, $x[n]$ is the nth sample, T is the sampling period, and with the substitution $z = e^{sT}$.

The inverse Z-transform is

$$x[n] = Z^{-1}\{X(z)\} = \frac{1}{2\pi j} \oint_C X(z) z^{n-1} dz \tag{9.16}$$

where C is a counterclockwise closed path encircling the origin and entirely in the region of convergence, which is the set of points in the complex plane for which the Z-transform summation converges. A special case of this contour integral occurs when C is the unit circle (and can be used when the region of convergence includes the unit circle). The inverse Z-transform simplifies to the inverse discrete-time Fourier transform:

$$x(n) = \frac{1}{2\pi} \int_{-\pi}^{+\pi} X(e^{j\omega}) e^{j\omega n} d\omega \tag{9.17}$$

Example of Finding the Difference Equation from the Transfer Function

Suppose we need a filter with two zeros at $z = -1$, and poles at $z = 1/2$ and $z = -3/4$. The transfer function is then

$$H(z) = \frac{(z+1)^2}{\left(z - \frac{1}{2}\right)\left(z + \frac{3}{4}\right)} \tag{9.18}$$

Expanding the above, we obtain

$$H(z) = \frac{z^2 + 2z + 1}{z^2 + \frac{1}{4}z - \frac{3}{8}} \tag{9.19}$$

Dividing by the highest power of z

$$H(z) = \frac{1 + 2z^{-1} + z^{-2}}{1 + \frac{1}{4}z^{-1} - \frac{3}{8}z^{-2}} = \frac{Y(z)}{X(z)} \tag{9.20}$$

The constants of the denominator a_j are the feed-backward coefficients and those in the numerator b_i are the feed-forward coefficients. Cross-multiplying in Equation 9.20 yields

$$\left(1 + \frac{1}{4}z^{-1} - \frac{3}{8}z^{-2}\right)Y(z) = \left(1 + 2z^{-1} + z^{-2}\right)X(z) \tag{9.21}$$

Taking the inverse Z-transform of the above results in

$$y[n] + \frac{1}{4}y[n-1] - \frac{3}{8}y[n-2] = x[n] + 2x[n-1] + x[n-2] \tag{9.22}$$

Solving for y[n], we obtain

$$y[n] = x[n] + 2x[n-1] + x[n-2] - \frac{1}{4}y[n-1] + \frac{3}{8}y[n-2] \tag{9.23}$$

Bilinear Transforms

The bilinear transform[13] is a useful approximation for converting continuous time filters (represented in Laplace s space) into discrete time filters (represented in z space), and vice versa. We start with the previously mentioned substitution:

$$z = e^{sT} = e^{(sT/2+sT/2)} = e^{sT/2}e^{sT/2} = \frac{e^{sT/2}}{e^{-sT/2}} \tag{9.24}$$

Using a series expansion for the exponential functions, we obtain

$$z = \frac{1 + \dfrac{sT}{2} + \dfrac{\left(\frac{sT}{2}\right)^2}{2!} + \cdots}{1 - \dfrac{sT}{2} + \dfrac{\left(\frac{sT}{2}\right)^2}{2!} + \cdots} \approx \frac{1 + \frac{sT}{2}}{1 - \frac{sT}{2}} \tag{9.25}$$

To reverse the mapping, we take the natural logarithm of the definition:

$$\ln(z) = \ln\left(e^{sT}\right) = sT \tag{9.26}$$

so that

$$s = \frac{1}{T}\ln(z) \tag{9.27}$$

Using a series expansion for the logarithm function, we find that

$$s = \frac{2}{T}\left[\frac{z-1}{z+1} + \frac{1}{3}\left(\frac{z-1}{z+1}\right)^3 + \frac{1}{5}\left(\frac{z-1}{z+1}\right)^5 + \cdots\right] \approx \frac{2}{T}\left(\frac{z-1}{z+1}\right) = \frac{2}{T}\left(\frac{1-z^{-1}}{1+z^{-1}}\right) \tag{9.28}$$

References

1. *Handbook of Chemistry and Physics*, 42nd Edition, Chemical Rubber Publishing Company, Cleveland, OH, 1960.
2. Rainville, E.D., *The Laplace Transform, An Introduction*, The Macmillan Company, New York, 1963.
3. Churchill, R.V., *Operational Mathematics*, Second Edition, McGraw-Hill Book Company, New York, Toronto, and London, 1958.

4. Electronics-Tutorials, http://www.electronics-tutorials.ws/.

5. Instrumentation for impact tests, SAE Recommended Practice J211, SAE International, 1988.

6. Butterworth, S., On the theory of filter amplifiers, *Wireless Engineer*, V. 7, 1930, pp. 536–541.

7. Antoniou, A., *Digital Filters: Analysis, Design, and Applications*, McGraw-Hill, New York, NY, 1993.

8. Gelfand, I.M. and Fomin, S.V., *Calculus of Variations*, translated from Russian by R.A. Silverman, Prentice-Hall, Englewood Cliffs, NJ, 1965.

9. Rabiner, L.R. and Gold, B., *Theory and Application of Digital Signal Processing*, Prentice-Hall, Englewood Cliffs, NJ, 1975.

10. Struble, D.E., Welsh, K.J., and Struble, J.D., Crush energy assessment in frontal underride/override crashes, SAE Paper 2009-01-0105, SAE International, 2009.

11. Smith, J.O. III, *Introduction to Digital Filters with Audio Applications*, W3K Publishing, http://books.w3k.org/, 2007.

12. Jury, E.I., *Sampled-Data Control Systems*, John Wiley & Sons, 1958.

13. Astrom, K.J., *Computer Controlled Systems, Theory and Design*, 2nd Edition, Prentice-Hall, Englewood Cliffs, NJ, 1990.

10

Digital Filters for Airbag Applications

Introduction

The detection of crash conditions, and the determination of whether and when an airbag deployment signal should be given, is a process that continues to grow in complexity as more and more "what if" conditions are considered. Thus digital electronics, with its inherent flexibility, has come to predominate in airbag control units, along with the replacement of mechanical sensors with electrical ones. The analysis of frequency content in the acceleration signals generated by crash sensors is an important element in the reliable detection of crash conditions and the accurate prediction of ultimate crash severities.

Example of Digital Filter in Airbag Sensor

The use of digital filters is one way of analyzing and responding to frequency content. In one instance, a bandpass filter with a bandpass of 20–150 Hz is employed for this purpose. The filter characteristics were identified by the manufacturer to be those of a second-order Butterworth filter,[1] and the sample rate was indicated to be 2 kHz (sample period of 1/2 ms). It was desired to develop a computer simulation of the filter, and the rest of the deployment algorithm, so that deployment decisions and timing could be identified for a variety of conditions.

The filter's responses to half-sine wave inputs were supplied in graphic form by the manufacturer, as shown in Figure 10.1.

In this case, the requisite transfer function was not available, since the zeros and poles were unknown. However, an Internet search revealed a web site for Dr. Anthony J. Fisher of the University of York in York, UK, that contains the mkfilter digital filter generation program.[2] This program designs an IIR digital filter from parameters specified by the user. Low-pass, high-pass, bandpass, and band-stop filters, with Butterworth, Bessel, or Chebyshev

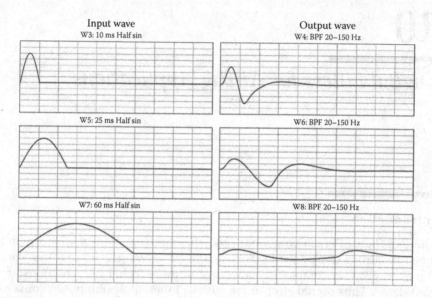

FIGURE 10.1
Half-sine pulses of 10, 25, and 60 ms duration.

characteristics, are designed using the bilinear transform or matched z-transform method. (The bilinear transform method is recommended for most applications.)

Output of the program is a set of s- and z-plane pole and zero positions, the filter difference equation (which Fisher calls the recurrence relation), a fragment of ANSI C code that implements the filter, and frequency-domain and time-domain response graphs.

For the problem at hand, the inputs were: filtertype = Butterworth, passtype = Bandpass, order = 2, samplerate = 2000, corner1 = 20, and corner2 = 150. Results indicated two s-plane zeros at the origin, and two sets of complex conjugate poles (four in all). Two sets of z-plane zeros were located at the intersection of the real axis with the unit circle (four in all). There were two sets of complex conjugate z-plane poles (four in all). Gain at the center was 30.84450009.

The difference equation was implemented in compiled BASIC as follows:

```
Program$ = "2ndFiltr.bas"
' Signal to be sampled is half-sine with duration of tDur
' Time t is in milliseconds
DEFInt i-k, n
Dim x(4), y(4)
Pi = 4.*Atn(1.0)
Fmt$ = "###.####"
tDur = 25.
```

```
' Sample rate is 2 kHz => dt = 0.5 msec
dt = 0.5
' Filter parameters
a1 = -3.3816513217 : a2 = 4.3386013793
a3 = -2.5177707758 : a4 = 0.5614923902
a0 = 1.0 : b0 = 1 : b1 = 0 : b2 = -2 : b3 = 0 : b4 = 1
Gain = 30.84450009

Open "E:\CaseWork\...\PBAS\25msCalc.txt" for Output as #3
Print #3, " Time Signal x0 x1 x2 x3";
Print #3, " x4 y1 y2 y3 y4"
FOR i = 0 to 4 : x(i) = 0.0 : y(i) = 0.0 : NEXT i
FOR t = -2.0 to 100. Step dt
        IF t < 0.0 OR t > tDur THEN Sig = 0. ELSE Sig = Sin(Pi*t/
tDur)
        FOR i = 1 to 4 : x(i-1) = x(i) : y(i-1) = y(i) : NEXT i
        x(4) = Sig/Gain
        y(4) = b0*x(4) + b1*x(3) + b2*x(2) + b3*x(1) + b4*x(0) - _
                    a1*y(3) - a2*y(2) - a3*y(1) - a4*y(0)
        y(4) = y(4)/a0
        Print #3, using "####.##"; t;
        Print #3, using Fmt$; Sig;
        FOR i = 0 to 4 : Print #3, using Fmt$; x(i); : NEXT i
        FOR i = 1 to 4 : Print #3, using Fmt$; y(i); : NEXT i
        Print #3,
NEXT t
Close #3
END
```

This particular run is for the 25-ms pulse of Figure 10.1. Note that the entire input signal is not put into an array, and then processed, since in a real-time application the only data that are available for processing are those from the beginning of the event up to the present time t. We see that at each value of t, there are four prior data points available, indexed 0 through 3. $x(4)$ is the present value of the independent variable (signal divided by gain), and $y(4)$ is the present value of the dependent variable. Before the present data point is calculated, a FOR loop shifts the values of x and y down by one index value, much as one would do with a moving average calculation.

Comparison of the filter output as calculated by the computer program, against the output as digitized from the plots of Figure 10.1, is shown in Figures 10.2 through 10.4.

It is readily seen that the computer code appears to reasonably replicate the results supplied in graphical form by the manufacturer, within the limits of accuracy imposed by the process of digitizing the provided curves.

A fourth-order bandpass filter designed to the same sample rate and corner frequencies has its odd-numbered feed-forward coefficients b_i equal to

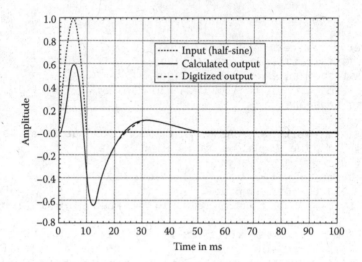

FIGURE 10.2
Bandpass filter response to 10 ms sine wave.

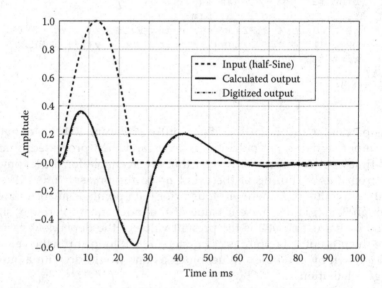

FIGURE 10.3
Bandpass filter response to 25 ms sine wave.

zero, as does the second-order filter. However, the "span" of the filter extends to $n-8$ instead of $n-4$. The pattern of the coefficients b_i in the second-order filter is 1, 0, −2, 0, 1, whereas in the fourth-order filter it is 1, 0, −4, 0, 6, 0, −4, 0, 1—patterns made easy to remember through the use of Pascal's Triangle,[3] or the coefficients used in binomial expansions.

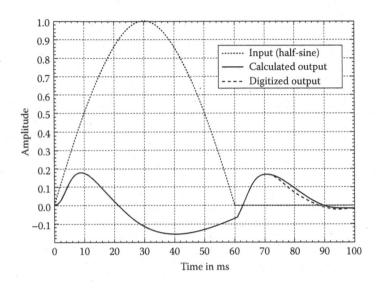

FIGURE 10.4
Bandpass filter response to 60 ms half-sine wave.

References

1. Butterworth, S., On the theory of filter amplifiers, *Wireless Engineer*, V. 7, 1930, pp. 536–541.
2. The mkfilter digital filter generation program, http://www-users.cs.york.ac.uk/~fisher/mkfilter.
3. Beckmann, P., *A History of Pi*, St. Martin's Press: New York, 1976.

11

Obtaining NHTSA Crash Test Data

Contemplating Vehicle Crashes

The terms "impact," "collision," and "crash" all refer to a dynamic event that happens during a very short (but finite) time frame, compared to most other events in the everyday usage of a vehicle or other object. Often the result of a crash is permanent damage to the vehicle, and perhaps injury to an occupant; thus it is something to be avoided, or at least mitigated. So why does one perform a crash test, what does one learn from it, and how does one go about gathering information? Because we want to know exactly what happened during the crash, why someone got a particular injury, or what we can do to make the vehicle a little more protective when the next crash like it occurs, as surely it will.

Clearly, one would run a test differently if the object being crashed were an egg, an airplane, a cube of jell-o, a golf ball, a clock, a bomb, or a passenger car. First of all, a road vehicle is not homogeneous like a cube of jell-o, nor filled with solid materials like a golf ball. Unlike a clock, it is designed to be accelerated to some speed, and to be decelerated to a stop. It is not intended to be blown to bits. The vehicle mass is not uniformly distributed. Some parts are so hard and dense that they can be thought of as solid and nondeformable, like an engine block, or perhaps a brake caliper. Some parts deform permanently in a crash and are made of relatively thin sheet metal that, as it turns out, is an excellent energy absorber. Some parts recover most of their original geometry, such as nonmetallic bumper blocks. Some subsystems mostly resist deformation in crashes up to some limit, and then deform only in certain areas, such as a passenger compartment.

The Crush Zone

The peculiar feature of a vehicle is that in a crash, some parts crush, and are designed to crush, a lot more than others. In severe crashes, the vehicle can be thought of as disposable impact packaging for its occupants. The part that is intended to crush is called the "crush zone."

In a perfect world, the crush zone would consist entirely of crushable materials that deform and absorb energy in such a way that crash forces are limited, and fed entirely into crush-resistant areas in the passenger compartment. But a vehicle is the result of many, many compromises in its design and manufacture. For example, the frontal crush zone in a high-speed barrier crash usually contains an engine block—a nondeformable component necessary to the vehicle's primary mission of providing mobility.

That hard mass moves in a different way (one hopes!) from the passenger compartment. As a vehicle crushes in a frontal barrier crash, the material in front of the engine "bottoms out," which is to say that it reaches the limit of its ability to deform, or that its crush stiffness becomes greatly increased. So the engine is brought to rest earlier than the compartment, after which it is no longer part of the mass undergoing deceleration. The mass being decelerated is suddenly decreased; we can expect that comparable forces in the crush zone will result in a rise in the compartment accelerations. Because it is brought to rest sooner, the engine can be expected to have a more severe, shorter acceleration pulse than does the passenger compartment.

Accelerometer Mount Strategy

This sort of thinking gives rise to the concept of a vehicle as a collection of hard masses—one of which may be the passenger compartment—interconnected in multiple ways by crushable elements, or "load paths." Such a concept was incorporated into what is known as a "lumped-mass" or "lumped-parameter" model, such as presented by Kamal[1] in the early 1970s. For each degree of freedom of each mass in a lumped-mass model, the differential equation of motion is derived using Newton's Second Law or a comparable method; the resulting system of equations is then solved by numerical means, for the purpose of simulating the vehicle (and perhaps occupant) response to a crash.

An occupant could be represented by one or several masses. These could be connected together by deformable elements so as to give a representation of a human being.

The velocity and displacement of each mass are often of interest. As discussed in Chapter 1, however, such quantities are difficult or impossible to measure in a crash environment. Therefore, it is common practice to use accelerometers, which are transducers that produce an electrical signal proportional to the acceleration. Subsequently, numerical integration and perhaps other mathematical calculations can be used to extract the velocity and displacement data from the acceleration signals.

One or more accelerometers may be mounted to a part of the vehicle that behaves like a hard mass, or to a structure that is sufficiently protected from being crushed or impacted by vehicle components that accelerometers and attached cables are not destroyed in a crash.

By definition of the crush zone, accelerometers located in that area are surrounded by crushable structures. Their signals are not generally representative of the acceleration environment experienced by occupants. Moreover, the structures can be subject to buckling, which results in the rotation of surfaces. Any accelerometer mount subject to rotation in a crash produces acceleration measurements that are not representative of what would be experienced had the accelerometer retained its orientation. It is also the case that the crush behavior of the structure should not be altered by the presence of an accelerometer mount. However, in some cases its location is representative of an actual or potential crash sensor location. In other cases, accelerometer mounts may be attached to hard masses in the crush zone for the purpose of obtaining inputs to lumped-mass models or other computer simulations. In either case, the attachment, or accelerometer mount, is on a sturdy structure, in a location protected from damage to the instrumentation. Generally, it is in an area deemed to be safe from buckling or localized rotations.

A single accelerometer can detect accelerations in a single direction, often parallel to the vehicle axis closest to the direction of the impact. In an angled collision, or where rotations are involved, an accelerometer mount may have two accelerometers at orthogonal directions, again generally parallel to two of the vehicle axes. This is called a bi-axial mount. A tri-axial mount, on the other hand, involves three accelerometers, again parallel to the axes of the vehicle. This may be used where accelerations in all three directions are anticipated, or where the resultant acceleration is of interest. The signal in each direction is independent (one hopes!) of the others; the resultant is calculated later, during post-test data processing.

In earlier discussion, it was seen that the acceleration–time history during a collision, or crash pulse, varies throughout the vehicle, as the various masses move in their various ways, and as the distribution of loads varies with the force–deflection characteristics and locations of the load paths. The crash pulses applicable to the vehicle as a whole are those obtained in or near the compartment, away from the crush zone. They also tend to be the pulses of most importance to occupant injury, unless intrusion is a factor (such as in side impacts, for example).

It is often the case that accelerometer mounts are located at various points around the vehicle, for various reasons. In frontal crashes, there is often one mount at the base of each B-pillar (latch pillar for the front door), which is where sturdy structure is located. There is often one or two mounts in the rear seat area, in an area often called the rear seat cross member, on the rear vertical lateral surface that rises from the rear seat foot well to the rear seat pan. If two such mounts are symmetrically located, their accelerations can be averaged at each step in time, thus eliminating the effect of yaw rotations on the calculation of longitudinal motions.

In side impacts, there are often a mount on the forward portion of the rocker panel opposite the impacted side, and another on the rear portion of that same rocker panel. There may also be mounts on the near-side doors, which may be of interest in examining their interactions with the occupants, or the effects of intrusion. There may also be mounts on the near-side rocker panels. The near-side mounts, however, are susceptible to concerns about transducers being in the crush zone, as mentioned earlier.

In rear impacts, mounts could be used at the base of the B-pillars, though such locations may be a little close to the crush zone for comfort. Locations at the base of the A-pillars (hinge pillars for the front doors) would be preferable if they are available.

Other Measurement Parameters and Transducers

Another quantity of interest is often force. For force measurement, transducers known as "load cells" are used. Tension force in some or all of the belt restraints is commonly measured. Also, in barrier crash tests run at 35 mph under the New Car Assessment Program (NCAP), it is common practice for compression forces against the barrier to be measured with a load cell array on the barrier face. Though other arrangements have been tried, many NCAP tests feature an array of four rows and nine columns. The analysis of such data will be discussed in Chapters 16 and 28.

Federal Motor Vehicle Safety Standard (FMVSS) 204[2] specifies a maximum rearward displacement of the steering assembly during a 30 mph frontal barrier crash test. Since displacement is at the root of the standard and involves a surface on a vehicle component, it does not suffice to measure acceleration on the component and perform a double integration. Displacement must be measured and recorded directly. This is accomplished using a scratch tube, in which the scratch on the tube provides a permanent record of the displacement. Consequently, a driver dummy is not used in this test, unlike almost all other crash tests.

Speaking of crash dummies, the measurement of compressive forces in the femurs has been in use for many years. In more recent times, crash test dummies have become increasingly sophisticated, and the quantities being measured have proliferated. Increased concern about lower leg injuries has led to the measurement of bending moments in the tibia and ankle, for example. It has become common to measure and report the neck forces and moments about all three axes, though older tests generally do not include such measurements. Another measure in use may be dummy chest compression.

Sign Conventions and Coordinate Systems

Every crash test report should make plain the sign convention used in the data presentation, which is to say that it should inform the reader what a positive value means for each of the quantities reported.

It is generally the case that an SAE coordinate system is used on each vehicle to report accelerations, velocities, and displacements. Thus, accelerations are positive forward in the vehicle, to the right in the vehicle, and down. Such a convention is usually indicated on a table or diagram. It is good practice to check to be sure, however.

In accordance with the above, crush values in a frontal impact may be reported as negative numbers. If they are reported as positive numbers, the convention used by the test agency may be to report structural deformation and force as positive in compression. This may also apply to barrier forces. Often accelerations are reported as positive, but plots of positive decelerations have also been seen. If the sign convention is not reported explicitly, the maximum values should make obvious what convention is being used, particularly if the force readings have had the maximum filtering applied (typically, Class 60). It is a good exercise to see what level of supposedly tensile readings are present at time zero, the instant of impact, in the results for barrier load cells (which cannot transmit tensile forces, of course), as a means of checking on any uncertainties in the load cell barrier data.

Processing NHTSA Crash Test Accelerometer Data

Many manufacturers and other entities, such as the Insurance Institute for Highway Safety (IIHS), conduct crash tests and presumably store the results in the form of reports, photographs, videos or films, and/or digital data files. Such information is not generally available to the public. However, there is a source that provides publicly available information: the U.S. Department of Transportation (DOT) and the NHTSA. While every agency has its own format for reports and other information, the discussion that follows will focus on the NHTSA.

Each test released by the NHTSA to the public is assigned a DOT number—a five-digit number assigned in chronological order as they are released. There appears to be no code embedded in the DOT number. Occasionally, a test sponsored by Transport Canada, the IIHS, or by a manufacturer will be released and given a DOT number. The DOT number does not correspond to the "NHTSA-Assigned Number," nor to the report number assigned by a contractor.

Data from NHTSA crash tests reside on the NHTSA web site, and exist in the form of one or more written reports, lists of pertinent summary information, videos, and digital files: one for each measured variable as a function of time. The reports usually include brief narratives, tables of pre- and posttest measurements, data plots, photographs, and dummy calibration data. Sometimes, the reports are broken into sections that are then stored in separate files. The following is a description of how to access the information and develop dynamic response information (accelerations, velocities, and displacements, as functions of time).

Summary of the Process

1. Find the desired test on NHTSA's web site.
2. Locate, download, and read the crash test report.
3. Digitize the desired time histories of acceleration, velocity, or displacement, or, alternatively, go through the following steps:
4. Identify the accelerometer channels to be downloaded.
5. Download the desired channels.
6. Run program(s) to read the files and process the data.
7. Prepare graphical presentation(s) of the digital data.

In many instances it suffices to obtain summary information only, such as maximum acceleration, minimum velocity, ΔV, maximum displacement, and so on, as opposed to a detailed analysis of one or more channels. When that is the case, one can download and run software from the NHTSA that provides many of the functions of steps 4 through 7 above, plus other tools for analysis. Go the NHTSA web site www.nhtsa.gov and click on the Research tab. Under the topic "Databases and Software" navigate to "Signal Analysis Software" and follow the instructions.

Perhaps the most generally useful of the applications is the "Signal Browser," the use of which will be discussed in Chapters 12 and 14.

Downloading Data from NHTSA's Web Site

To proceed with a data download, the first step is to create a destination, if one does not already exist, for the downloaded files, such as a sub-directory titled "DownLoad," under the project name being worked on. (There may be a number of files to store and analyze.)

Crash test data from NHTSA can be downloaded from http://www-nrd. nhtsa.dot.gov. From the browse topics, select *Databases and Software*. Among the links under "Vehicle Crash Test Database" are "Query by test parameters and Query by vehicle parameters such as make, model, and year. One can also print catalogs as follows: Sorted by vehicle MAKE, MODE., and YEAR, Sorted by NHTSA assigned test number (TSTNO), and Event Data Recorder Reports." Notable by its absence is the ability to select or sort tests by crash mode and vehicle parameters simultaneously. For example, one might want to find all frontal tests of Toyota Corollas from model years 2002 to 2006. To do so, it will be necessary to do some filtering by hand: for example, one might query by vehicle parameters to find all 2002–2006 Toyota Corollas, and then pick through the results by hand to weed out everything but the frontals.

On the other hand, if one wants a particular test and knows its DOT number, which can be accessed through a query by test parameters, one of which is DOT number. In any case, fill out the query criteria table and click on "Submit."

Identifying the Accelerometer Channels to be Downloaded

Crash test data from NHTSA can be downloaded from http://www-nrd. nhtsa.dot.gov. This table lists the documents retrievable for each crash test. The number and type of documents vary from test to test. Generally, there is a column of links called *Reports*, each of which will lead to one or more report documents. (Sometimes, the photographs and data traces are in separate documents.) Clicking on the link for the desired test will cause a pdf file containing the report to be downloaded. The file may take considerable time to download, but if one needs the report anyway, it can be consulted to find vehicle and test parameters, the sign convention in use, crush measurements, accelerometer locations, and so on. Because there are usually numerous instrumentation channels for the dummies—many more than for the vehicle—it is generally not a good idea to try to dope out the vehicle data channel numbers exclusively from the test report.

A better approach involves a column of links called *Test No*. Clicking on the link for the desired test brings up another table titled "Vehicle Database Query Results—Vehicle Information" (http://www-nrd.nhtsa.dot.gov), which contains a small bit of information about the test. Included in this information may be the vehicle make and model, crash test mode, and crash test speed. This can be very helpful in identifying the tests of particular interest.

Off to the right of this table is a dialog box titled "Download Test: XXXXX," where XXXXX is the DOT Number. Select "NHTSA EV5 ASCII X-Y" in this box and click on "Go." Select "Open" in the ensuing dialog box.

File vXXXXXascii.zip will be downloaded and can be opened by a zip program (WinZip, for example). The result is a list of zipped files that can be extracted. The first of these is called vXXXXX.EV5, which is an ASCII file. It can be opened with a text editor such as EditPad Lite, the DOS Edit command, or even Excel (which puts everything in the first column). A fragment of such a file, with the rows truncated so as to fit on a single line, appears below.

```
# Source: NHTSA Vehicle Database - Test Number: 05404
# Date: <5/7/2009>
- - - EV5 - - -
- - - TEST - - -
V5|56 KPH 2005 TOYOTA COROLLA INTO A LCB|INVESTIGATE FRONT AND R
- - - VEHICLE - - -
1|16|02|2005||4S|JTDBR32E750054251|4CTF|1.8|AF|1566|2600|4525|17
- - - BARRIER - - -
R|LCB|0|0|NO COMMENTS
- - - OCCUPANT - - -
1|03|H3|0|F|0|0|H3|05|FIRST TECHNOLOGY, S/N: 324||ADDITIONAL INS
1|04|H3|0|M|0|0|H3|50|VECTOR, S/N: 110||ADDIT. INSTR.: LWR NECK
- - - RESTRAINT - - -
1|03|1|3PT|BC|UN|
1|04|1|3PT|BC|UN|
- - - INSTRUMENTATION - - -
1|1|AC|04|HDCG|XL|SEC|G'S|1650|MFG: ENDEVCO, MODEL: 7264, S/N: J
1|2|AC|04|HDCG|YL|SEC|G'S|1650|MFG: ENDEVCO, MODEL: 7264, S/N: J
1|3|AC|04|HDCG|ZL|SEC|G'S|1650|MFG: ENDEVCO, MODEL: 7264, S/N: J
1|4|LC|04|NEKU|XL|SEC|NWT|1650|MFG: DENTON, MODEL: 1716A, S/N: 1
1|5|LC|04|NEKU|YL|SEC|NWT|1650|MFG: DENTON, MODEL: 1716A, S/N: 1
1|6|LC|04|NEKU|ZL|SEC|NWT|1650|MFG: DENTON, MODEL: 1716A, S/N: 1
1|7|LC|04|NEKU|XL|SEC|NWM|1650|MFG: DENTON, MODEL: 1716A, S/N: 1
1|8|LC|04|NEKU|YL|SEC|NWM|1650|MFG: DENTON, MODEL: 1716A, S/N: 1
1|9|LC|04|NEKU|ZL|SEC|NWM|1650|MFG: DENTON, MODEL: 1716A, S/N: 1
1|10|LC|04|NEKL|XL|SEC|NWT|1650|MFG: DENTON, MODEL: 1794A, S/N:
1|11|LC|04|NEKL|YL|SEC|NWT|1650|MFG: DENTON, MODEL: 1794A, S/N:
1|12|LC|04|NEKL|ZL|SEC|NWT|1650|MFG: DENTON, MODEL: 1794A, S/N:
1|13|LC|04|NEKL|XL|SEC|NWM|1650|MFG: DENTON, MODEL: 1794A, S/N:
1|14|LC|04|NEKL|YL|SEC|NWM|1650|MFG: DENTON, MODEL: 1794A, S/N:
1|15|LC|04|NEKL|ZL|SEC|NWM|1650|MFG: DENTON, MODEL: 1794A, S/N:
1|16|AC|04|CHST|XL|SEC|G'S|1650|MFG: ENDEVCO, MODEL: 7264, S/N:
1|17|AC|04|CHST|YL|SEC|G'S|1650|MFG: ENDEVCO, MODEL: 7264, S/N:
1|18|AC|04|CHST|ZL|SEC|G'S|1650|MFG: ENDEVCO, MODEL: 7264, S/N:
1|19|DS|04|CHST|XL|SEC|MM|1650|MFG: SERVO, MODEL: 2897, S/N: CST
1|20|AC|04|PVCN|XL|SEC|G'S|1650|MFG: ENDEVCO, MODEL: 7264, S/N:
1|21|AC|04|PVCN|YL|SEC|G'S|1650|MFG: ENDEVCO, MODEL: 7264, S/N:
1|22|AC|04|PVCN|ZL|SEC|G'S|1650|MFG: ENDEVCO, MODEL: 7264, S/N:
1|23|LC|04|FMRL|ZL|SEC|NWT|1650|MFG: DENTON, MODEL: 2430, S/N: 7
1|24|LC|04|FMRR|ZL|SEC|NWT|1650|MFG: DENTON, MODEL: 2430, S/N: 9
```

```
1|49|AC|NA|SELR|XG|SEC|G'S|1650|MFG: ENDEVCO, MODEL: 7264C-2 K-2-
1|50|AC|NA|SELR|ZG|SEC|G'S|1650|MFG: ENDEVCO, MODEL: 7264C-2 K-2-
1|51|AC|NA|SERR|XG|SEC|G'S|1650|MFG: ENDEVCO, MODEL: 7264C-2 K-2-
1|52|AC|NA|SERR|ZG|SEC|G'S|1650|MFG: ENDEVCO, MODEL: 7264C-2 K-2-
1|53|AC|NA|ENGN|XG|SEC|G'S|1650|MFG: ENDEVCO, MODEL: 7264C-2 K-2-
1|54|AC|NA|ENGN|XG|SEC|G'S|1650|MFG: ENDEVCO, MODEL: 7264C-2 K-2-
1|55|AC|NA|BRCR|XG|SEC|G'S|1650|MFG: ENDEVCO, MODEL: 7264C-2 K-2-
1|56|AC|NA|BRCL|XG|SEC|G'S|1650|MFG: ENDEVCO, MODEL: 7264C-2 K-2-
1|57|AC|NA|DPLC|XG|SEC|G'S|1650|MFG: ENDEVCO, MODEL: 7264C-2 K-2-
1|58|LC|01|LPBO|NA|SEC|NWT|1650|MFG: LEBOW, MODEL: 3419T, S/N: 8
1|59|LC|01|SHBT|NA|SEC|NWT|1650|MFG: LEBOW, MODEL: 3419T, S/N: 8
1|60|DS|01|SHBT|NA|SEC|MM|1650|MFG: CELESCO, MODEL: PT-1010-0050
1|61|LC|04|LPBO|NA|SEC|NWT|1650|MFG: LEBOW, MODEL: 3419T, S/N: 8
1|62|LC|04|SHBT|NA|SEC|NWT|1650|MFG: LEBOW, MODEL: 3419T, S/N: 8
1|63|DS|04|SHBT|NA|SEC|MM|1650|MFG: CELESCO, MODEL: PT-1010-0050
0|64|LC|NA|LCA1|NA|SEC|NWT|1650|MFG: INTERFACE, MODEL: 1220TX-50
0|65|LC|NA|LCA2|NA|SEC|NWT|1650|MFG: INTERFACE, MODEL: 1220TX-50
0|66|LC|NA|LCA3|NA|SEC|NWT|1650|MFG: INTERFACE, MODEL: 1220TX-50

0|72|LC|NA|LCA9|NA|SEC|NWT|1650|MFG: INTERFACE, MODEL: 1220TX-50
0|73|LC|NA|LCB1|NA|SEC|NWT|1650|MFG: INTERFACE, MODEL: 1220TX-50
0|74|LC|NA|LCB2|NA|SEC|NWT|1650|MFG: INTERFACE, MODEL: 1220TX-50

0|81|LC|NA|LCB9|NA|SEC|NWT|1650|MFG: INTERFACE, MODEL: 1220TX-50
0|82|LC|NA|LCC1|NA|SEC|NWT|1650|MFG: INTERFACE, MODEL: 1220TX-50
0|83|LC|NA|LCC2|NA|SEC|NWT|1650|MFG: INTERFACE, MODEL: 1220TX-50

0|90|LC|NA|LCC9|NA|SEC|NWT|1650|MFG: INTERFACE, MODEL: 1220TX-50
0|91|LC|NA|LCD1|NA|SEC|NWT|1650|MFG: INTERFACE, MODEL: 1220TX-50
0|92|LC|NA|LCD2|NA|SEC|NWT|1650|MFG: INTERFACE, MODEL: 1220TX-50

0|99|LC|NA|LCD9|NA|SEC|NWT|1650|MFG: INTERFACE, MODEL: 1220TX-50
- - - END - - -
```

Blank lines indicate where records have been deleted from the down-loaded file. Channel numbers are indicated in the second column. Data for channels 25 through 48 pertain to dummy occupant 03 (column 4); these lines were removed for brevity.

Column 3 indicates the type of transducer (AC for accelerometer, LC for load cell, and DS for displacement), and column 8 indicates the measurement units. Column 6 indicates the direction of mounted accelerometers or load cells. Column 7 indicates the units of the independent variable (time).

Column 5 is a mnemonic indicating where the transducer is mounted. Load cell barrier transducers are indicated by "LC" followed by the location A1 through D9, and are thus found in channels 64 through 99 in this particular test.

From the mnemonics, channels 61 through 63 appear to be the lap belt (outer) load, shoulder belt load, and shoulder belt spool-out, in that order, for dummy 03. Channels 58 through 60 are the same quantities, probably for dummy 04 (though 01 is indicated). That leaves us with channels 49 through 57 as the vehicle channels. "ENGN" (column 5) indicates engine. BRCR and BRCL are the right and left brake calipers. DPLC is the dash panel center.

Since this test was a frontal impact, the vehicle accelerometer locations of interest are in the left and right rear. In this test, they were on the rear seat cross-member ("SELR" and SERR"), where bi-axial accelerometer mounts (X and Z) were used. The desired X-direction accelerations are in channels 49 and 51.

The right end of each row (here truncated) is a comments field, so one can scroll to the right ends of the records and see whether any data anomalies were recorded. For example, it is common practice for the NHTSA to use two engine accelerometers—one on top and one on the bottom—as was the case in this test. One cannot tell from the above file fragment which is which. However, in the comments field for channel 53, the notation is "Engine top; exceeded full scale at approximately 46 ms." For channel 54, the comment is "Engine bottom."

Additionally, one can check out the "Anomalies" page in the test report, and perhaps best of all, look at the data plots for acceleration (and velocity, if the data have been integrated). Leaving the zip file open, one can then go back to the zip file window and highlight the appropriate channel number files to be extracted.

To verify whether the mnemonics have been interpreted correctly, go back to the previous table, titled "Vehicle Database Query Results—Test Parameters," or "Vehicle Database Query Results—Vehicle Information," as the case may be. Another column in this table contains links titled *Instrumentation Information*. Clicking on the link for a particular test will bring up a list of data channels. This list shows such variables as Vehicle No, Sensor Type, Sensor Location, Sensor Attachment, Sensor Direction, and Data Measurement Units. The desired channel numbers are easily identified using this table.

Downloading the Desired Channels

The zip file vXXXXXascii.zip contains a list of all the zipped files, including vXXXXX.EV5, as discussed above. It also contains the data files, which have names of the form vXXXXX.NNN, where XXXXX is the DOT number, and NNN is the three-digit channel number. For this particular test (DOT 05404), the files to be extracted are v05404.049 and v05404.051. In the zip program window, highlight the desired files, click on the "Extract" button, select the destination drive and directory, and the selected files will be extracted to that location. Again, it is recommended that the files be placed in their own subdirectory, such as "DownLoad," under the name of the project being worked on.

Parsing the Data File

Below is the beginning of a downloaded file called v05404.049, which contains data from channel number 049 from test DOT 05404. The file is shown as it would appear in WordPerfect with Reveal Codes turned on. As we can see, it is tab-delimited. The left-hand column is time in seconds, and the right-hand column is acceleration in Gs. The time values run in uniform increments; in this test the time increment is 0.000080 s. The reciprocal of this number gives us the sample rate of 12.5 kHz. We also notice that in this test, data recording starts before barrier contact (zero event).

```
-0.020000[TAB]-0.099386[HRt]
-0.019920[TAB]-0.654563[HRt]
-0.019840[TAB]-0.901308[HRt]
-0.019760[TAB]-0.716249[HRt]
-0.019680[TAB]-0.346131[HRt]
-0.019600[TAB]-0.037700[HRt]
-0.019520[TAB]-0.006857[HRt]
-0.019440[TAB]-0.222759[HRt]
-0.019360[TAB]-0.469504[HRt]
-0.019280[TAB]-0.469504[HRt]
-0.019200[TAB]-0.253602[HRt]
-0.019120[TAB]0.147359[HRt]
-0.019040[TAB]0.394104[HRt]
-0.018960[TAB]0.455790[HRt]
-0.018880[TAB]0.332418[HRt]
-0.018800[TAB]0.085673[HRt]
-0.018720[TAB]-0.099386[HRt]
-0.018640[TAB]0.085673[HRt]
-0.018560[TAB]0.332418[HRt]
-0.018480[TAB]0.363261[HRt]
-0.018400[TAB]0.085673[HRt]
-0.018320[TAB]-0.531190[HRt]
-0.018240[TAB]-0.962994[HRt]
-0.018160[TAB]-1.271430[HRt]
-0.018080[TAB]-1.271430[HRt]
-0.018000[TAB]-1.086370
```

Such a file cannot be read by programs like NotePad, although it can be read by programs such as Word, WordPad, EditPad Lite, or Excel, using a fixed data width. However, consider that in the above test, the data extends to 310 msec, for a total time interval of 330 msec. With a sampling rate of 12.5 kHz, the data file for every channel contains 4126 records. Thus, a standard spread sheet program can be used to look at or print out the data, but is not suitable for data analysis, which instead requires the use of a special-built program.

On the other hand, tab-delimited files cannot be read using the generic input commands that come with certain programming languages, such as Power BASIC. If a program is being used that has such limitations, the lines have to be parsed using a procedure that developed by this author called GetLine(), explained below.

First, a null character variable FirstWord$ is defined. The line in the data file is read one character at a time, and the character is concatenated onto FirstWord$. When the procedure reaches the Tab character [Chr$(9)], it knows it has completed reading the time value. FirstWord$ is then converted from a character variable to a number using function like the PBas function Val(). The process continues by starting with another null character variable SecondWord$ and concatenating onto it until the Line Feed character [Chr$(10)] is encountered. This marks the end of the record. SecondWord$ is then converted to a number, which for this channel is the acceleration in Gs. The procedure GetLine() has been incorporated into various programs that read and process the NHTSA data files.

Although the data plots in crash test reports are filtered, the NHTSA data files contain just a time series of the raw data. Any filtering that is required has to be performed by the user. This is discussed further in Chapter 13.

Filtering the Data

The NHTSA and its contractors use second-order Butterworth filters,[3] which are discussed in Chapter 10, to filter crash test data. The FORTRAN source code for the filter is publicly available, and is included in the test procedure for performing and reporting NCAP tests.[4]

First, the time interval between samples is used to calculate the filter coefficients, which are also functions of the corner frequency for the class of filter being employed. Then, the values of the independent variable (time) and the dependent variable (acceleration, force, etc.) are read one record at a time, in the forward direction, and placed in arrays. At the same time, the dependent variable array value is replaced by a weighted average that is function of the filter coefficients and the previous two weighted averages. Next, the dependent variable array is read backwards, from last to first, and the weighted averaging process is repeated for the dependent variable values. Finally, the arrays can be cycled through in a forward direction, and the values of the independent and dependent variables written to a file that will now contain the independent variable and the filtered dependent variable.

This double-filter procedure, forward and backwards, is employed for the purpose of eliminating any phase shift in the filtered data. Of course, the fact that the filtering is done after all the data are collected is what allows the use of a backwards procedure. A Power Basic implementation of the NHTSA filter is provided in Chapter 13.

The result of the filtering process is no reduction of the number of data points. The only change is that the filtered values of the dependent variable reflect a weighted average of the surrounding values.

References

1. Kamal, M.M., Analysis and simulation of vehicle to barrier impact, SAE Paper 700414, SAE International, 1970.
2. U.S. Department of Transportation, Steering assembly rear displacement, *Code of Federal Regulations*, Title 49, Part 571-204, Government Printing Office, Washington, DC.
3. Butterworth, S., On the theory of filter amplifiers, *Wireless Engineer*, V. 7, 1930, pp. 536–541.
4. *Laboratory Indicant Test Procedure: New Car Assessment Program*, Appendix IV, U.S. Department of Transportation, National Highway Traffic Safety Administration, 1990.

12

Processing NHTSA Crash Test Acceleration Data

Background

The National Highway Traffic Safety Administration (NHTSA) has various contractors conducting full-scale crash tests as part of its standards enforcement, public information, and research activities. For each crash test type (frontal, side, rear, etc.), the NHTSA has published test procedures to ensure uniformity across all test labs. In particular, the NHTSA has developed computer software for data acquisition and reduction, the use of which is required.[1]

Analog data are prefiltered to CFC (channel frequency class) 1000 per SAE Recommended Practice J211[2] and digitized at a minimum rate of 8 kHz. (Rates of 8, 8⅓, 10, 12½, 13⅓, and 20 kHz have been encountered.) Digital data files are provided to NHTSA and are subsequently made available to the public through its web site (http://www-nrd.nhtsa.dot.gov). While the crash pulses and other traces in NHTSA crash test reports show the variables filtered per SAE J211, the data files downloaded from the NHTSA have not been processed through a digital filter. Filtering and certain processing procedures can be accomplished via the NHTSA signal browser—discussed later in this chapter—or performed by the user.

Integrating the Accelerations

Since the time step of integration is very small (<⅛ ms), it would probably be sufficient to assume a constant acceleration in each interval. However, for consistency with calculations used in other programs, and to ensure maximum accuracy, we assume that the crash pulse is adequately represented by a piecewise linear function that connects all the data points. In any interval between data points, the function is linear, as shown in Figure 12.1. In the time interval of interest, we can then write

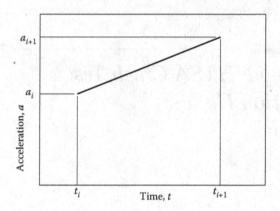

FIGURE 12.1
Piecewise linear acceleration–time relationship.

$$a = a_i + \frac{a_{i+1} - a_i}{t_{i+1} - t_i}(t - t_i) \qquad t_i \le t < t_{i+1} \tag{12.1}$$

Integrating this function is made much easier by incorporating the following variable substitution:

$$\tau = t - t_i \qquad t_i \le t < t_{i+1} \tag{12.2}$$

Then when $t = t_i$, $\tau = 0$, and when $t = t_{i+1}$, $\tau = t_{i+1} - t_i = \Delta t$, and $d\tau = dt$. We thus have

$$\frac{dV}{dt} = a = a_i + \frac{a_{i+1} - a_i}{\Delta t}\tau \tag{12.3}$$

Noting that the interval Δt is the same for all segments, it thus does not carry a subscript. Integrating the above, we obtain

$$V_{i+1} = V_i + \int_{\tau=0}^{\tau=\Delta t} a\,d\tau = V_i + \int_0^{\Delta t}\left[a_i + \frac{a_{i+1} - a_i}{\Delta t}\tau\right]d\tau \tag{12.4}$$

Performing the indicated integration and evaluating between limits produces

$$V_{i+1} = V_i + \frac{a_i + a_{i+1}}{2}\Delta t \tag{12.5}$$

We see that the incremental change in velocity given by Equation 12.5 is simply the trapezoidal area under the acceleration–time curve. Performing a second integration on the right-hand side of Equation 12.4 gives us the displacements, as follows:

$$x_{i+1} = x_i + \int_0^{\Delta t} \left[V_i + a_i \tau + \frac{1}{2}\left(\frac{a_{i+1} - a_i}{\Delta t} \right) \tau^2 \right] d\tau \tag{12.6}$$

Performing the indicated integration and evaluating between limits results in

$$x_{i+1} = x_i + V_i \Delta t + (2a_i + a_{i+1}) \frac{(\Delta t)^2}{6} \tag{12.7}$$

If $a_i = a_{i+1}$, Equation 12.7 reduces to the classical formula for constant acceleration.

Equations 12.5 and 12.7 are utilized for in all calculations that involve integrating downloaded acceleration data. Assuming that rotational motions of the accelerometer are small over the time frame of interest, velocity values at the recorded time values may be computed by integrating the accelerations, *starting at the time zero* and proceeding in a time-forward fashion, using Equation 12.5. A similar statement can be made for displacements. Equation 12.7 can be used to obtain the displacement at each time value, starting at time zero and proceeding in a time-forward direction, as long as there was no significant vehicle rotation during the crash phase.

Filtering the Data

NHTSA's filtering algorithm incorporates a second-order Butterworth filter.[3] The algorithm is written in FORTRAN, but it can be just as easily implemented in other languages, such as compiled BASIC. This has been done in a program called NHTFiltr.bas. This program is designed for reading t versus Y tab-delimited acceleration files such as those in NHTSA's crash test library, which have names of the form vXXXXX.NNN, where XXXXX is the five-digit DOT number, and NNN is the three-digit channel number.

The first two records in the data file are read, in order to determine the sample rate. Then the file is then closed and re-opened, in order to relocate the file pointer to the beginning of the file for subsequent calls to the Filter(j) subroutine. Depending on the sample rate, filter coefficients are calculated, using a bi-linear transform for various filter classes.

Filter(*j*) Subroutine

The filter algorithm is contained in a subroutine called `Filter(`*j*`)`, where *j* is an index for the filter class being used. ($j = 1 \Rightarrow$ CFC 1000; $j = 2 \Rightarrow$ CFC 600; $j = 3 \Rightarrow$ CFC 180; $j = 4 \Rightarrow$ CFC 60.) The corresponding corner frequencies are 1650, 1000, 300, and 100 Hz. The appropriate corner frequency and the time between samples are used to calculate the filter coefficients, by way of a bilinear transformation. The subroutine is shown below:

```
SUB Filter(j)
  ' Calculates parameters of the j-th filter
  Shared Pi, DT, NPts, Fcor(), CFC%(), b0(), b1(), b2(), a1(),
    a2(),_
  AFact, Fmt1$, Fmt2$, InFile$, InFileCh$, TempOutFile$(),_
  Time(), Y(), MinAcc(), MaxAcc(), MinATime(), MaxATime(),
    MaxPts%

  F6 dB(j) = 1.25*Fcor(j)  ' Freq. at which filter
    response = -6 dB, Hz
  Wd = 2.*Pi*F6 dB(j)
  Wa = Tan(Wd*DT/2.)
  b0(j) = Wa⊥2/(1. + Sqr(2.)*Wa + Wa^2)
  b1(j) = 2.*b0(j)
  b2(j) = b0(j)
  a1(j) = -2.*(Wa^2 - 1.)/(1. + Sqr(2.)*Wa + Wa^2)
  a2(j) = (-1. + Sqr(2.)*Wa - Wa^2)/(1. + Sqr(2.)*Wa + Wa^2)
  ' Filter forward
  Open InFile$ for Input as #1
  Y1 = 0.0
  FOR n = 1 to 10
      Call GetLine (1, n, Time(), Y()) : Y1 = Y1 + Y(n)
  NEXT n
  Y1 = Y1/10.
  X2 = 0.0
  X1 = Y(1)
  X0 = Y(2)
  Y(1) = Y1
  Y(2) = Y1

  DO UNTIL EOF(1) OR n > MaxPts%
    Call GetLine (1, n, Time(), Y())
    X2 = X1
    X1 = X0
    X0 = Y(n)
    Y(n) = b0(j)*X0 + b1(j)*X1 + b2(j)*X2 + A1(j)*Y(n-1) + A2(j)*Y
      (n-2)
  Incr n
  LOOP
  NPts = n-1
  Close #1
```

```
' Filter backwards
Y1 = 0.
FOR n = NPts to NPts-9 Step -1: Y1 = Y1 + Y(n)  : NEXT n
Y1 = Y1/10.
X2 = 0.0
X1 = Y(NPts)
X0 = Y(NPts-1)
Y(NPts) = Y1
Y(NPts-1) = Y1

MinAcc(j) = 1E9  : MaxAcc(j) = -1E9
FOR n = NPts-2 to 1 Step -1
   X2 = X1
   X1 = X0
   X0 = Y(n)
   Y(n) = b0(j)*X0 + b1(j)*X1 + b2(j)*X2 + A1(j)
      *Y(n+1) + A2(j)*Y(n+2)
   IF Y(n) < MinAcc(j) THEN MinAcc(j) = Y(n)  :
      MinATime(j) = Time(n)
   IF Y(n) > MaxAcc(j) THEN MaxAcc(j) = Y(n)  :
      MaxATime(j) = Time(n)
NEXT n

TempOutFile$(j) = "DataTrac\" + InFileCh$ + "-" + FnP$(3, CFC%(j))_
      + ".tmp"
Open TempOutFile$(j) for output as #3
Print #3, "Time Accel" FnP$(3,CFC%(j))
FOR n = 1 to NPts
   Print #3, using Fmt2$; Time(n)*1000., Y(n)
NEXT n
Close #3
END SUB
```

Again, the feed-forward coefficients b_i are in the ratio 1, 2, 1, except in this case a multiplier is used that depends on two circular frequencies ω_d and ω_a, which themselves are functions of the corner frequency. Also, the feedback coefficients a_j are calculated explicitly and are also functions of the two circular frequencies ω_d and ω_a.

Parsing the Data File

As can be seen in the code, a subroutine GetLine() is called for the purpose of reading a line of data. This is done because the NHTSA data files are tab-delimited, whereas the Power BASIC input commands are intended for space- or comma-delimited data, and thus cannot be used for this application. A description of the routine is provided in Chapter 11.

NHTFiltr.bas Program Output

Segments from a typical output file are shown below.

Time	Accel1000	Accel0600	Accel0180	Accel0060	Veloc	Displ
-20.0000	-3.3453	-2.2112	-1.9819	-4.1158	M	M
-19.9200	-3.3923	-1.9817	-2.0200	-4.1561	M	M
-19.8400	-2.9219	-1.5650	-2.0770	-4.1922	M	M
-19.7600	-2.4044	-1.1997	-2.1707	-4.2248	M	M
-19.6800	-1.6517	-1.0638	-2.3001	-4.2526	M	M
-19.6000	-1.1813	-1.1648	-2.4455	-4.2727	M	M
-19.5200	-0.9461	-1.4587	-2.5860	-4.2820	M	M
-19.4400	-0.8990	-1.8664	-2.7045	-4.2786	M	M
-0.1600	5.8753	3.3276	0.9889	1.3383	M	M
-0.0800	-1.0401	0.6925	1.1050	1.3580	M	M
0.0000	-6.7795	-1.3885	1.2107	1.3765	43.93	0.00
0.0800	-7.8615	-2.0461	1.3164	1.3937	43.93	0.06
0.1600	-3.8628	-1.1103	1.4250	1.4094	43.94	0.12
0.2400	2.5352	0.8367	1.5306	1.4233	43.94	0.19
0.3200	7.2396	2.8340	1.6221	1.4351	43.94	0.25
0.4000	7.9923	4.0868	1.6886	1.4443	43.94	0.31

Time is in milliseconds, accelerations are in Gs, velocity is in miles per hour, and displacement is in inches. Prior to the zero event (and prior to the start of integrations), Ms are written to the velocity and displacement columns as placeholders. When the file is imported to the graphics program (via the script file Import Filtered Data.axs), the Ms signify missing data.

Averaging Two Acceleration Channels

Theoretically, it would be desirable to have an accelerometer mount at the center of mass, or CG. However, many practical considerations render such a location inadvisable or impossible. The CG may not be located on sufficiently stout structure (or may even be in thin air!), it may be in the potential path of dummy motions, or it may not be accessible, for example. Even mounting an accelerometer on the vehicle centerline for a frontal barrier impact might not be the best choice. It is common, therefore, to use accelerometer mounts symmetrically located about a vehicle axis. In that situation, calculating the average of the two acceleration traces cancels out the effects of vehicle rotation about the CG, leaving just the translational motion.

To enable such a process, a program called AccelAvg.bas was developed to:

a. Read *t* versus *Y* tab-delimited acceleration files such as those in NHTSA's crash test library.

b. Average the accelerations.

c. Perform the necessary integrations.

d. Filter the resulting acceleration, velocity, and displacement signals.

e. Place the results in an ASCII text file.

The input files are assumed to have been downloaded, per Chapter 11, into a subdirectory called "DownLoad," under the project directory. The user can specify the channel number of one acceleration channel for analysis, or the numbers of two to be averaged and then analyzed.

The acceleration data from the first channel are parsed using `GetLine()`, as discussed in Chapter 11. Because of DOS memory limitations and the large number of data, the data are also written to a space-delimited temporary file, as are the data from the second acceleration channel, if it is used. The space-delimited data from the first channel are filtered to the maximum extent (CFC 60), to obtain an approximation of the overall maximum and minimum values. These values are subsequently displayed to help the user determine the sign convention used on accelerations. (This can be ascertained by looking at plots of filtered data, but in the absence of such plots, searching 4000 records of unfiltered data by hand can be tedious and error-prone. Sign conventions have been encountered with either acceleration or deceleration as positive, although NHTSA data generally have acceleration as positive.) The user selects acceleration or deceleration as being positive, so that the data may be integrated correctly. For the same reason, the user is also asked to specify the initial velocity (in kph, consistent with current NHTSA practice).

Then the first space-delimited file is reopened (along with the second file, if it is used), and the average accelerations are calculated and stored in the average acceleration (space-delimited) temporary file. Also, the average accelerations are double-integrated, using the initial velocity specified by the user, and the results are placed in velocity and displacement temporary files.

Next, the average accelerations are filtered to CFC 60, and the velocities and displacements are both filtered to CFC 180. The results are placed in three more temporary files. Then the filtered accelerations, velocities, and displacements are read and printed out to a permanent file containing all three variables. A second permanent file is also created, containing run information such as the names of the NHTSA files used, the number of records, the sample rate, the filter coefficients, and so on. Other summary information, such as the maximum and minimum average accelerations, the maximum and minimum velocities, the maximum displacement, the total velocity change, and the CG rebound coefficient, are also printed to this file. Finally, the various temporary files are deleted.

Using the NHTSA Signal Browser

The NHTSA Signal Analysis Software package includes several applications: Ankle Rotation, Auto Injury Criteria, Injury Assessment Values, Load Cell Analysis, MHD Filter, Roof Crush, Signal Browser, Toe Pan Intrusion, and Vehicle Checker. For a standardized package, it is very useful in that tool bar options allow the user a good deal of flexibility in the choice of channels to be analyzed, the types of analysis to be performed, and the presentation of the results. The load cell analysis application deals with the measurement of forces in instrumented crash barrier faces, and will be discussed further in Chapter 14.

To use the Signal Browser, one runs the Signal Analysis program (after having downloaded and installed it in one's computer, of course) and selects the Signal Browser option. A dialog box is used to specify the source (e.g., NHTSA web site), the database (e.g., vehicle), and the test, which is selected in a fashion similar to that described for the NHTSA crash test database in Chapter 11 (although the process is hierarchical with this software). A structured list of channel identifiers is then presented, with check boxes that allow the user to select which ones to analyze. The data are immediately graphed (at CFC 1000). The user can click on an icon and select a lower channel class if desired.

An edit tab allows the user to customize the graph, including line colors, line styles, and many other features. The data or the graph can also be copied to the Windows clipboard, and the graph can be saved to a variety of formats, including MetaFile, JPG, and so on. The title or a legend shows which channels are being displayed. If only one channel is graphed, a subtitle shows the minimum and maximum values of the variables, and the times at which they occurred. When the cursor is moved over the plot area, its coordinates are displayed, which allows the user to point at a particular location on the curve(s) and digitize its coordinates.

Another tab, called "Operate," contains some very useful functions: bias removal, filter, integrate, derivative, FIR 100, resample, scale x axis, scale y axis, truncate, average, resultant (if appropriate), sum, and reload. The filter option allows one to specify the cutoff frequency. The truncate function operates on the time variable, and allows one to look in more detail at a portion of the curve, by specifying the beginning and end time values.

Operations can be performed in series, such as average, filter, and integrate. However, if multiple curves have been combined (e.g., sum or average), the default title shows information for only the first channel in the series. The software allows the user to specify his or her own title, however, so it is a good idea to specify the channel numbers and the process in the title. If only one curve results from the process, the subtitle still indicates minimum and maximum values, and the associated time values.

References

1. *Laboratory Indicant Test Procedure: New Car Assessment Program*, Appendix IV, U.S. Department of Transportation, National Highway Traffic Safety Administration, 1990.
2. Instrumentation for impact tests, SAE Recommended Practice J211, SAE International, 1988.
3. Butterworth, S, On the theory of filter amplifiers, *Wireless Engineer*, V. 7, 1930, pp. 536–541.

13

Analyzing Crash Pulse Data

Data from NHTSA

Crash pulse data (acceleration vs. time) from NHTSA are usually presented graphically as part of the hardcopy test report. The data presentation reflects filtering according to the applicable test procedure, which, in turn, references SAE J211.[1] The plots generally indicate what filter class was used to present the data. Sometimes, they do not indicate the minimum and maximum acceleration values. Sometimes, the plots have notations showing the minimum and maximum values, which may or may not reflect filtering. If the notations do not comport with the plotted curve, it is probable that the minimum and maximum reflect unfiltered rather than filtered values.

In crash test reports, sometimes further processing of acceleration data is presented in the form of velocity–time traces, displacement–time traces, and resultant accelerations. The latter are obtained from tri-axial acceleration packs, which contain three individual accelerometers, all located very close together, and oriented in mutually orthogonal x, y, and z directions.

The analysis of hardcopy plots entails the creation of data file(s) by a digitization process, which, in turn, involves digitizing the coordinate axes for scaling purposes, and then digitizing selected points on the plot. Generally speaking, hardcopy plots show data after the appropriate filtering has been applied. Minimum and maximum values, if indicated on the plots, usually—but not always—come from the filtering process; therefore, the reported peaks and valleys do not correspond with what is seen in the digital data for the same transducer. Since the degree of filtering significantly affects the peaks and valleys, any comparisons based on maxima and minima must reflect consistent filtering, and reports of the results should indicate the nature and extent of filtering used.

Crash pulse data can also be downloaded as digital files, one per channel, in which a time series of the transduced values is presented. That is to say, each file contains uniformly spaced numeric time values in one column, and corresponding acceleration values in a second column, as discussed in Chapter 12.

Since the crash pulse typically has so much high-frequency content—that is, it is so jagged—the acceleration signal is generally not the best way to understand a particular crash or to compare it to others. The peaks and valleys are heavily dependent, as we shall see, on the filtering applied to the data. Early in the crash, before the occupants contact the vehicle interior and restraint systems, they are decoupled from the crash pulse entirely (and are not exposed to those peaks and valleys); when they contact the vehicle interior, it is the contact velocity that assumes importance. Later on, they are coupled to the crash pulse, which is then filtered (mechanically) by their restraint systems and other components in the vehicle interior with which they may interact.

So the occupant's injury exposure is dependent mostly on the vehicle's velocity–time curve. Velocity is derived from acceleration by an integration process, of course, and that process inherently smooths the accelerations. That truth can be understood mathematically, and it can also be understood by simply comparing an acceleration–time plot with a velocity–time plot.

To demonstrate this point, data were examined for crash test DOT 5683, in which a 2006 Ford F-250 (Vehicle 1, the "bullet" vehicle) traveling 24.0 mph was crashed head-on into a 2002 Ford Focus (Vehicle 2, the "target" vehicle) traveling 43.9 mph. The ratio of impact speeds was almost exactly the reciprocal of the ratio of the vehicle masses, so the test was apparently designed for a stationary system center of mass. The test was performed at Transportation Research Center of Ohio (TRC), data were sampled at 12.5 kHz, and high-quality hardcopy plots (see Appendix B of the test report) were produced having maximum and minimum values indicated. Integrations were not provided for the left and right rear seat crossmember accelerometers. However, a tri-axial accelerometer pack was located at the center of mass, and integrations were provided for all three channels. A plot of the resultant acceleration was provided as well. X-axis acceleration, velocity, and displacements are shown in Figures 13.1 through 13.3, and Figures 1.4 through 1.6.

Note that accelerations were filtered to CFC 60, whereas the velocities and displacements are smoother curves even though the data were filtered less heavily—CFC 180. This filtering is in accordance with SAE J211, which seems to have been written with an eye to the smoothing nature of the acceleration process.

Digital data for selected channels were downloaded from the NHTSA web site, using the procedures outlined in Chapter 11. Table 13.1 lists the channels of interest.

For Channel 280, the summary results are shown in Table 13.2.

Clearly, the reported values printed on the data plot are consistent with the acceleration–time curve, and thus reflect the filtering (at Class 60) that was applied to the curve. The raw digital data do not reflect that filtering. If the crash pulse were to be described by its maximum and minimum accelerations, using the raw data would lead to a very different characterization of the crash.

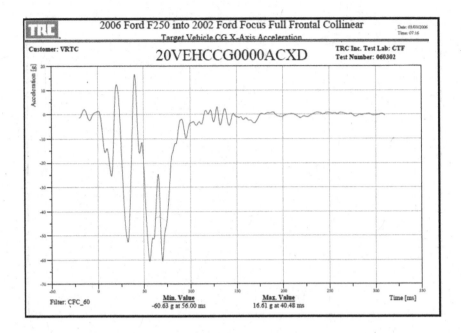

FIGURE 13.1
(See color insert.) CG *x*-axis acceleration, vehicle 2 (target vehicle).

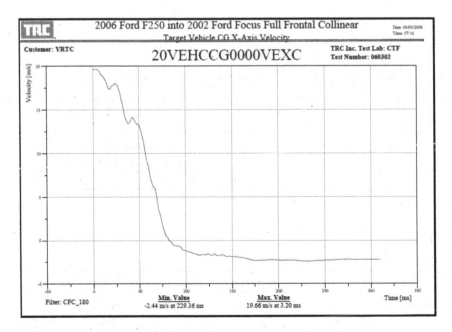

FIGURE 13.2
(See color insert.) CG *x*-axis velocity, vehicle 2 (target vehicle).

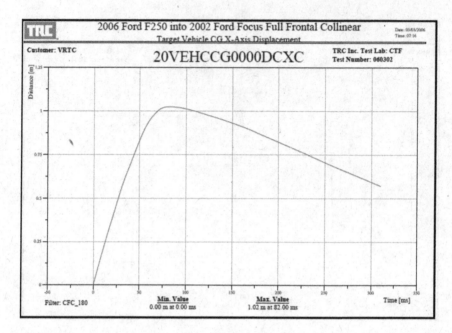

FIGURE 13.3
(**See color insert.**) CG *x*-axis displacement, vehicle 2 (target vehicle).

TABLE 13.1

Channels Selected for Analyzing Test 5883

Channel No.	Vehicle	Quantity	Location
100	1	x Accel	Left rear seat cross member
101	1	x Accel	Right rear seat cross member
111	1	x Accel	CG
269	2	x Accel	Left rear seat cross member
270	2	x Accel	Right rear seat cross member
280	2	x Accel	CG

TABLE 13.2

Results Found for Channel 280, Test 5683

Quantity	Reported on Data Plot	Found in Digital Data File
Minimum acceleration, g	−60.63	−183.29
Maximum acceleration, g	16.61	160.32

Repeatability of Digitizing Hardcopy Plots

The process for digitizing hardcopy plots of crash pulses was incorporated by this author in a Power BASIC program called PulseInt.bas. Of course, this software requires that the coordinate axes be digitized so that the appropriate coordinate rotations and scaling factors can be applied. Once the axis calibration is done, the process need not be repeated if there are multiple curves plotted on the same graph.

The plotted hardcopy crash pulse (Figure 13.1) was digitized three times. The first set of digitized data was identified as CGX1. Afterward, the plot was removed from the digitizing tablet and then reattached; the digitizing process (which includes calibrating the axes and digitizing the points) was repeated, and the data were given the identifier CGX2. Finally, the curve was digitized again, on the same axis calibration; these data were identified as CGX3. The results are summarized in Table 13.3.

The ΔV ranged from 49.10 to 49.55 mph (as compared to the reported value of 49.44 mph); the variation of 0.45 mph was 0.91% of the total ΔV. The maximum displacement ranged from 40.07 to 40.24 in. (as compared to the reported value of 40.16 in.); the variation of 0.17 in. was 0.42% of the total displacement.

In the last two runs, variability associated with the axis calibration process is removed, leaving only the variability associated with digitizing the curve itself. In these two runs, the ΔV ranged from 49.54 to 49.55 mph; the variation of 0.01 mph was 0.02% of the total ΔV. The maximum displacement varied by 0.12 in., which was 0.30% of the total displacement. For the crash pulse analyzed, eliminating the axis calibration variability reduced the variability of the calculated results.

Generally, it appears that the repeatability of digitizing hardcopy plots is within about 1%, at least for plots of the quality of Figure 13.1.

TABLE 13.3

Results of Digitizing the Hardcopy Plot of Accelerations, Test 5683

	Parameter and Time of Occurrence					
	Maximum Speed		Minimum Speed		Maximum Displacement	
Data Set Name	Value (mph)	At Time (msec)	Value (mph)	At Time (msec)	Value (in.)	At Time (msec)
CGX1	43.97	2.76	−5.13	230.65	40.24	82.81
CGX2	43.95	2.26	−5.60	231.66	40.07	81.16
CGX3	43.96	2.51	−5.58	230.40	40.19	82.16

Effects of Plotted Curve Quality

In digitizing a plotted crash pulse, it is reasonable to expect the quality of
the plot to affect the quality of the results. Presumably, such deficiencies as
axis skewness, poor resolution resulting from nonoptimal axis scaling, poor
line quality, multiple generations of copies, and general clutter can adversely
influence the ability to extract good-quality data. An example of this opposite end of the spectrum from Figures 13.1 to 13.3 is shown in Figure 13.4, an
exact copy of the plots provided by the manufacturer. Despite its very poor
quality, the reconstructionist was expected to determine the ΔV of the test.

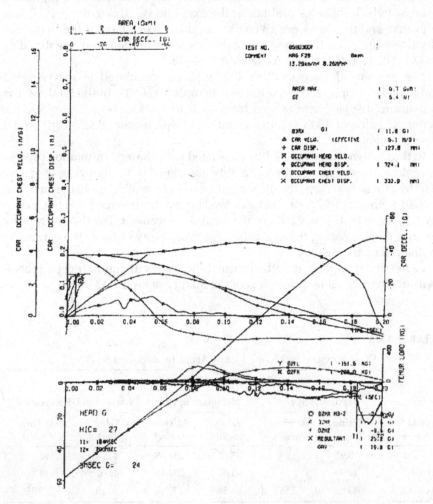

FIGURE 13.4
Right-side accelerometer data from crash test report produced by the manufacturer.

TABLE 13.4

Results of Digitizing Poor-Quality Hard Copy

	Parameter and Time of Occurrence				
	Minimum Speed		Velocity	Maximum Displacement	
Data Set Name	Value (mph)	At Time (msec)	Change (mph)	Value (in.)	At Time (msec)
031RX1	−3.38	181.82	11.64	5.02	55.13
031RX2	−3.60	107.63	11.86	4.94	53.79
031RX3	−3.56	107.40	11.82	4.96	54.74

No maximum and minimum values for acceleration were reported, so there could be no comparison of digitized to reported values. To evaluate process repeatability as before, the crash pulse was digitized three times: twice with the entire digitizing process repeated (031RX1 and 031RX2), and a third time on the same axis calibration as the second digitization (031RX3). The results are shown in Table 13.4.

Note that these results are from the right-side accelerometer only, and not the average of the left and right sides. The ΔV for the right-side transducer ranged from 11.64 to 11.86 mph; the variation of 0.22 mph was 1.87% of the average of the three right-side values. The maximum displacement for the right side ranged from 4.938 to 5.017 in., compared to a reported value of 5.031 in. Taking the reported value as "gospel," the discrepancy ranged from −0.093 to −0.014 in., or −1.85% to −0.28%.

Comparing the last two digitizations, the variation in ΔV was 0.04 mph, or 0.31% of the average. Variation in the maximum displacement was 0.025 in., or 0.50% of the reported value. Again, the contribution of axis calibration to the overall variability can be seen.

The variability in the analysis of this data channel, while small, is not as small as that seen for the target vehicle CG channel in DOT 5683. This appears to be a consequence of the relatively poor quality of the traces shown in Figure 13.4.

Accuracy of the Integration Process

Channel 280 of Test 5683 was deliberately chosen for study because velocity and displacement plots for that channel were included in the crash test report, thus permitting a check on the integration process. When the downloaded digital data were integrated using AccelAvg.bas, the results in Table 13.5 were calculated.

The initial velocity used in the calculations was the value published in the crash test report; namely, 70.7 kph. Presumably, the same value was

TABLE 13.5

Results of Integrations for Channel 280 of Test 5683

| | Parameter and Time of Occurrence | | | | | |
| | Maximum Speed | | Minimum Speed | | Maximum Displacement | |
How Obtained	Value (mph)	At Time (msec)	Value (mph)	At Time (msec)	Value (in.)	At Time (msec)
Calculated	44.00	3.20	−5.48	229.36	40.41	81.92
Reported	43.98	3.20	−5.46	229.36	40.16	82.00

used in the preparation of TRC's crash report (which is worth mentioning because all of the integrations depend on it). In any case, the reported ΔV was 43.98 + 5.46 = 49.44 mph, whereas the calculated ΔV was 49.48 mph, a discrepancy of 0.08%. The maximum displacement had a discrepancy of 0.62%, so it would appear that the integration process, when applied directly to the NHTSA data, produces accurate results. The same integration equations are used for hand-digitized crash pulse plots. Even though the integration intervals are far larger, the results are probably still accurate to within 1%, as long as the hardcopy plots are of good quality, and the digitizing is done carefully.

Accuracy of the Filtering Process

Probably the best way of quantifying the accuracy of the filtering process comes from looking at the peak values in the filtered acceleration data, as compared to the filtered velocities and displacements, which involve integration in addition to filtering. For Channel 280, the comparisons are shown in Table 13.6.

The timing of the peaks and valleys is identical. The reported range of accelerations, when filtered to CFC 60, was 77.24 Gs, whereas the calculated range at CFC 60 was 77.31 Gs, a discrepancy of 0.09%.

TABLE 13.6

Accuracy of Filter Process, Channel 280 of Test 5683

| | Parameter and Time of Occurrence | | | |
| | Maximum Acceleration, CFC 60 | | Minimum Acceleration, CFC 60 | |
How Obtained	Value (Gs)	At Time (msec)	Value (Gs)	At Time (msec)
Calculated	16.66	40.48	−60.65	56.00
Reported	16.61	40.48	−60.63	56.00

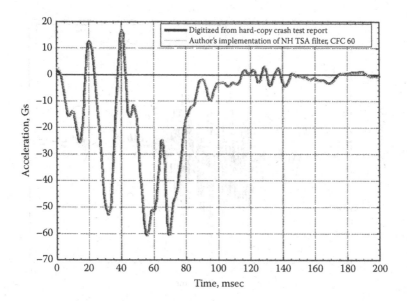

FIGURE 13.5
Comparison between CFC 60 filters for CG *x*-axis accelerations, Channel 280.

To judge the accuracy of the filtering process over the entire crash pulse, the best we can do is to digitize the hardcopy plot in the crash test report and then replot the data, so that we can overlay the results of the present filter calculations on it. Doing this produces Figure 13.5.

Close examination reveals only the slightest differences along the time axis, even though the gray line has been thickened so as to not to be obscured by the black line laid over it. Lest one concludes that such differences are due to a slight phase shift, however, it is important to note that the time values for the maximum and minimum accelerations are identical, suggesting no shift at all. Rather, it is more likely that the differences are due to the limits of digitizing resolution. Note, for example, that the digitized trace shows a slight bend to the left at about 75 msec. This bend is not seen on the original plot, indicating that the point digitized at 75.6 msec has a slightly erroneous time coordinate.

For this data set, at least, it appears that the present implementation of the filter algorithm provides a faithful replica of the NHTSA CFC 60 filter.

Effects of Filtering on Acceleration and Velocity Data

Since the acceleration signal has a significant high-frequency content, we can expect that filtering to CFC 60 will have a dramatic effect, not only on its

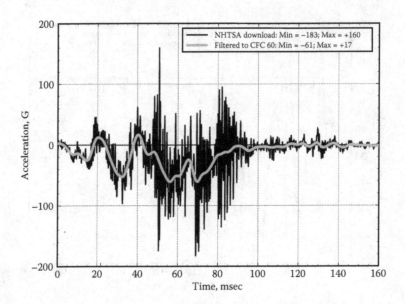

FIGURE 13.6
Channel 280 accelerations: raw versus filtered to CFC 60.

peaks and valleys, but on the overall appearance of the trace. This is easily seen in Figure 13.6.

As expected, the effects on the minimum and maximum values are dramatic. Clearly, the raw data are not suitable for characterizing the crash pulse.

It was argued earlier that since integration is a smoothing process, the results of integration should not be much affected by the extent of filtering applied to the original data. Figure 13.7 shows that it takes only one integration cycle to remove most of the noise.

At first glance, it appears that these curves are virtually identical; overall differences due to filtering, even at CFC 180, are barely perceptible. A closer look, as seen in Figure 13.8, shows that differences do exist.

In particular, rapid fluctuations in the velocity around 50 msec would hardly seem representative of actual physical behavior. One has to wonder whether the actual compartment accelerations behaved like this, or whether this is yet another example of "ringing" that can result when the effort to locate an accelerometer at the center of mass leads to a mount that is not as rugged as one might like. In any case, the effect is removed by the CFC 180 filter.

A second integration can be expected to produce even more smoothing for the displacements. Nevertheless, it is standard practice, as required by SAE J211, to apply a CFC 180 filter to the displacement curves as well.

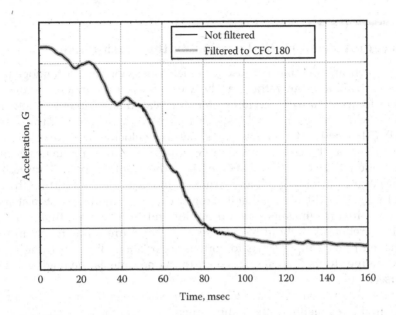

FIGURE 13.7
Velocities from Channel 280: raw versus filtered to CFC 180.

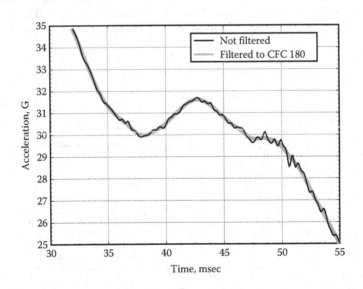

FIGURE 13.8
Detail from Channel 280 velocity data.

Effect of Accelerometer Location on the Crash Pulse

It has been argued that because a vehicle does not have a homogeneous mass distribution, but rather can be viewed more accurately as a collection of lumped masses, accelerometers are placed at various locations in a crash-tested vehicle. Figure 13.9 shows four *x*-axis accelerometer traces from NHTSA Test 5404, and Figure 13.10 shows the velocity–time curves derived from the same channels. We see in Figure 13.9 that the engine has a major acceleration spike of about –130 Gs early in the event, at about 30 msec. This is the engine bottoming out any structure in front of it, and hitting the barrier. In Figure 13.10, we see that it actually comes virtually to a stop at about 40 msec, then resumes its forward motion until it reaches a final stop and starts its rebound at about 55–60 msec. The structure immediately in front of the left brake caliper starts to bottom out at about 35 msec, bringing the brake caliper to its final stop at about the same time as the engine—about 55 msec.

These frontal elements have less structure between them and the barrier, compared to elements in the compartment, such as the dash panel and the right rear crossmember. The more rearward elements have lower deceleration spikes, gentler velocity–time curves, and later deceleration peaks, and do not reach a stop against the barrier face until about 75–80 msec. After that point, all four elements are in rebound, and all velocities are negative.

The dash panel acceleration tracks the engine bottom acceleration from about 18 to about 23 msec, then abruptly does an about face. This reversal may be due to the two front seat occupants engaging their airbag restraint

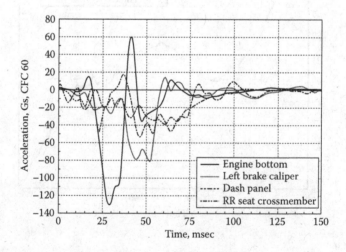

FIGURE 13.9
Acceleration traces from test 5404.

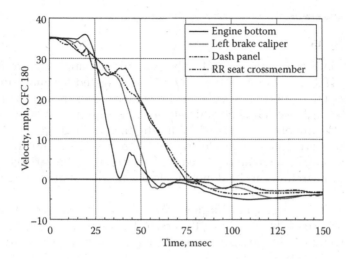

FIGURE 13.10
Velocity–time traces from test 5404.

systems, which then develop (compressive) restraint forces that are passed forward from the steering column on the left, and from the restraint mount on the right, into the dash panel. These forward-directed forces cause a hesitation in the dash panel's velocity–time curve, which is not seen in the right rear cross member.

As we move rearward in the vehicle—and out of the crush zone—we find fewer and fewer differences in the velocity–time traces. In a sense, they are converging on a single trace that is applicable to the compartment. For Test 5404, that trace is the average of the two rear seat crossmember traces.

We can think of the four locations presented here as having experienced four different crashes—crashes that are different in important respects. Attempts to summarize the crashes will result in conclusions that are also different in important respects, as seen in Table 13.7.

TABLE 13.7

Quantities Evaluated at Various Accelerometer Locations

Accelerometer Location	Peak Deceleration (Gs)	Velocity Change (mph)	CG Rebound Coefficient
Engine bottom	130.8	41.1	0.140
Left brake caliper	81.5	39.7	0.134
Dash panel	54.0	38.3	0.094
RR seat cross member	39.0	38.7	0.107

Conclusions

The NHTSA filtering algorithm has been accurately duplicated in the program NHTFiltr.bas. Of all of the methods investigated for analyzing crash pulse data, the numerical processing of raw data comes the closest to reproducing the peak accelerations and velocity change reported by the test lab (within 0.1% on both parameters). Hand-digitizing a high-quality Class 60 plot is not as accurate in reproducing the ΔV, but is still within 1.0%. If the crash pulse plot being hand digitized is of poor quality, the variability of the results of integration can be doubled. However, the variation of ΔV is still within 2%.

One should be careful in choosing which accelerometer traces to analyze for the purpose of characterizing the crash. The location(s) selected for that purpose should be in the compartment, as far away from the crush zone as possible, and on sturdy structure.

Reference

1. Instrumentation for impact tests, SAE Recommended Practice J211, SAE International, 1988.

14

Downloading and Analyzing NHTSA Load Cell Barrier Data

The Load Cell Barrier Face

A load cell is a force-measuring device: a transducer that converts a force or load into an electrical signal. Even a small vehicle with an engine compartment 2 ft high, and a width of 5 ft, for example, has a cross-section area of at least 10 ft^2 that comes into contact with the barrier face. Even if the forces were uniformly distributed, this is too large an area to be handled by a single load cell. Moreover, the forces are distributed far from uniformly, and it is this very lack of evenness that is of interest for a variety of reasons, ranging from vehicle-to-vehicle compatibility issues to the development and validation of computer models.

Consequently, the test protocol for the New Car Assessment Program (NCAP)[1] entails the use of load cells on the barrier face. Usually, the NHTSA contractors employ a 4 × 9 array of load cells. The columns are labeled 1 through 9; the rows are labeled, bottom to top, A through D. Figure 14.1 shows the standard arrangement and dimensions.

If no data-acquisition problems occur, 36 channels of force-versus-time data are available for each NCAP test, from which can be constructed a time-varying contact force distribution across the front of a vehicle. Sometimes, crash test reports contain data plots or data summaries from individual load cells; sometimes, they present results for the sum of all 36 load cells, or groups of load cells, such as those shown in Figure 14.1. However, digital data files are available only for individual load cell channels.

Unfortunately, barrier load cell arrays are rarely used in government-sponsored tests other than the NCAP tests at 35 mph.

Various load cell arrangements have been proposed and used. For example, high-resolution load cell arrays, in conjunction with deformable barrier faces, have been utilized for research purposes, primarily in Europe. One arrangement utilizes an 8 × 16 matrix of load cells 4.9 in. (125 mm) in size.[2] However, consider the following: a data channel covering 300 msec

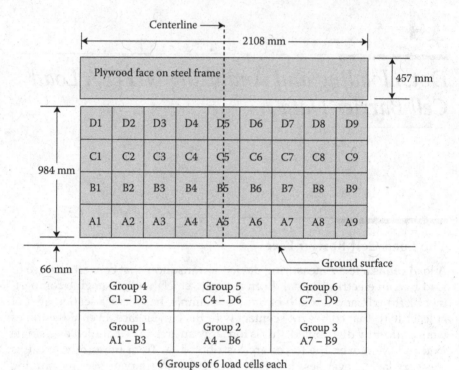

FIGURE 14.1
NHTSA load cell barrier configuration.

of duration and sampled at 12.5 kHz produces a data file containing 12,500 × 0.3, or 3750 data points. Processing 36 load cell channels plus 2 acceleration channels (to be integrated for velocity and displacement) results in analyzing over 142,000 data points—the equivalent of about 647 years of stock market closing price data! It hardly seems necessary to add to the computational burden. In the discussion that follows, a 4 × 9 matrix will be assumed.

Downloading NHTSA Load Cell Barrier Data

To download and analyze load cell barrier data from an NHTSA crash test, one needs to identify the desired test (preferably by DOT number), discern the channel numbers of the load cells and the requisite accelerometers, and download the corresponding data files into a suitable location on the hard drive. In view of the multiplicity of channels per test, it is good practice to place downloaded data files in a separate subdirectory called Download. The process of locating the test of interest on the NHTSA web site, and identifying the acceleration channels to be downloaded, is described in detail in Chapter 11. In contrast, this chapter focuses on the load cell channels.

As mentioned in Chapter 11, important instrumentation information is contained in a file called vXXXXX.EV5 that can be downloaded from the NHTSA web site, where XXXXX is the five-digit DOT number. A fragment of such a file, with the rows truncated so as to fit on the page, is shown below.

```
# Source: NHTSA Vehicle Database - Test Number: 05404

- - - INSTRUMENTATION - - -
1|1|AC|04|HDCG|XL|SEC|G'S|1650|MFG: ENDEVCO, MODEL: 7264, S/N: J36743|

1|63|DS|04|SHBT|NA|SEC|MM|1650|MFG: CELESCO, MODEL: PT-1010-0050-111, S/N: A60897|
0|64|LC|NA|LCA1|NA|SEC|NWT|1650|MFG: INTERFACE, MODEL: 1220TX-50 K, S/N: 62467|
0|65|LC|NA|LCA2|NA|SEC|NWT|1650|MFG: INTERFACE, MODEL: 1220TX-50 K, S/N: 122042A|

0|72|LC|NA|LCA9|NA|SEC|NWT|1650|MFG: INTERFACE, MODEL: 1220TX-50 K, S/N: 68045|
0|73|LC|NA|LCB1|NA|SEC|NWT|1650|MFG: INTERFACE, MODEL: 1220TX-50 K, S/N: 68027|

0|81|LC|NA|LCB9|NA|SEC|NWT|1650|MFG: INTERFACE, MODEL: 1220TX-50 K, S/N: 68047|
0|82|LC|NA|LCC1|NA|SEC|NWT|1650|MFG: INTERFACE, MODEL: 1220TX-50 K, S/N: 62466|

0|90|LC|NA|LCC9|NA|SEC|NWT|1650|MFG: INTERFACE, MODEL: 1220TX-50 K, S/N: 62453|
0|91|LC|NA|LCD1|NA|SEC|NWT|1650|MFG: INTERFACE, MODEL: 1220TX-50 K, S/N: 68050|

0|99|LC|NA|LCD9|NA|SEC|NWT|1650|MFG: INTERFACE, MODEL: 1220TX-50 K, S/N: 68033|
- - - END - - -
```

Blank lines indicate rows that have been deleted from the file fragment in the interest of saving space.

The first column seems to indicate the crash partner: 1 for the vehicle and 0 for the barrier. The second column is the channel number. We see that in this test, channels 1 through 63 pertain to the vehicle; channels 64 through 99 are for the barrier load cells. In all NHTSA load cell barrier tests reviewed as of 2012, the load cell channel numbers have been assigned in a contiguous block, as in this example.

The third column is the type of transducer: AC for accelerometer, DS for displacement transducer, and LC for load cell. The fifth column indicates where the transducer is located. The barrier transducers are denoted LCA1 (for load cell A1) through LCD9 (for load cell D9). This is the sequence seen in all the NHTSA load cell barrier tests reviewed to date. The sixth column indicates acceleration direction (NA for load cells). The seventh column is the units of the independent variable (the first column in the data file), which is SEC.

The eighth column, most important for data analysis purposes, is the units of the dependent variable (the second column in the data file): G'S for accelerometers, MM for displacement transducers, NWM (newton-meters) for moment transducers, and NWT (newtons) for load cells.

Crash Test Data Files

The names of the files available for download from the NHTSA web site have the form vXXXXX.NNN, where XXXXX is the five-digit DOT number, and NNN is the three-digit data channel number (with leading zeros in the file extension as necessary). As discussed in Chapter 11, they are tab-delimited ASCII sequential files that contain two columns of data. The left-most column is an evenly spaced time series that includes, and usually begins before, time zero. Of course, the sample rate is simply the inverse of the time interval between data points. The second column is the value of the dependent variable (acceleration, force, etc.) at the corresponding times. The files do not include headers, but the physical quantities being measured, and the units of measurement, are indicated in the .EV5 file, as discussed above.

All electronic data for a given crash test are sampled simultaneously, and thus reflect the same start time, stop time, and sample rate. Consequently, all files for that test have the same number of records (data points), allowing various dependent variables to be cross-correlated.

Grouping Load Cell Data Channels

It may be a tedious, and perhaps uninformative, process to examine 36 load cell traces from a particular test. Therefore, load cells are often grouped together as indicated in Figure 14.1. Sometimes, other groupings are used. Data plots may (or may not) appear in crash test reports for various load cell groups. However, even when groupings are employed to present the data in crash test reports, the underlying digital data are not available. The digital data files available from NHTSA's web site reflect individual data channels only, and no digital filters have been applied. Thus any combining of load cell data, and any filtering, has to be performed by the user, or through the use of the NHTSA data browser. A method for performing row-by-row analysis is the main subject of this chapter.

If the barrier force readings are to be correlated against acceleration, velocity, or displacement, then one or more acceleration channels will have to be processed for that purpose, which usually means averaging, integrating once or twice, and filtering to the appropriate channel frequency class (CFC). See Chapter 12 for discussions of these subjects.

Computational Burden of Load Cell Data Analysis

As indicated in Chapter 12, data from NHTSA crash tests must be sampled at a rate of at least 8 kHz,* and that sample rates ranging from 8 to 20 kHz have been found. For example, consider Test 05404: a 35 mph fixed barrier NCAP test of a 2004 Toyota Corolla. The test was conducted at Transportation Research Center of Ohio. Data were sampled at 12.5 kHz. Since the total duration of each transducer signal was 330 msec, each data file contains 4126 records. Analyzing 38 channels of data involves processing almost 157,000 records per crash test. This is a significant computational, programming, and computer storage burden.

Another issue arises because the data are stored in sequential (not random-access) files, which means that the records have to be read sequentially. Multiple files can be open simultaneously, and other computer instructions can be executed between the reading of sequential records. However, those other instructions cannot include the reading of a separate, already opened, file. All read instructions for one file have to be completed before any read instructions are executed for another file.

This issue can be dealt with by reading the entire file at once, and storing the data in an array in time-series order, vertically, if you will, before proceeding to the other files and repeating the process. Then all of the load cell values can be read at a given time value—horizontally, if you will—and the necessary calculations performed. But with the number of records in the data files, the arrays become enormous. Imagine an array with 157,000 cells! This is beyond the reach of many computer systems.

One way of reducing this burden is to retain only some of the data in an array, which can then be much smaller. Since the data are to be filtered anyway, this is potentially a feasible approach. However, aliasing errors must be avoided. After all, sample rates and large data files are the results of avoiding aliasing errors in the first place. Aliasing is a sufficiently important topic to warrant a little discussion before going farther.

Aliasing

Aliasing refers to the ability, or rather the inability, to reconstruct a signal or image from a digitized sample. For example, a Moiré interference pattern

---

* NCAP test procedure.

can sometimes be seen in the television image of a patterned tie or a striped shirt or suit because of the spatial resolution in use. Or a wagon-wheel effect may make a spoked wheel appear to rotate backwards in a movie or video because of the limited frame rate. The presence of aliasing means that the reproduced signal or image may not be faithful to the original.

The term aliasing comes from the idea that because of uncertainties, the sampling of different signals can render them indistinguishable (or aliases of one another). To the extent that aliasing is present in a data sample, there can be multiple interpretations of what the original signal was. Of course, the more detail in the image, the higher the resolution that is required to faithfully capture and replicate that detail. Similarly, the higher the frequencies that make up the signal, the higher the sample rate that is required to document the signal and avoid misinterpreting the digital sample. This is undoubtedly the reason behind NHTSA's sample rate requirement. On the other hand, the smoother the signal is in the first place, the coarser the sampling that can be used without misrepresenting it.

Consider, for example, a signal that has been digitized at a 500 Hz sample rate and plotted in Figure 14.2 with the solid squares. The usual way of representing the stored data is to draw straight lines through the points. If the actual signal was a 50 Hz sinusoid, such a piecewise linear representation will result in minimal error when the signal is evaluated at intermediate time values. The straight lines will make for a convincing graph. However, the graph will be very misleading if in fact the actual signal was a 450 Hz

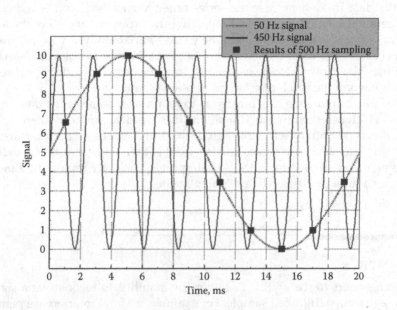

FIGURE 14.2
Two different curves fitted to one set of data sampled at 500 Hz.

sinusoid sampled at 500 Hz. In that case, the 50 Hz sinusoid would be called an alias of the actual 450 Hz wave form.

Aliasing is an important topic in instrumentation engineering. Suffice it to say that: "It can be shown that if the spectrum of a data waveform is confined to the region below some maximum frequency f_m, this waveform can be completely described by samples spaced $1/2f_m$ sec apart. In practice, data spectra are rarely truncated abruptly at some maximum frequency, but rather they decrease more or less gradually at the higher frequencies. Under these conditions, it can be shown that a source of error called aliasing is encountered when a portion of the data spectrum overlaps into the region above half the sampling frequency. To avoid aliasing errors, one of two procedures may be followed. First, the data may be sharply filtered at the input to the sampling instrument to eliminate all spectrum components above half the sampling frequency. Alternatively, the sampling frequency can be raised to four or five times greater than the maximum data frequency of significance. The drawback of sharp filtering is that it normally produces phase distortion. The drawback of the higher sampling rate is that it vigorously consumes digital storage space"[3] (p. 130).

It has been this author's rule of thumb that a pulse (or perhaps a half-sine wave) can be adequately represented if it is sampled 8 or more times at uniform intervals. This is consistent with NHTSA's highest frequency filter class of 1000 and a minimum sample rate of 8 kHz.

So let us get on to the problem at hand. The channels for Test 5404 were sampled at 12.5 kHz. Every tenth data point would represent a sampling rate of 1250 Hz. Aliasing errors could then be expected for all frequencies above 625 Hz. But barrier load cell channels are filtered at CFC 60, for which the corner frequency is 100 Hz (see Chapter 12). The required roll-off characteristic is between –9 and –24 dB per octave; second-order filtering produces a roll-off characteristic of about –12 dB per octave. This degree of filtering means that any signal components above 625 Hz would be attenuated by about 32 dB[*] (i.e., reduced to essentially zero), and thus considered as unwanted noise.

Therefore, after the noise is filtered out of the load cell barrier channels, taking each 10th data point and thereby using an effective sampling rate of 1250 Hz will dramatically reduce the sample sizes (by a factor of 10), while still avoiding the introduction of aliasing errors. The time intervals (of 0.8 msec) will still be sufficiently small to permit the analysis of loads and graphing of the results.

The analysis software has been written so that the interval may be easily changed by the user. This may be necessitated by the use of a lower original sampling rate, or the analysis of data that have higher frequency filter requirements.

---

[*] $625/100 = 2^n$; $n = \ln(6.25)/\ln(2) = 2.6438$; Gain $= -12(2.6438) = -31.73$ dB.

Example of Load Cell Barrier Data Analysis

As an example of processing load cell barrier data, the load cells will be grouped by rows, which is an arrangement not to be found in NHTSA crash test reports. The total force will be found for each row by summing the force values for the nine load cells in that row at each instant of time. The row force will be correlated against the vehicle displacement, as obtained by double-integrating the accelerometer data, to provide an individual force–deflection characteristic for each row. See Chapter 17 for a discussion of force–deflection characteristics. Since the crush energy is the area under the force–deflection characteristic, the crush energy will then be calculated for each row, by using the row force sum and the vehicle displacement as obtained by integrating accelerometer data, again at each instant of time. The results will be applied to the analysis of underride/override collisions, as discussed in Chapter 27.

For the purposes of this example, it is assumed that a "displacement file" already exists. It has the name XXXX-Avg.txt, where XXXX is the four-digit DOT number (without the leading zero). The file is assumed to contain the variables time, acceleration in Gs filtered to CFC 60, velocity in mph filtered to CFC 180, and displacement in in. likewise filtered to CFC 180. A program like AccelAvg.bas, discussed in Chapter 12, can generate such a file.

To compute the time variation of the total force in each load cell row, and perform crush energy calculations, the program LCAnal.bas was created. After doing the necessary preliminaries like identifying the DOT number of the test and locating the downloaded files, the program looks for a file of the name XXXX-Avg.txt. Finding one, the program then obtains the channel number of the load cell A1. The 4×9 array of load cell channels is assumed to occupy a contiguous block of 36 numbers, starting from A1 and going through D9. As of 2012, all load cell barrier tests that have been analyzed have adhered to this pattern, using filenames like v0XXXX.NNN, where NNN is the three-digit channel number (with leading zeros as necessary).

Each of the load cell data files is then subjected to an implementation of the NHTSA CFC 60 filter, which runs both forward and backward so as to eliminate any phase shift. The filtering is done by processing each of the 36 files separately, and is discussed in Chapter 12. For each channel, after the filtering is completed both forward and backward, the filtered data are written to a file having a name of the form XXXXRC.txt, where again XXXX is the four-digit DOT number, R is the row designation (A through D), and C is the one-digit column number (1 through 9). Since the data are written at intervals, the number of records for the filtered data is reduced by a factor of 10 or more. (In this case, the raw data files have 4126 records, whereas the filtered data start at time zero and are written at intervals of 10, so the filtered data files contain only 388 records.) Now, the mountain of data has been reduced to a digestible size.

Once the relatively sparse files of filtered data have been written, the nine files for each row can be read into an array that consists of the 388 rows, one for each time value. Each row contains the time value, plus the nine values of the load cell forces at that particular time. At each time value (0.8 msec apart in this example), the load cell forces are summed, and that row total is written to a file, each record of which contains the time value and the row force total. Four such files are written, one for each row.

The next step is to read the file that contains the displacements. The creation of this file is discussed in Chapter 13. The displacement file is fully populated, in that it does not have a reduced number of records. Therefore, its values must also be thinned in precisely the same way as the downloaded load cell data channels: the displacement values are loaded into an array in intervals.

Finally, the row forces, the total barrier force, the displacement, the row crush energies, and the total crush energy are written to a detail output file, a fragment of which is shown below. Some of the columns have been excised so that the fragment could fit on the page without making the print too small. Time is in msec, force is in lb, displacement is in inches, and energy is in ft-lb. The results pertain to Test 5404. The nonzero displacement at time zero is a result of filtering the displacements to CFC 180. The method of calculating the crush energy is discussed in Chapter 22.

Time	ForceA	ForceB..	TotForce	Displ	EnergyA..	EnergyD	TotEnergy
0.00	-47	711..	661	0.341	0..	0	0
0.80	-122	2037..	2046	0.614	-2..	-0	31
1.60	-224	4028..	4203	0.979	-7..	-2	126
2.40	-329	6506..	7004	1.468	-18..	-5	354
3.20	-408	9210..	10186	1.969	-34..	-12	713
4.00	-464	11990..	13543	2.463	-52..	-19	1201
4.80	-479	14761..	16964	2.952	-71..	-26	1822
5.60	-417	17397..	20293	3.436	-89..	-31	2574
6.40	-266	19674..	23260	3.915	-103..	-31	3443
7.20	-35	21296..	25556	4.390	-109..	-25	4411
8.00	256	21997..	26939	4.865	-104..	-13	5448
8.80	589	21538..	27177	5.337	-88..	3	6513
9.60	938	19827..	26148	5.808	-58..	20	7560
10.40	1239	17521..	24491	6.278	-15..	37	8552
11.20	1338	16421..	23860	6.749	35..	55	9499

For Test 5404, the total barrier force as a function of time is shown in Figure 14.3.

Of course, variables can be cross-plotted. A common format is the force–deflection characteristic, in which force is plotted versus displacement. For Test 5404, the result is shown in Figure 14.4. This subject is taken up in Chapter 17.

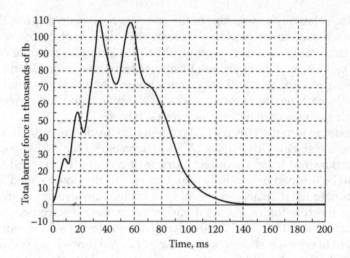

FIGURE 14.3
Total barrier force versus time, test 5404.

We see that the use of displacement as an independent variable dramatically changes the appearance of the plot. Of course, time marches on at a steady pace, while the vehicle slows to a stop and then changes direction. Once the displacement starts to decrease, the vehicle is in rebound, the structure unloads, the force drops dramatically, and some of the crush energy is recovered. When the force drops to zero, the displacement is almost 26 in., and the structure is

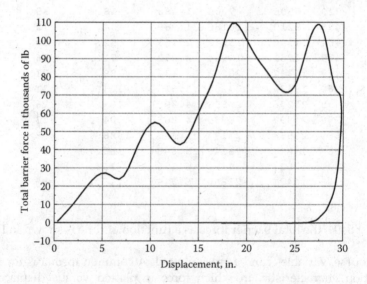

FIGURE 14.4
Force–displacement characteristic, test 5404.

completely unloaded (by definition). This displacement does not match the residual, or static, crush because the structural relaxation is not completed as of the end of the electronic data-gathering process. On the other hand, the vehicle as a whole is moving backwards when its front separates from the barrier. At that point, a gap starts to open up between the vehicle and the barrier; subsequently, the vehicle displacement is no longer related to structural crush.

Using the NHTSA Load Cell Analysis Software

One part of the NHTSA Automated Signal Analysis package discussed in Chapter 11 is Load Cell Analysis. It is designed specifically for analyzing load cell barrier data. Like the Signal Browser, it walks the user through a short series of dialog boxes so as to select the test to be analyzed. The opening screen shows a graph of total barrier force as a function of time for the selected test.

The main choices in this software come through the selection of Plot Type. There are 26 choices, among which are: acceleration, average height of force, contour peak force, displacement, force by columns, force by rows, force–displacement, group force, individual forces, initial stiffness, total force, and velocity. In group force plots, the user can choose the load cells to be included in the group.

All of these plots have time as the independent variable, except for force displacement. Acceleration, velocity, and displacement data come from the rear seat cross member channels, or their integrals. The filter class is not identified, but all data appear to be filtered to CFC 60. Unlike the Signal Browser, the subtitles in the graphs do not show the minimum and maximum values.

This software is particularly useful for the study of compatibility of vehicle front structures with vehicles struck in the front, side, or rear. The grouping of forces by rows is a subject we come back to in Chapter 26—"Underride/Override Collisions."

References

1. Laboratory Indicant Test Procedure, New Car Assessment Program, U.S. Department of Transportation, National Highway Traffic Safety Administration, 1990.
2. Edwards, M., Happian-Smith, J., Davies, H., Byard, N., and Hobbs, A., The essential requirements for compatible cars in frontal collisions, *17th International Technical Conference on the Enhanced Safety of Vehicles*, U.S. Department of Transportation, 2001.
3. Keast, D.N., *Measurements in Mechanical Dynamics*, Bolt Beranek and Newman Inc., McGraw Hill Book Company, New York, St. Louis, San Francisco, Toronto, London, and Sydney, 1967.

15

Rollover Forensics

Introduction

A rollover accident requires different kinds of analysis, and different kinds of inspection techniques, than do crashes into objects or other vehicles. Its severity is usually not expressed in terms of ΔV. Except for crashes involving a quarter turn from its wheels to its side, it does not consist of a single collision. Instead, it entails a whole series of crashes that occur every time the vehicle touches down on the ground. As a whole, the string of crashes in a rollover is a random, chaotic sequence. It is almost completely lacking in repeatability, which is a primary reason why staged rollover tests have never been prescribed as part of vehicle safety standards. The contact velocity or ΔV of each individual impact that occurred is virtually impossible to determine. While a computer simulation would contain such measures as part of its calculations, rollovers of actual vehicles are unpredictable. For example, in one rollover test, a roof rail may impact the ground and be deformed; in a subsequent test under identical conditions with an identical vehicle, the same roof rail may miss the ground by a fraction of an inch and be essentially unscathed.

A crash with another vehicle or a fixed object is often (but not always) a single event, whereas a rollover is usually a series of events. It is usually important to know what the sequence was, even if their individual impact speeds or ΔVs cannot be determined. However, if the sequence can be determined, it is usually possible to know the approximate translational and rotational speeds the vehicle possessed when they occurred. This may be important information, especially if an occupant is ejected.

Crush in a rollover is distributed over various parts of the vehicle, particularly the "greenhouse"—the portion of the vehicle that extends above the base of the windshield and other glass areas. The demarcation between the greenhouse and the lower portions of the vehicle is often referred to as the "belt line"—probably because at one time it tended to be at about the same height off the ground as the belt on a man's pants.

In any case, the deformation of either the greenhouse or other structures can, and often does, come from multiple impacts, and these impacts may occur to some portion of the vehicle remote from where the crush is measured ("induced damage"). Therefore, measurement of the crush does not lead to a quantification of the rollover initiation speed or any other speed measure. Greenhouse deformation may or may not have significance to the vehicle occupants. Its significance has been hotly debated for decades, and continues to be controversial at the time of this writing.[1,2]

Measurement of greenhouse crush is a three-dimensional exercise, which renders it fundamentally different from crush measurements for coplanar collisions with vehicles and fixed objects. For the latter, it generally suffices to take measurements in the X–Y plane, but for rollovers the Z component is often even more important than the X and Y components. Therefore, the crush measurement has additional requirements that may dictate the use of entirely different equipment. On the other hand, crush measurement may not be required at all because the numbers tell little about speed.

Measurements of Severity

Speed is always the bogeyman when it comes to vehicle safety. We have already encountered one speed measure—ΔV—as the primary measure of severity in vehicle-to-vehicle and vehicle-to-fixed-object collisions. Speed at impact can also be important in these crashes. In rollovers, however, different severity measures are commonly used.

One has to think of how to define the initiation of rollover, because it is usually in the rollover where the injuries occur. Everything that happens before the rollover may be important in determining why the rollover occurred, but the single most significant determinant of the risk to injury is the speed at rollover initiation. Initiation may be defined as a certain roll angle, or it may be defined as lift-off of one or two wheels. These measures are often used in testing. For the reconstructionist, however, initiation is usually expressed in terms of scene evidence. The evidence may be in the form of some surface feature, such as a curb, that the vehicle tripped on. The evidence may be in the form of the cessation of tire marks on the pavement. See, for example, Figure 15.1. It may be in the form of a furrow in the ground, as in Figure 15.2, or less often, a gouge in the pavement, that comes to an end at the point of roll initiation. Depending on the mechanism of roll initiation, and the scene evidence left behind, the roll may be classified as a tripped rollover or an on-road rollover. The reconstructionist should be careful in his use of language, because to use the expression "trip" for roll initiation is to imply a tripped roll to some persons. Perhaps one of the parties contends the accident was an on-road roll. In that case, the reconstructionist can step into the middle of a controversy that is outside his

FIGURE 15.1
Cessation of tire marks at roll initiation.

or her area of expertise (rollover resistance designed into the vehicle, for example).

In any case, vehicle speed at roll initiation (or at trip, if you like) is almost always of interest. It is analogous to impact speed in a coplanar collision. The primary evidence of initiation speed is the distance on the ground the

FIGURE 15.2
End of furrow at roll initiation.

vehicle rolled between initiation and rest. As with coplanar collisions, roll-over test data have been analyzed with an eye to determining an average drag factor that the reconstructionist could use to find the initiation speed, as a function of rollover distance.[3–5] A drag factor value of 0.4–0.5 is widely used for this purpose.

Another measurement of rollover severity is the total angular rotation of the vehicle. For analytical purposes, the reconstructionist can express this in degrees, which may extend well beyond 360. A more common measure among lay persons, however, is the number of rolls, expressed as a proper fraction. (E.g., 3½ rolls implies that the vehicle went from being on its wheels to being on its roof, after having rotated through 3 complete revolutions, plus 1/2 more. 2¼ rolls usually means that the vehicle wounded up on its side.) In accident databases, the angular rotation may be expressed in quarter turns, where four quarter turns would be one complete revolution through 360°.

There is no hard and fast formula connecting rollover distance to number of rolls because of the randomness of rollovers, but certainly broad trends exist. For one thing, one could take a cloth measuring tape and wrap it around the vehicle laterally, including the wheels. This vehicle circumference would relate to the distance consumed in one roll by the vehicle, if it were always in contact with the ground and never sliding. A vehicle 5 ft high and 6 ft wide could be expected to have a lateral circumference of 22 ft or a little less (due to "tumblehome," which is the inward slope of the sides of the greenhouse closer to the roof). Of course, actual rolling vehicles are often airborne at some stages and sliding at others. Often in the early stages of a rollover, the translational velocity is the highest, while the roll rate is still building, so sliding is at its maximum. Later in the sequence, the roll rate has built to a peak while there is less translational speed and less sliding. These trends can often be observed in scene evidence. On average, though, many rollovers consume about 20–35 ft per roll, as seen in Figure 15.3.[6–9,4,11] Since the roll angle at roll initiation is approximately known, and the vehicle attitude at rest is also known, the question of number of rolls often comes down to an integer. If a vehicle is on its wheels at rest, it rolled either zero or some integer number of rolls. If a vehicle is on its roof, the net number of rolls was either 1/2 or 1½ or 2½ rolls, and so forth. Or perhaps in its final roll, the vehicle rolled up onto its side, stopped, and then rolled backwards onto its roof. This is always possible, and evidence on the vehicle may well shed light on this issue. In any case, Figure 15.3, even with all its scatter, can often provide a key piece of information, by way of ruling out integer variations in the net number of rolls.

In summary, usually the first questions asked of the reconstructionist are: Rollover distance? Number of rolls? After some analysis, the question often is: Speed at initiation? And still later on, the questions might be: Average roll rate? Peak roll rate? Vehicle orientation and location when ejection occurred (if it did)?

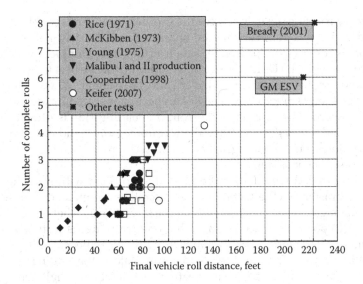

FIGURE 15.3
Trend between rollover distance and number of rolls.

Evidence on the Vehicle

With these questions in mind, let us get on to the vehicle inspection. The inspection should be prepared for carefully, under the assumption that there will be only one opportunity to see the vehicle. So the inspection should be all-inclusive. Existing photographs, particularly at-scene photos, should be collected and studied carefully beforehand. After deciding what sort of inspection to conduct and selecting the necessary inspection equipment, packing it, and transporting it to the inspection site, and gaining access to the vehicle and permission to inspect it, the first thing to do is placard the photographs (i.e., take a picture of a placard or a piece of paper that identifies the case name, the vehicle being inspected, and the date and place of inspection). So take every conceivable photograph and record every conceivable detail, because inevitably there is always one photograph that "got away." The number of lost photo opportunities should be minimized.

The next thing to do is a photographic walk-around to document the vehicle as found, before any changes are made, however inadvertent. Front, rear, side, and quarter views should be photographed. Then, the inspection can start in earnest.

The first bit of analysis to determine is the direction of roll. It is not good to call the direction "clockwise" or "counterclockwise," because it will be clockwise if viewed from one end of the vehicle, and counterclockwise if viewed from the other. Most investigators refer to the side of the vehicle that, during the roll sequence, first faces the ground. That is referred to the "leading side." Thus, a rollover may be characterized as "driver side leading," "driver side down," or "passenger side leading," for example. If the accident occurred outside the United States, it would probably be best to use the terms "left" and "right," rather than "driver" and "passenger." "Left" and "right" are understood to refer to the view of an occupant seated inside the vehicle, facing forward. The side opposite from the leading side is often called the "trailing" side. It is good to have the terminology in hand at the outset, so that any notes, either photographed, hand-written, or dictated, will be consistent and nonconfusing.

The sheet metal surfaces can be liberally sprinkled with scratches in the paint or abrasions of the metal, but often scratches in metal can be searched in vain for signs of the directionality of the scratch (e.g., whether it was made from top to bottom or vice versa). What one really needs to find is a situation where the material has been melted by the heat, because in such situations it is forced to flow, and tends to accumulate at the trailing edge of the abrasion. Plastic surfaces, such as weather seals, emblems, light lenses, bumper covers and body cladding, are much more fruitful places to look. If they were abraded, the directionality is mostly unmistakable. See, for example, Figure 15.4, which shows where a seal around a radio antenna mount was abraded, and a sheet metal drip rail was folded over, due to contact with the ground. Since this component was located on the driver's side roof rail, the flow occurred toward the passenger side, while the vehicle was upside down. Therefore, this evidence clearly establishes the fact that the vehicle was involved in a passenger-side leading roll.

This type of scrutiny applies to other types of collisions. Even though it was not involved in a rollover, the vehicle in Figure 15.5 came into contact with the tire of another vehicle. The plastic flow on its bumper cover, and the

Flow direction

FIGURE 15.4
Flow direction shows roll direction.

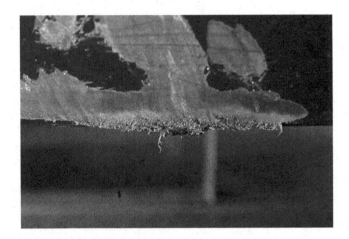

FIGURE 15.5
Material flow on a plastic bumper cover.

resulting deposition of melted material at the trailing edge of the abrasion, shows in which direction the tire was rotating.

It is a good idea to come to the inspection armed with outline drawings or sketches of the vehicle as seen from the top, both ends, both sides, and perhaps the bottom. Therefore, any evidence appearing on any surface of the vehicle can be sketched, or written down if necessary, including such things as broken, rolled up, or rolled down glazing (glass, usually), broken or missing lights or light lenses, outside rear view mirrors, radio antennas, tires, wheels, bumper covers, emblems, and so on. If the roll direction has been ascertained, the drawings can be annotated with scratch, scrape, or gouge angular orientations that indicate whether the roll was "tail-leading" or "nose-leading." Nose-leading means that the crab angle was <90°; the nose of the vehicle was pointing somewhat toward where the vehicle was going; tail-leading means that the tail was pointing somewhat in that direction. It has been this author's practice to characterize each area of evidence marks by how many degrees they lie from a pure lateral roll, or "barrel roll." "15° nose-leading" means, for example, that the direction is 15° away from a pure lateral roll, nose-leading. It might also be called "+15°," as distinct from "−10°," which would be tail-leading. If the roll direction (driver-side leading, or passenger-side leading) has already been determined, then the inspection notes can be internally consistent. An example of such an inspection sketch is shown in Figure 15.6.

Notice in this inspection sketch that the scratch patterns have been color-coded. This does not imply that the reconstructionist comes to the inspection knowing how to group the patterns, or what color to use. The opposite is mostly true. In fact, many investigators use small lengths of inch tape on the vehicle to mark the scratched areas and scratch directions, so that they

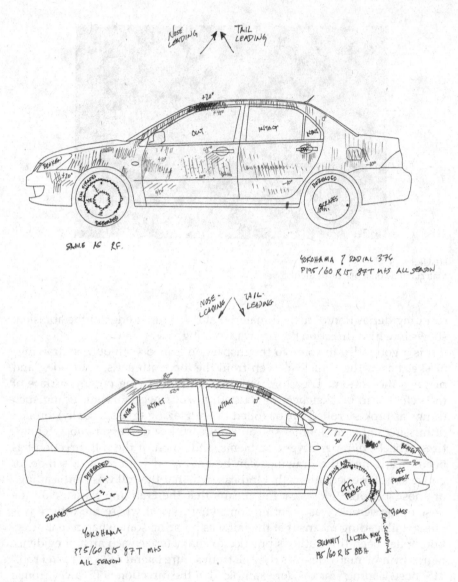

FIGURE 15.6
(See color insert.) Vehicle inspection sketch.

can be seen from a distance. However, this author just uses 1/2-inch masking tape—blue for light-colored vehicles, and white for dark ones. A visual estimate of angle is written directly on the tape. (Exactitude is not important, because the exact angle will probably vary somewhat from one location to another.) Later, when all the scratched areas are marked, the patterns can be grouped by angle and marked with a number. Then a color can be

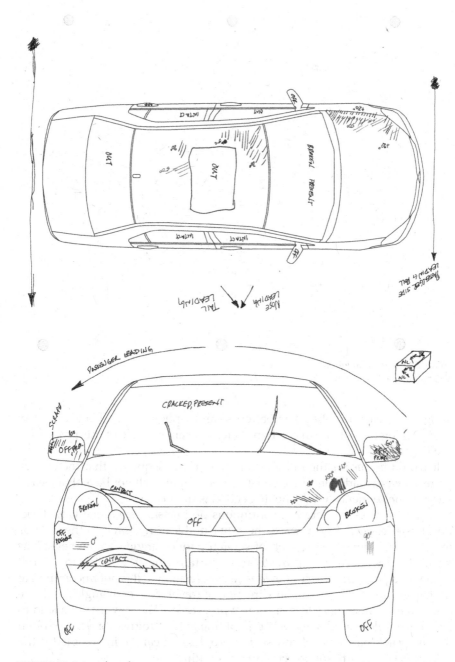

FIGURE 15.6 (continued)
(See color insert.) Vehicle inspection sketch.

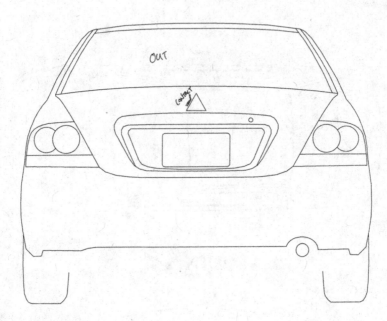

FIGURE 15.6 (continued)
Vehicle inspection sketch.

assigned to each number. The patterns can be photographed, and they can be drawn on the vehicle inspection sketch. For an example of this procedure, see Figure 15.7.

It would be nice if the numbering were in the sequence that the scratch patterns were made, but the reconstructionist may well not have that clear a view of the sequence while the inspection is underway.

In most rollovers, there is a continuous progression from one yaw direction (nose-leading, say) toward the other (tail-leading, say), so there tends to be a continuous progression of scratch pattern angles on the vehicle as well. The directions may provide an important clue as to when in the roll sequence a particular piece of evidence was generated. It is not uncommon to find a vehicle heading along its direction of travel (zero crab angle), with the crab angle increasing as loss of control progresses. The yaw motions tend to continue while the vehicle is rolling, leading to the progression seen so often in the scratch patterns. Of course, the progression can be interrupted if the vehicle hits a significant object while it is rolling.

Figure 15.8 is a mosaic of two photographs showing how the various scratch patterns have been marked with blue tape. One of them, a nose-leading pattern at about +15°, is highlighted in red. Figure 15.9 shows that a second pattern has been highlighted in magenta.

FIGURE 15.7
Example of scratch pattern demarcation.

Figure 15.10 shows that a third pattern has been highlighted in green.
Figure 15.11 shows that a fourth pattern has been highlighted in light blue.
Clearly, the top surface of the vehicle has been on the ground four times. Therefore, the vehicle has rolled at least 3½ times. The term "at least" is used because the vehicle could have been upside down for one or more rolls

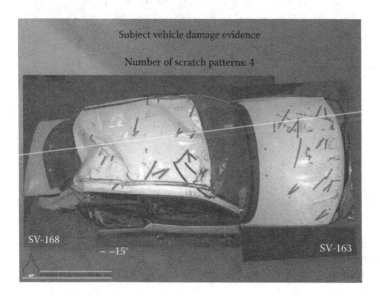

FIGURE 15.8
(**See color insert.**) Scratch pattern highlighted in red.

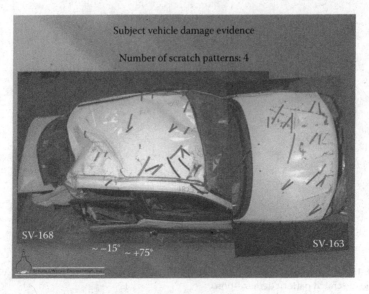

FIGURE 15.9
(See color insert.) Scratch patterns highlighted in red and magenta.

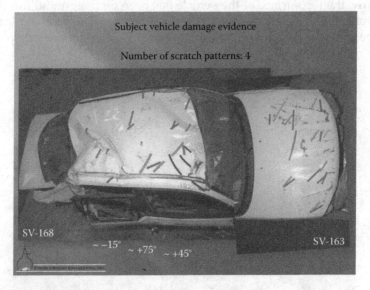

FIGURE 15.10
(See color insert.) Scratch patterns highlighted in red, magenta, and green.

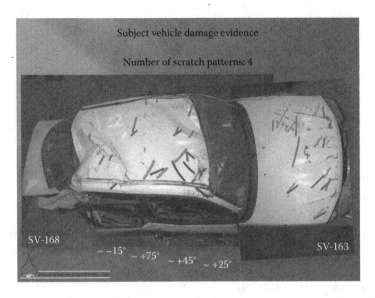

FIGURE 15.11
(**See color insert.**) All four scratch patterns highlighted.

without touching the ground. The reconstructionist can say that the vehicle itself indicates at least 3½ rolls; the scene (particularly the rollover distance) may indicate something more.

As to the sequence of patterns, further examination is required to reveal the order in which the four patterns were made.

We do know that protruding surfaces are more subject to ground contact than receding ones. An airplane, flying at a constant altitude, is more likely to hit a high mountain than it is a deep canyon. Therefore, a scratch pattern in a valley was most likely made before the valley was a valley, before vehicle damage created the peaks surrounding the valley.

Since a thin material will buckle when in-plane compressive stresses are comparatively low, ridges in sheet metal tend to form perpendicular to the principal compressive stresses imposed by damage. A look at Figure 15.11 with an eye to the compressive sheet metal ridges reveals that in-plane loading to this vehicle's roof panel must have occurred from a forward-directed contact to the backlight (rear window) header. Indeed, examination of this area showed wood fibers embedded in the weather seal. See Figure 15.12. Therefore, we know the vehicle went backwards, perhaps with the nose pitched up, into a tree, creating the ridges in the roof panel, prior to its rolling over onto its roof and having the scratch marks created on the tops of the ridges.

Longitudinally directed scratch marks can raise the question of whether the vehicle experienced a lateral, or barrel-type roll, or whether it pitched over end over end ("did an end-o," in the vernacular). Consider, though,

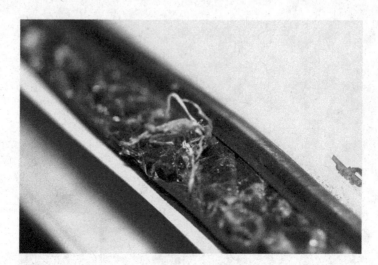

FIGURE 15.12
Wood fibers trapped in weather seal.

that almost all vehicles are much longer than they are wide. Therefore, their pitch moment of inertia is much higher than their roll moment of inertia, so dynamically they have much more resistance to pitch rotations than roll motions. In addition, when the vehicle pitches, the reaction from the ground acts over a much longer moment arm from the center of mass. They require a far higher impulse to pitch over than to roll over; consequently, pitch-over kinematics are quite rare. In this author's experience, a pitch-over has only occurred when steep, rugged terrain is being traversed downhill. Because of the high impulse required for pitch-overs, one should look for evidence of a substantial impact in the front, or the front undercarriage, of the vehicle. For example, see Figure 15.13, in which heavy damage was done to the stout front suspension of a truck-based SUV when its left front wheel hit a large hole while the vehicle was careening out of control down a steep slope.

Examination of the vehicle showed left front tire deflation and debeading, bending of the longitudinal stabilizer strut, damage to the suspension attachments, bending of the lateral suspension I beam, damage to the antiroll bar mounts, and mud deposits all over the left front suspension. Damage was also found to the rear lift gate where it went into a tree as the vehicle was upside down and backwards to its direction of travel (see Figure 15.14).

The suspension, tires, and wheels are important evidence in any rollover, not just pitch-overs. The wheels are hard elements that tend to leave marks wherever they hit. The tires are relatively soft and black, and leave tell-tale smears whenever they come into contact with a surface like the road (see Figure 15.15). And they both protrude outside the rest of the body element, which increases the likelihood of touching something. Consider, for

FIGURE 15.13
Suspension damage due to pitch-over impulse.

example, Figure 15.16, which shows an alloy wheel after a vehicle rolled over several times.

Obviously, the tire has been deflated and debeaded. The hard, brittle alloy material has fractured. Moreover, the material has fractured in two distinct areas, indicating that the wheel touched down at least two separate times.

FIGURE 15.14
Rear lift gate damage from tree impact.

FIGURE 15.15
Black tire marks on roadway.

FIGURE 15.16
Fractured wheel with two impacts.

Evidence at the Scene

Surely, there would be matching evidence on a paved surface, such as seen in Figure 15.17 or Figure 15.18. The latter has been identified by the police as "u," marked with paint.

A mark like the one labeled "u" would be expected to remain long after the accident, so the investigator should go looking for it on the roadway,

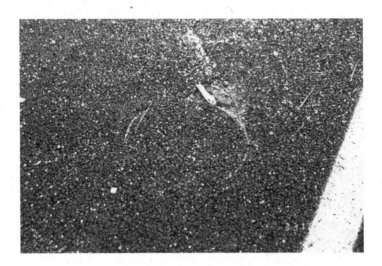

FIGURE 15.17
Classic "Rim-Down" mark.

FIGURE 15.18
Classic "Crescent Moon" mark.

particularly if the roadway is asphalt (asphaltic concrete, formally). In this case, this author did find the mark 2 years later, shown in Figure 15.19 labeled in a manner consistent with the police investigation. The area had been restriped but not resurfaced.

Of course, if the road had been repaved in the interim, the evidence would have been destroyed. Therefore, if the reconstructionist finds gouges or other marks that he or she intends to rely on, and that would have been destroyed

FIGURE 15.19
Crescent mark visible at scene inspection.

by resurfacing, it behooves him or her to find out whether such has occurred. Sometimes, this question can be answered by comparing details in the present paving and striping to the details seen in at-scene photos; other times, it may be necessary to obtain information from the local transportation agency to find out if the road has been resurfaced post-accident.

Often, the nature of the surface a vehicle has rolled on will be revealed by materials captured in weather seals, sheet metal joints, or tire seating beads on wheel rims. Figure 15.20 shows asphalt deposits on a wheel rim that rolled over on asphalt, whereas Figure 15.21 shows a rim with plant material deposits that came from rolling over on dirt.

FIGURE 15.20
Asphalt deposits in wheel rim.

FIGURE 15.21
Plant materials in wheel rim.

The nature of the ground surface—whether it has been paved or not—may have a significant effect on the appearance of the marks made on the vehicle that contacts it. Therefore, a transition at the scene—such as the edge of pavement—can be expected to show up on the vehicle if the vehicle contacts that area during the rollover. See, for example, Figures 15.22 and 15.23.

If a vehicle rolls on grass, the scratches tend to be much different, and much more subtle. In fact, they might not even be called "scratch marks" at all, but rather "brush marks," as seen in Figure 15.24. These marks may be sufficiently hard to see, especially in a dark inspection or storage facility, that one becomes concerned with the "dirtiness" of the vehicle, which is to say

FIGURE 15.22
Effects of pavement edge on vehicle.

FIGURE 15.23
Details of edge-of-pavement scratches.

the coating of dirt or dust that has occurred post-accident, and which may obscure evidence of interest to the reconstructionist. The more subtle the brush marks, the easier it is to confuse them with artifacts that are not accident related. However, a word of caution to the investigator before "cleaning" the surface. Make sure the custodian is aware of what is being done, and that approval has been given first. It may be a good idea to take some 'before" and "after" photos. No one wants to be accused—especially falsely—of tampering with the evidence.

One needs to be especially wary of "tarp marks," those made by transporting the vehicle when covered by a tarp. It is almost impossible to keep a tarp

FIGURE 15.24
Brush marks.

from fluttering when exposed to the slip stream, and when that happens, the tarp often leaves abrasion marks on the vehicle paint. These may be difficult to distinguish from brush marks. Vehicle owners, custodians, and attorneys should be admonished, in the strongest terms, to avoid covering rollover vehicles with a tarp when transporting rollover vehicles in the open air.

Window glazing is often an area of high interest in a rollover, particularly if an ejection has occurred. In that case, the ejected occupant will often leave clothing and tissue smears behind as he or she scrapes against the periphery of the ejection portal on the way out of the vehicle. See, for example, Figure 15.25.

There were scrapes in the inner door trim at the base of the window. Sometimes, there are disturbances of the weather seal as well. In this case, the window was broken out, and pieces of glass were trapped in the weather seal. The type of glazing should be noted (tempered glass in this case). The thickness should also be measured with a micrometer caliper, and recorded. When a vehicle rolls over off the road, glass deposits will often remain on the ground. Over time they may be disturbed by mowing, but usually they will stay concentrated in the same general area. See, for example, Figure 15.26 that was taken long after the accident.

The type and thickness of glass found on the ground can be determined, and compared to the pieces still trapped in the weather seal. If there is a match, important information may be gained as to what opening the glass came from, and/or what point in the rollover the glass fractured. This may shed light on just when an ejection could have occurred. A map of the accident scene, with the areas of different glass sources marked out, will often paint a clear picture of how the vehicle was oriented as the rollover occurred. Parts broken off the vehicle, such as exterior rear view mirrors, will add definition to the picture.

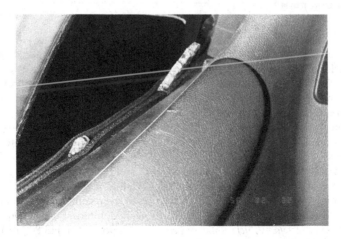

FIGURE 15.25
Ejection portal evidence.

FIGURE 15.26
Glass deposit long post-accident.

If the ejection occurred on the leading side of the vehicle, the occupant would have been projected in front of the rolling vehicle (and probably downward), and subjected to the possibility of being run over. If the vehicle rest position is ahead that of the occupant, that is very well what may have happened. On the other hand, if there is evidence that the occupant was ejected upwards, one should look for an ejection portal on the trailing side, with the vehicle positioned so that the ejection portal is rotating up from the ground surface.

References

1. Cooperrider, N.K., Thomas, T.M., and Hammoud, S.A., Testing and analysis of vehicle rollover behavior, SAE Paper 900366, SAE International, 1990.
2. Bahling, G.S., The influence of increased roof strength on belted and unbelted dummies in rollover and drop tests, *The Journal of Trauma: Injury, Infection, and Critical Care*, April 1995;38(4):557–563.
3. Orlowski, K.F., Bundorf, R.T., and Moffatt, E.A., Rollover crash tests—The influence of roof strength on injury mechanics, SAE Paper 851734, SAE International, 1985.
4. Orlowski, K.R., Moffatt, E.A., Bundorf, R.T., and Holcomb, M.P., Reconstruction of rollover collisions, SAE Paper 890857, SAE International, 1989.
5. Martinez, J.E. and Schlueter, R.J., A primer on the reconstruction and presentation of rollover accidents, SAE Paper 960647, SAE International, 1996.
6. Rice, R.S., *Test for Vehicle Rollover Procedure*, Report DOT HS 800 615, Federal Aviation Administration, National Aviation Facilities Experimental Center, November 1971.

7. McKibben, J.S., Clark, G.S., and Carlson, L.E., *Development of Techniques to Prevent Occupant Ejection during Rollover*, Vol. 1. Executive Summary, Report PB 231 563, Agbabian Associates, October 1973.
8. Young, R. and Scheuerman, H., *Test for Vehicle Rollover Procedure*, Report DOT HS-801 776, Federal Aviation Administration, National Aviation Facilities Experimental Center, December 1975.
9. Bahling, G.S., Bundorf, R.T., Kaspzyk, G.S., Moffatt, E.A., Orlowski, K.R., and Stocke, J.E., Rollover and drop tests—The influence of roof strength on injury mechanics using belted dummies, SAE Paper 902314, *Proceedings, 34th Stapp Car Crash Conference*, SAE International, 1990.
10. Cooperrider, N.K., Hammoud, S.A., and Colwell, J., Characteristics of soil-tripped rollovers, SAE Paper 980022, SAE International, 1998.
11. Keifer, O.P., Richardson, W.C., Layson, P.D., Reckamp, B.C., and Heilmann, T.C., Vehicle linear and rotational acceleration, velocity and displacement during staged rollover collisions, SAE Paper 2007-01-0732, SAE International, 2007.

16

Rollover Analysis

Introduction

As discussed in Chapter 15, rollovers are chaotic events, often involving multiple impacts. While each impact follows the laws of physics (of course), rollovers are so nonrepeatable on the whole as to take on a somewhat random nature. Even carefully controlled rollover tests using nominally identical vehicles will exhibit considerable scatter, as seen in the Malibu tests of Figure 15.3.

If the rollover of a vehicle is a stochastic process, what hope is there of predicting all the details? Worse, if all one has to go by is the evidence remaining a few years later, how can the details be reconstructed? In a 1989 paper,[1] it was argued that a computer simulation adapted from an occupant model could be used to replicate (after very considerable effort and "tuning," no doubt) the vehicle motions seen on video during an actual rollover test, if only for the first second or so. Even if the simulation effort were successful, one can virtually guarantee that it would not predict the motions of an identical vehicle, if an identical test were to be run the following day. Claims that a computer model can "be an effective tool in reconstructing the motion of the vehicle"[2] would seem to be preposterous on their face.

The many complexities of modeling rollovers with a computer simulation are outlined briefly by Day and Garvey,[3] not the least of which is that the deformation seen in any given area of the vehicle can be the result of multiple impacts. These deformations change the shape of the vehicle (by definition), and thus change the behavior of the vehicle during subsequent rolls, meaning that any subsequent impacts must begin with a deformed vehicle. About the best that can be said for computer models is that if they can offer a suitable representation of the early phases of a rollover test, and they can allow parametric studies of the effects of vehicle parameters during those early phases, as opposed to running multiple tests.

As a result of such considerations, the reconstructionist must take a broader-brush approach. This is not to imply that details in the evidence as discussed in Chapter 15 can be overlooked. Indeed, they provide a solid underpinning to the developments of this chapter. While much can be discovered, some claims cannot be asserted without jeopardizing credibility.

Use of an Overall Drag Factor

The usual approach to rollover reconstructions is to work out the vehicle kinematics from initiation to rest, using the sort of investigation techniques discussed in Chapter 15. Then the entire sequence is analyzed as a whole, rather than analyzing each ground contact, each airborne segment, and so on. The approach is to apply a single drag factor to the entire rollover trajectory, so that the speed at roll initiation, time duration, and average roll rate can be determined, in much the same way that separation speed can be calculated from a vehicle's run-out trajectory. It is thus common practice to use reverse trajectory methods to analyze the rollover accidents.

This approach has been delineated in papers by Orlowski et al.[4] and others.[5-7] Here, the essential task is to use a reasonable drag factor. If the overall drag factor is known, the speed at initiation is related to the final roll distance by the simple formula

$$V = \sqrt{29.938(DragF)S} \approx \sqrt{30(DragF)S} \qquad (16.1)$$

where V is in mph and S is the final roll distance in feet. The ability of this equation to describe the experimental data is shown in Figure 16.1. Of course, there is scatter in the data, as might be anticipated by now. However, using a drag factor of 0.38–0.50, Equation 16.1 aptly describes most of the tests at initiation speeds at 30 mph and above. This range of drag factors is consistent with Orlowski, who found an average value of 0.43 for the Malibu rollovers.

It might be tempting to think that some lower drag factor might apply below 30 mph. However, it is good to remember that Equation 16.1 is a special case of Equation 4.3; that is, it is a statement of energy conservation, in the absence of potential energy changes. In the Cooperrider tests,[8] the trip speeds were below 20 mph and the vehicles rolled 1½ times or less, so the loss of potential energy (from the CG height at trip to the CG height at rest) assumed more significance relative to the other losses in those events. The tripping mechanisms in these tests were curbs and soil furrowing. The analysis of soil trips is a separate topic, which will be discussed further later in this chapter.

The above considerations beg the question: "Shouldn't different drag factors be used for different surfaces?" The short answer is that there is presently no evidence to support such a thing. Warner et al.[9] indicate a friction coefficient of about 0.3–0.4 for sheet metal sliding on a hard surface, which is not inconsistent with the rollover data. And while rollovers have airborne segments, they also have hard touch downs that result in gouges and tire marks, so what is gained in some areas seems to be lost in others. The same can be said for rollovers on soil, in which considerable displacement of soil can occur.

Keifer et al.[10] have suggested a possible inverse relationship between the number of rolls and the overall drag factor. However, the totality of test

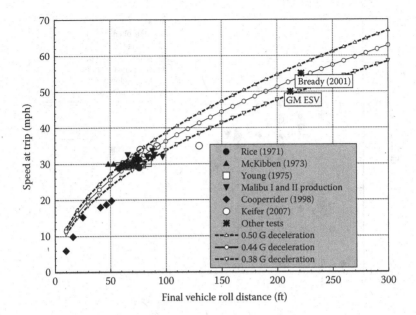

FIGURE 16.1
Speed versus roll distance in tests.

results, shown in Figure 16.2, does not suggest that a statistically significant correlation exists.

Once the rollover distance is known, the speed at roll initiation is found using Equation 16.1, using an overall drag factor between 0.38 and 0.50. While it may not be obvious at first glance, this is a reverse trajectory approach of the simplest form (since one works backward toward the unknown initiation speed). However, this approach reveals nothing about other parameters of interest, such as the velocity–time history, roll rates, and so forth. On the other hand, it is often the case that more is known about the rollover trajectory than just the total distance. Might not a more comprehensive reverse trajectory analysis, such as discussed in Chapter 4, utilize this additional information and provide more information about the rollover accident?

Laying Out the Rollover Trajectory

To use a full reverse-trajectory technique, we start by laying out the vehicle trajectory from roll initiation to rest, using the scene measurements (or survey), just as we did with the vehicle run-outs in Chapter 4. The first step is to

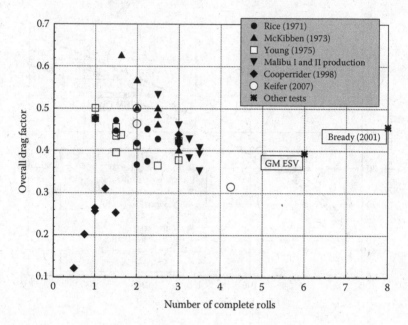

FIGURE 16.2
Drag factor as a function of rolls.

postulate the number of rolls, based on the final rollover distance in Figure 15.3, combined with the number of rolls that may be indicated by evidence on the vehicle. Then using a scale scene drawing that shows the scene evidence, we place scale drawings of the vehicle on the drawing so that evidence on the vehicle is matched up with evidence on the scene, consistent with the postulated number of rolls. Sometimes, this step will lead to a change in the postulated number of rolls one way or the other.

Figure 16.3 is a portion of the scene drawing that shows vehicle placements such that three of the evidence marks on the roadway (U, V, and W, for example) are matched with damage on the vehicle (damage to the wheel rims). Mark U is seen in Figures 15.18 and 15.19; Marks V and W appear in the background of Figure 15.18. Gouges V and W are sufficient to define the yaw position of the vehicle at what will become key position 5, which is labeled on the drawing. For the sake of convenience, the postulated number of rolls at that position (1½) is also indicated. The yaw angle at key position 4 is not known, but must be inferred from the yaw angles at adjacent positions. The subsequent analysis will show whether that yaw position needs to be adjusted one way or the other (mostly, by looking at the yaw rate history). Part of the vehicle at key position 6 is also shown; the "X" drawn on it indicates that the vehicle is upside down.

Note that to carry out this step of the process, scale drawings of the vehicle must exist from various perspectives. A good way to make these vehicle

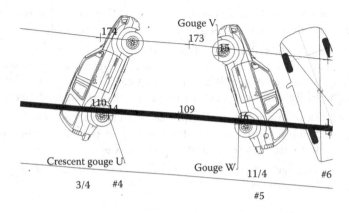

FIGURE 16.3
Vehicle placement on scene evidence.

drawings is to photograph an exemplar vehicle, and trace over the photo using a drawing package like AutoCAD. Alternatively, one may use photographs of a scale model vehicle. This is particularly convenient for the ⅛ positions. Vehicle drawings may also be purchased from a supplier.

The object of the vehicle placement exercise is to explain the physical evidence, and to locate the vehicle CGs when the vehicles are in the key positions. Therefore, the vehicle drawings need to show (at least approximately) the CG locations in the vehicle. The CG coordinates can then be scaled off the drawing.

At this point, one might ask, "Why not do a reverse trajectory analysis something like we did in Chapter 4?" Why not, indeed? We can expand on the reverse trajectory analysis commonly used in rollover reconstructions, add key points based on the physical evidence, subdivide the trajectory using a spline curve that passes through the key points as in Chapter 4, and work through the trajectory backwards, thereby obtaining time-based information on the vehicle kinematics.

Setting Up a Reverse Trajectory Spreadsheet

A reverse roll trajectory spreadsheet is an analog of the reverse trajectory spreadsheets discussed in Chapter 4. However, there are some important differences:

1. The vehicle is not rolling on its wheels. Therefore, the tire properties do not come into play, and the drag factor does not depend on the coefficient of friction and the crab angle.

2. A single drag factor is assumed to pertain throughout the trajectory. A commonly used value is 0.43, as discussed previously.

3. Since the coefficient of friction and the crab angles are not variables that figure in the calculations, they could be eliminated in the spreadsheet. However, they will be input and calculated, respectively, for compatibility with the previous reverse trajectory spreadsheets.

4. An additional variable—roll angle θ—is to be tracked. This variable will need to be represented in the key values (input) section and in the splined variables section of the spreadsheet.

5. Once a time base is established, the roll rate, a variable of key interest, can be calculated.

With these differences in mind, the input section of the spreadsheet appears as shown in Figure 16.4. A description field has been added (column C), as well as a key roll angle variable (column G). The user input fields are shaded, and the entered values are in bold type. The input cells have been prepared for 15 key points, which is probably more than necessary. However, more key points could be used if desired. As before, we have created an arc length variable Key S that represents the distance traveled along straight lines connecting the key variables. The length of each such straight line is Δ Key S (column I), calculated by the Pythagorean theorem. The cumulative sum Key S of the line lengths is in column J and becomes our independent variable for splining purposes. The length of travel from roll initiation to rest was a little over 152 ft.

The CG locations at key positions 4 and 5, shown in Figure 16.3, are key points 4 and 5, shown in rows 8 and 9 of Figure 16.4. The key point coordinates were picked off the scene drawing. Their true precision is less than that indicated by their precision on the spreadsheet, but two decimal places were retained for verification against the drawing. While the scene was surveyed and Z coordinates of the vehicle CG could therefore be obtained (by accounting for the CG heights above the ground), the accident site was flat and level, and the Z coordinates were not deemed to affect the analysis.

The remaining inputs are in cells F34 through F36. The friction coefficient does not figure in the analysis, as explained above, but is represented in input cell F35 for compatibility purposes. Since the drag factor is uniform, a single input cell F36 was used to fill the entire column in the output section of the spreadsheet, a portion of which is shown in Figure 16.5.

Except for roll initiation, it can be seen that the key roll angles (column G) correspond exactly to the cumulative number of rolls (column B). However, the roadway evidence—except at rest—was created while the vehicle was rotating through a range of roll angles (and ranges of values for X, Y, Z, and ψ, for that matter). Therefore, there is some degree of latitude in assigning a specific roll angle to a specific key position. That latitude can be exercised if it helps smooth out unrealistic variations in the roll rate. While keeping

	A	B	C	D	E	F	G	H	I	J
1										
2			**Key Trajectory Points**							
3	Key	# of		X	Y	Z	Roll θ	Yaw ψ	Δ Key S	Key S
4	Pt #	Rolls	Description	feet	feet	feet	deg	deg	feet	feet
5	1	0	End of tire marks	176.90	−20.60	0.00	25.0	54.0		0.000
6	2	¼	Scrape T1	195.59	−21.59	0.00	90.0	66.9	18.719	18.719
7	3	½	Scrapes T2 & T3	211.19	−22.28	0.00	180.0	77.0	15.614	34.333
8	4	¾	Gouge U	231.06	−22.99	0.00	270.0	92.6	19.881	54.214
9	5	1¼	Gouges V & W	249.10	−24.02	0.00	450.0	106.8	18.069	72.283
10	6	1½	Scratches on Left Roof	258.18	−24.51	0.00	540.0	109.9	9.098	81.381
11	7	1¾	Gouge X & Rim Down Y	265.35	−24.61	0.00	630.0	112.9	7.170	88.550
12	8	2¼	Print on RF fender	284.58	−23.90	0.00	810.0	137.0	19.242	107.792
13	9	2¾	Scrape a and Gouge b	306.24	−19.89	0.00	990.0	159.2	22.027	129.820
14	10	3¼	Rest	328.60	−19.30	0.00	1170.0	180.0	22.367	152.186
15										
16										
17										
18										
19										
20										
21										
22										
23										
24										
25										
26										
27										
28										
29										
30										
31										
32										
33					**Final Conditions**					
34				Final Speed =	0.00	mph				
35				Friction Coefficient =	0.70					
36				Rollover Drag Factor =	0.43					

FIGURE 16.4
Input section of RollTraj spreadsheet.

in mind that some variations are to be expected, such latitude was not exercised in this case.

To set up the calculation section of the spreadsheet, the first task is to set up the subdivision of the trajectory (in columns K and L in Figure 16.5). The number of subdivisions is quite arbitrary. In this case, 75 spline points (and 74 segments) would provide plenty of points in a time history plot of any of the variables, and would create trajectory segment lengths of just over 2 ft. See column S. Given that decision, the trick is to set up fractions 0, 1/74,

			ROLLOVER TRAJECTORY ANALYSIS							Trajectory Analysis										
K	L	M	N	O	P	Q	R	S	T	U	V	W	X	Y	Z	AA	AB	AC	AD	AE
Spline Pt #	Arc len Ŝ	Key Pt #	Spline X	Spline Y	Spline Z	Roll θ	Spline ψ	Seg Len	Cum S	Direc	Crab Ang	Mu	LockF	DragF	Veloc	Slope	Δ time	Roll Rate	Yaw Rate	Cum time
	feet		feet	feet	feet	deg	deg	feet	feet	deg	deg				mph		sec	deg/sec	deg/sec	sec
1	0.000	1	176.900	-20.600	0.000	25.000	54.000	0.000	0.000						44.27					0.000
2	2.057		178.954	-20.711	0.000	30.395	55.455	2.057	2.057	-3.1	57.82	0.700	1.000	0.430	43.97	0.000	0.032	169.74	45.79	0.032
3	4.113		181.007	-20.822	0.000	35.917	56.908	2.057	4.113	-3.1	59.27	0.700	1.000	0.430	43.67	0.000	0.032	172.58	45.39	0.064
4	6.170		183.061	-20.932	0.000	41.696	58.355	2.057	6.170	-3.1	60.71	0.700	1.000	0.430	43.36	0.000	0.032	179.33	44.90	0.096
5	8.226		185.114	-21.042	0.000	47.859	59.793	2.057	8.226	-3.1	62.14	0.700	1.000	0.430	43.06	0.000	0.032	189.91	44.33	0.128
6	10.283		187.168	-21.151	0.000	54.534	61.221	2.057	10.283	-3.0	63.55	0.700	1.000	0.430	42.75	0.000	0.033	204.23	43.67	0.161
7	12.339		189.222	-21.259	0.000	61.849	62.634	2.057	12.339	-3.0	64.94	0.700	1.000	0.430	42.44	0.000	0.033	222.20	42.93	0.194
8	14.396		191.276	-21.366	0.000	69.933	64.031	2.057	14.396	-3.0	66.31	0.700	1.000	0.430	42.12	0.000	0.033	243.74	42.12	0.227
9	16.453		193.329	-21.472	0.000	78.913	65.408	2.057	16.453	-2.9	67.66	0.700	1.000	0.430	41.81	0.000	0.033	268.75	41.22	0.261
10	18.719	2	195.593	-21.586	0.000	90.000	66.900	2.266	18.719	-2.9	69.05	0.700	1.000	0.430	41.46	0.000	0.037	298.72	40.19	0.298
11	20.566		197.437	-21.678	0.000	100.017	68.094	1.847	20.566	-2.8	70.34	0.700	1.000	0.430	41.17	0.000	0.030	328.67	39.19	0.328
12	22.622		199.492	-21.778	0.000	111.979	69.409	2.057	22.622	-2.8	71.54	0.700	1.000	0.430	40.85	0.000	0.034	349.84	38.43	0.362
13	24.679		201.546	-21.875	0.000	124.465	70.714	2.057	24.679	-2.7	72.77	0.700	1.000	0.430	40.52	0.000	0.034	362.28	37.89	0.397
14	26.735		203.600	-21.969	0.000	137.137	72.021	2.057	26.735	-2.6	73.99	0.700	1.000	0.430	40.19	0.000	0.035	364.73	37.60	0.432
15	28.792		205.655	-22.059	0.000	149.657	73.337	2.057	28.792	-2.5	75.20	0.700	1.000	0.430	39.86	0.000	0.035	357.41	37.57	0.467
16	30.849		207.710	-22.146	0.000	161.687	74.671	2.057	30.849	-2.4	76.41	0.700	1.000	0.430	39.53	0.000	0.035	340.58	37.78	0.502
17	32.905		209.765	-22.228	0.000	172.890	76.034	2.057	32.905	-2.3	77.64	0.700	1.000	0.430	39.19	0.000	0.036	314.47	38.24	0.538
18	34.333	3	211.191	-22.282	0.000	180.000	77.000	1.427	34.333	-2.2	78.69	0.700	1.000	0.430	38.96	0.000	0.025	285.48	38.80	0.563
19	37.018		213.875	-22.377	0.000	191.736	78.873	2.686	37.019	-2.0	79.97	0.700	1.000	0.430	38.51	0.000	0.047	248.23	39.63	0.610
20	39.075		215.931	-22.446	0.000	199.668	80.357	2.057	39.075	-1.9	81.53	0.700	1.000	0.430	38.17	0.000	0.037	216.88	40.57	0.646
21	41.131		217.986	-22.513	0.000	207.132	81.884	2.057	41.132	-1.9	82.99	0.700	1.000	0.430	37.82	0.000	0.037	202.22	41.36	0.683

FIGURE 16.5

First 21 spline points of calculation in RollTraj spreadsheet.

2/74, ..., 74/74 so that the splined arc length S (column L) can be calculated from the length of the straight line segments connecting the key points (cell J14, 152.186 ft, in this case). To do this, column K was filled with integers from 1 to 75, ending in cell K79. Then, the formula for splined arc length S (cell L5) is

$$= (K5 - 1) * (\$J\$14 - \$J\$5)/(\$K\$79 - 1) \tag{16.2}$$

This evaluates to

$$= (0) * (152.186)/(74) \tag{16.3}$$

for that cell. The absolute reference operator $ is used because we intend to copy the formula into the rest of column L. When we do that, we get our independent variable Arc len S, which is the total distance of 152.186 ft, carved up into 74 segments. They would be of equal length if we made no further adjustments of some of the splined trajectory points, as discussed later.

Now we are ready to set up the spline operations. As discussed in Chapter 4, to set up the spline on X, the formula in Cell N5 is

$$= Spline(\$J\$5:\$J\$14, \$D\$5:\$D\$14, \$L5, TRUE) \tag{16.4}$$

where $J$5:$J$14 is the key independent variable Key S, $D$5:$D$14 is the key dependent variable X, $L5 is the splined independent variable Arc len S, and TRUE enables extrapolation. Again, the absolute reference operator $ is used because we intend to copy the formula.

To spline Y, we need to set up cell O5 to reference the dependent key variable Y (column E). Formula 16.4 is changed to

$$= Spline(\$J\$5:\$J\$14, \$D\$5:\$D\$14, \$L5, TRUE) \tag{16.5}$$

Similar formulas are set up in cells P5, Q5, and R5. When these formulas are copied into the other 74 rows, the splining operation is complete. The values can be compared against the inputs in row 14 of Figure 16.4 to verify that the splines were set up properly.

If a different number of key points were to be used, the cell ranges of the splined variables would have to be changed accordingly. In Equations 16.4 and 16.5, and the formulas in cells P5, Q5, and R5, the references to row 14 would have to be changed to whatever row now contains the last key point.

The next step does not affect the results that are to be obtained, but it does produce calculations that coincide with the key points. In column L, the splined independent variable Arc len S was examined to identify the cells having values closest to the key independent variable key S. In each such cell, the value was set equal to the corresponding Key S cell value. (Be sure to remember from Chapter 4 that the progression of values in Arc len S has to be monotonic.) This action causes an immediate recalculation of all the splines in that row, and shows that the spline curves pass exactly through the key points. Key point numbers were entered by hand in column M, and the rows were highlighted in gray to facilitate locating the calculated values corresponding to the key points.

Columns U and V are carry-overs from the reverse trajectory spreadsheet, and were retained for compatibility. For the friction coefficient μ, Cell W6 contains the formula = F$35, which copied into the remaining cells of column W. Similarly, the cells in column Y reference Cell F$36. Since the vehicle is sliding rather than rolling on wheels, a lock fraction value of 1.000 was copied into all the cells of column X. Columns U through Y all refer to trajectory segments, so there are no entries in row 5. (Segment i exists between point i and point $i + 1$.)

The variables Veloc, Slope, Δ time, Yaw Rate, and Cum time are calculated just as they were in the reverse trajectory spreadsheet, as explained in Chapter 4. Roll rate is a new variable. Its calculation is analogous to that of the variable Yaw Rate. Thus, adapting Equation 4.5, we have

$$\dot{\theta}_i = \frac{\theta_{i+1} - \theta_i}{\Delta t_i}$$

(16.6)

where the over dot indicates differentiation with respect to time (rate). Again, if the roll angles are in degrees, the roll rates will be in deg/sec; if they are in radians, the rates will be in rad/sec. The formulas for roll rate and yaw rate are copied into all except the first cells in columns AC and AD, respectively. It is to be noted that the last entries in the roll rate and yaw rate columns—AC79 and AD79 in this case—are not zero, since they represent the averages in the last segment, not the values at the point of rest.

At this point, the spreadsheet is complete. We see that the velocity at roll initiation turns out to be a little over 44 mph, for this case. The value in cell Z5 is displayed in bold face because of the importance of the result.

Examining the Yaw and Roll Rates

In the usual reverse trajectory analysis, knowledge of the roll distance and overall drag factor leads to initial speed and elapsed time, which lead in turn to average roll rate. A reconstruction that describes roll rates only in terms of the average value was most assuredly done this way.

If there is no physical evidence at the scene between initiation and rest, all that can be said about yaw and roll rates are their average values. Further information is hidden from the reconstructionist's view. Often, however,

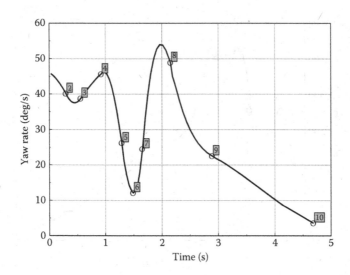

FIGURE 16.6
Yaw rate history for sample accident.

clues are left at the scene that enable the discovery of additional information. The current analysis, using key positions, is the means by which the discovery can be made.

Figure 16.6 shows the average yaw rate in each of the trajectory subdivisions, calculated from the reverse trajectory spreadsheet. Unreasonable yaw rates may be a hint that one or more of the key yaw angles are incorrect. Within the limits of physical evidence, certain of the key yaw angles may need adjustment in some cases so as to bring the yaw rates into a reasonable range.

The roll rate history that results from reconstructing the example rollover using the spreadsheet analysis is shown in Figure 16.7. We see that the roll rate diminishes toward zero, as it should at the end. It is not exactly zero, because the roll rate is the average in each (subdivided) segment, including the last one. The roll rate history has peaks and valleys that have been revealed by analyzing the copious physical evidence at the scene. The peak roll rate is in the neighborhood of 550 deg/sec.

The key points are numbered, and plotted with the open circles. In general, the key points cannot be expected to coincide with the peaks and valleys in the roll rate history.

One might question whether such a roll rate history is representative of what has been seen in tests. It is the case that dynamic measurements of roll rates are difficult to obtain, because the simultaneous motions about all three vehicle axes impose a significant data processing burden. Hence, the rollover tests with dynamic roll rate data are few in number. However, such measurements have been reported in Orlowski's Malibu Test 5,[11] and three of the tests reported by Keifer.[10] A comparison of the reconstructed roll rates with those

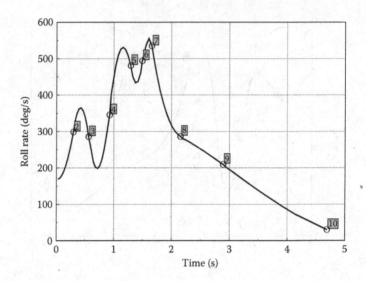

FIGURE 16.7
Roll rate history for sample accident.

measured in the four tests is shown in Figure 16.8. The reconstructed roll rate of Figure 16.7 is repeated with the dashed line in Figure 16.8, and the test curves are shown as solid lines.

As might be expected with chaotic events like rollovers, the detailed roll rate histories vary all over the place. Of course, all of the vehicles are dissimilar. The tests involve the use of an inclined, decelerating dolly from which the vehicles were launched sideways, at speeds ranging from 32 to 35 mph, whereas the reconstruction involves an on-road rollover at about 44 mph. (The Orlowski test used a launch angle at the standard 23°, while the Keifer tests started at 34° and used a lower dolly deceleration.) As opposed to the tests, the reconstruction has a nonzero initial roll rate because rolling motions have already started at time zero, when the leading tires reach the ends of their marks. Nevertheless, the durations range between 4½ and 6½ sec, and the peak roll rates were banded between about 350 and 550°/sec. In this regard at least, the reconstructed roll rates are not unrealistic.

Of course, the reconstructed roll rate curve is much smoother than the test curves. Therefore, it must not be interpreted as exact, but as a reasonable representation. However, it serves another purpose as well: namely, to check on the postulated number of rolls. Because the roll position of the vehicle at rest is known, the postulated number of rolls can be varied only by integer amounts. In this particular case, the abundance of physical evidence does not readily allow the insertion or removal of an additional roll anywhere in the trajectory. In other cases, the physical evidence may not be so constraining (such as a rollover into a plowed field, for example), in which case postulated roll(s) may be added or subtracted. This will obviously have a

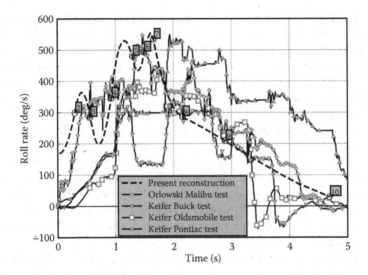

FIGURE 16.8
Comparison of reconstructed with actual roll rates.

significant effect on the computed roll rate histories, which can be compared to Figure 16.8. It will probably be immediately apparent from the calculated roll rates what the correct number of rolls is for a given crash.

Since the spreadsheet for rollover calculations is a super-set of that used for reverse trajectory run-out calculations, the pre-roll and post-roll trajectories could be combined into a single spreadsheet if desired. This would allow the roll angle to start at zero, during the pre-roll phase, before the rollover has initiated (i.e., before the CG is positioned above the tire–road interfaces of the leading tires). An even larger spreadsheet is thereby created, however, which was not deemed helpful for instructional purposes. It is also the case that key positions tend to be more closely spaced in the rollover phase than during the pre-roll maneuvers. The use of separate spreadsheets allows the subdivided trajectory segments to be different before and after roll initiation, which may be useful in fitting splines through the key points. Also, the rollover trajectory spreadsheet may simply start during the pre-roll maneuver, before roll initiation, in order to allow the roll angle to build from zero.

Scratch Angle Directions

As discussed in Chapter 15, scratch pattern observation and scratch angle measurement are important in the reconstruction of vehicle yaw angles during a rollover event. Of course, it is true that the crab angle does not affect the translational speed or roll rate calculations for a rolling vehicle, since the lock fraction is assumed to be unity (unlike a run-out calculation). Nevertheless, the crab angle may very well be significant in analyzing the occupant motions during ground contacts, and the vehicle heading can have an important influence on the airborne trajectories of thrown objects or ejected occupants. To determine these angles, various investigators[4-7,11] have indicated that one can align the vehicle scratches with its direction of travel at the time of contact.

Thus, it is puzzling indeed to encounter scratch patterns that are largely longitudinal, often mixed in with others that indicate a more or less lateral, or "barrel" roll. As discussed in Chapter 15, end-over-end tumbling of a vehicle is relatively rare, and would create obvious damage on the vehicle were it to occur. In the absence of such evidence, what is one to make of these scratch patterns?

This apparent conundrum is taken up in a 2001 paper by Bready et al.,[12] who correctly point out that scratches and abrasions are caused by the relative motion between the contacted vehicle surface and other objects (more specifically, the ground). For the vehicle scratch angles to match the crab angle, the vehicle must be in a state of rigid-body vehicle translation (without rotation). Yet it is rollovers that we are analyzing, so by definition such an angle-matching procedure must entail faulty logic, strictly speaking.

Like any three-dimensional rigid body, a vehicle has three principal mass moments of inertia. Rotations about the axis of minimum inertia will be stable. For vehicles, this is virtually always the roll axis, which is roughly parallel to the geometric x-axis. Thus, we assume that a longitudinal axis through the CG is the one about which roll occurs. (Of course, pitch and roll can also occur due to ground contacts, giving rise to alternating contacts at the vehicle corners, or "football" motions, but such detail is far beyond the scope of the present discussion.)

If the vehicle rolls like a wheel that rolls without slip, the instantaneous center of its rotation is at the contact point, and there is no relative velocity. Relative velocity at the contact point can only be created when the instantaneous center is elsewhere; that is, when the translational velocity of the CG is not equal and opposite to the tangential velocity at the point of contact.

If a vehicle is in a pure barrel roll (with a crab angle of 90°) where a relative velocity exists—as it almost always does—at the point of contact, the resulting scratches and abrasions will be purely lateral. In other words, the roll will be neither nose-leading nor tail-leading. In keeping with the discussion in Chapter 15, the scratch angle may be said to be zero. Of course, the roll rate has no effect on the scratch angle.

On the other hand, vehicles often exhibit some crab angle other than 90° while rolling. In such cases, the roll axis is not perpendicular to the path of travel. The translational and tangential velocity vectors at the point of contact are not only unequal; they are also not parallel. Moreover, the relative velocity at that point depends on the difference between these two vectors. Because a subtraction is involved, the angle of the relative velocity vector (and thus the scratch angles) can be fairly sensitive to small changes if the vectors are roughly equal (or are close to zero). In these circumstances, the scratch angles can deviate significantly from the perpendicular to the crab angle. The more the crab angle deviates from the perpendicular, the greater effect the roll rate will have on the scratch angle. (Be aware that while they are defined relative to a pure barrel roll in this discussion; in the Bready paper, the scratch angles are defined relative to the vehicle's longitudinal axis.)

In the Bready paper, there is a presentation of typical translational and tangential velocity histories. The tangential velocity rises from zero near roll initiation to a peak value, and then decreases to zero at rest. The translational velocity starts off at its maximum value, generally remains above the tangential velocity, and then decreases to zero in the same time frame. It is usually late in the event, when both of these velocities are heading toward zero, that there can be the greatest swing in scratch angles. Of course, uncertainties in these velocities will lead to uncertainties in the scratch angles as well.

For calculation purposes, we adopt the convention that for a roll in the positive direction about the longitudinal axis (passenger-side leading), a positive scratch angle means nose leading; a negative angle means tail-leading. If the roll direction is reversed, the interpretation of the results is reversed also.

If a vehicle is not rolling, but is sliding on the ground with velocity V and at some crab angle (Crab), then the relative velocity of the ground with respect to the vehicle is also V. The component of V parallel to the vehicle's x-axis is $V \cos(Crab)$, and the component in the circumferential direction is $V \sin(Crab)$. If at the same time the vehicle is rolling about its x-axis with a velocity ω, then there will be a relative velocity $-r\omega$ in the circumferential direction due to the roll, where r is the radius between the CG and the vehicle surface being contacted. With respect to the vehicle's x-axis, the resultant relative velocity will be the (vector) sum of the two relative velocity vectors. The angle of this resultant velocity vector will be the scratch angle SA, which can be described in terms of how much the scratch directions deviate from the circumferential direction, as discussed above. The tangent of this angle is the ratio of the total velocity component parallel to the x-axis, divided by the total velocity component in the circumferential direction. In mathematical terms,

$$SA = Arc\tan\left[\frac{V\cos(Crab)}{V\sin(Crab) - r\omega}\right]$$

(16.7)

where, of course, the quantities are all in consistent units.

This sort of equation could be incorporated in the spreadsheet of Figure 16.5. However, for much of the rollover event, scratches would not be generated because the vehicle was airborne, contacting with its tires, and so on. So the only place one would need such a calculation would be the key positions at which a comparison between observed and calculated scratch angles is desired. For instance, none of the key positions in Figure 16.5 had scratch angle measurements recorded for them (though perhaps positions 2 and 3 should have had them!). Cell AF22 could then have contained the formula

$$= \text{Degrees} \left\{ A\text{Tan2} \begin{bmatrix} Z22 * Sin(Radians\{V22\}) \\ -F\$37/12 * Radians\{AC22\}/1.46667, \\ Z22 * Cos(Radians\{V22\}) \end{bmatrix} \right\} \quad (16.8)$$

where the vehicle radius (distance from the CG to the contact point in inches) has been placed in Cell F37, and where the absolute row reference symbol \$ has been utilized to facilitate copying the formula into other rows. The Excel two-argument Arctangent function ATan2(x,y) is used to keep track of the quadrant it is working in, and returns a value between $-\pi$ and $+\pi$ radians (excluding $-\pi$). Here, the crab angle and roll rate are expressed in degrees, so the Radians() function is used to convert them to radians. The factor of 12 converts the radius into feet. Since the translational velocity is already in mph, the factor 1.46667 is used to convert the circumferential velocity from ft/sec into mph. The Degrees() function converts the results back into degrees. The various kinds of brackets are not recognized by Excel, but are used here to assist in keeping track of how the functions are nested.

In this case, an effective radius is contemplated for the vehicle. An effective radius can be calculated by finding the radius of a circle having the same circumference as a rectangular prism of the same overall dimensions as the vehicle. In this case,

$$r_{eff} = \frac{OAW + OAH}{\pi} \quad (16.9)$$

To be more exact about it, a different radius, obtained by measurement, could be used for each contact point. An even finer point could be put on the calculations by accounting for the (progressive) deformation of the vehicle with each contact. However, such exactitude would seem unwarranted in view of uncertainties regarding just how the deformations occurred.

If accident vehicle observations permit the use of scratch angle comparisons using Formula 16.8, the "Goal Seek" feature of Excel can be used to great advantage. For a specific example, one could ask Excel to set the calculated scratch angle at key position 3 in Cell AF22 to the observed value, by changing the vehicle yaw angle input for key position 3 in cell H7 (Figure 16.4). Thus, the vehicle heading angle is quickly set to a value that is consistent with the scratch angle measurement. The process does not always converge, but usually it does so quickly. The splines, velocities, roll and yaw rates, and all the other calculations are automatically updated.

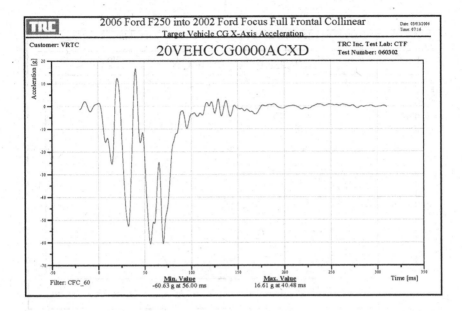

FIGURE 1.4
CG *X*-axis acceleration, target vehicle, test DOT 5683.

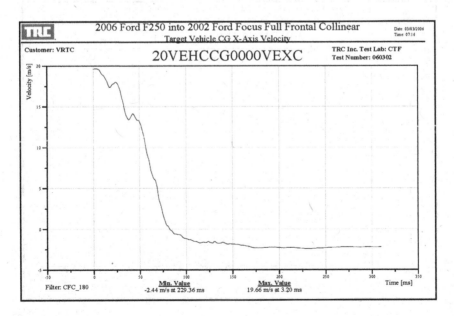

FIGURE 1.5
CG *X*-axis velocity, target vehicle, test DOT 5683.

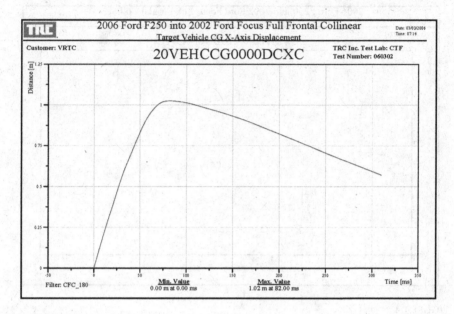

FIGURE 1.6
CG X-axis displacement, target vehicle, test DOT 5683.

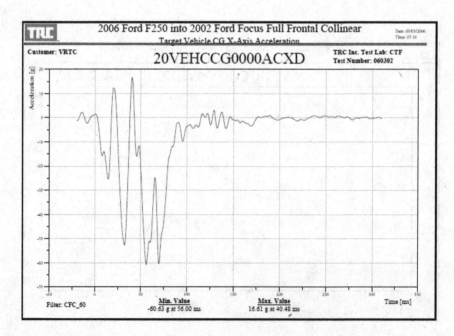

FIGURE 13.1
CG x-axis acceleration, vehicle 2 (target vehicle).

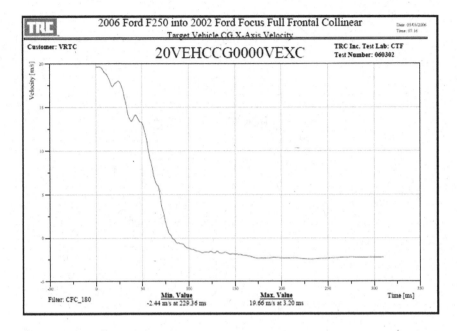

FIGURE 13.2
CG x-axis velocity, vehicle 2 (target vehicle).

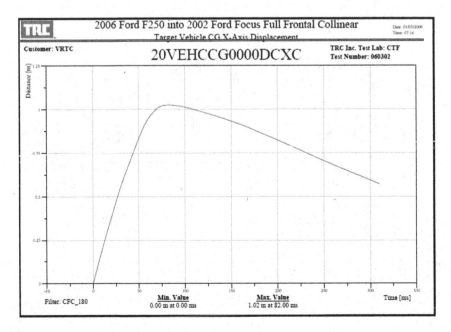

FIGURE 13.3
CG x-axis displacement, vehicle 2 (target vehicle).

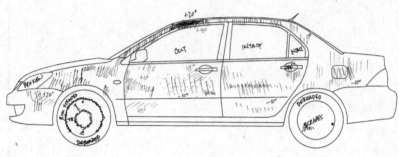

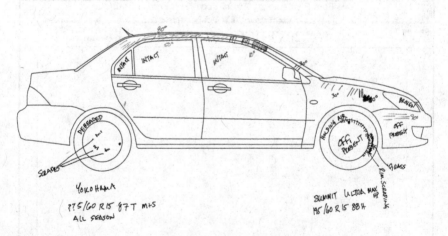

FIGURE 15.6
Vehicle inspection sketch.

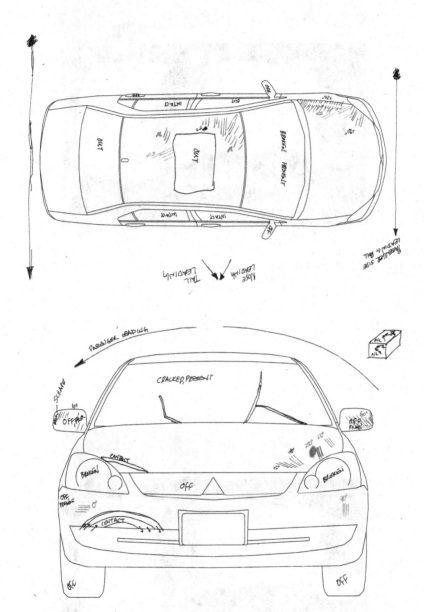

FIGURE 15.6 (continued)
Vehicle inspection sketch.

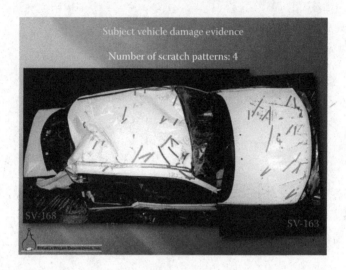

FIGURE 15.8
Scratch pattern highlighted in red.

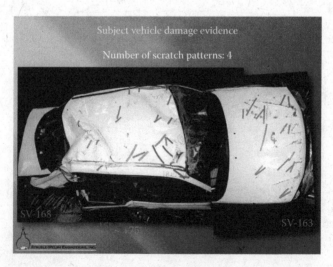

FIGURE 15.9
Scratch patterns highlighted in red and magenta.

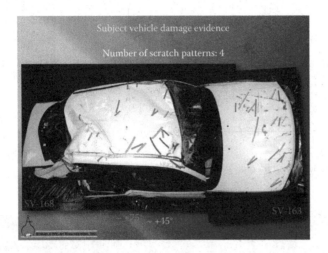

FIGURE 15.10
Scratch patterns highlighted in red, magenta, and green.

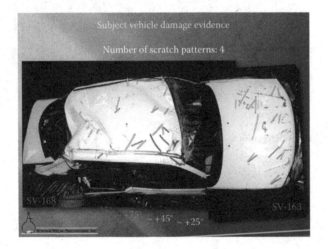

FIGURE 15.11
All four scratch patterns highlighted.

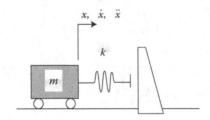

FIGURE 17.6
Simple barrier crash model.

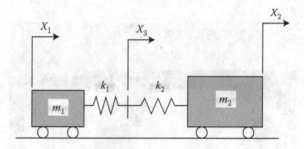

FIGURE 17.10
Vehicle-to-vehicle crash model.

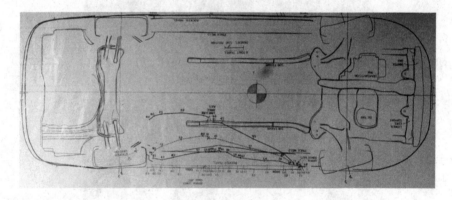

FIGURE 22.3
Damage map for accident vehicle.

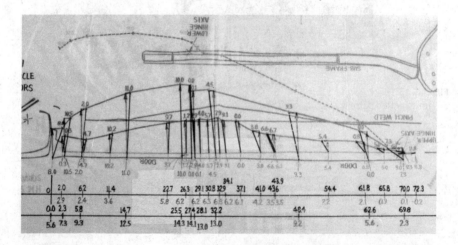

FIGURE 22.4
Crush profiles.

Soil and Curb Trips

Much effort has been expended over the years to develop a test procedure that would reduce the nonrepeatability of rollover tests. Some approaches have focused on controlling the vehicle so that a ground impact occurs in a specified, repeatable manner,[13-15] but that impact is only one of many that may or may not occur in the whole rollover event. The fact remains that a vehicle that is tripped and allowed to roll without constraining its motion, as in actual field crashes, will do so in a random, chaotic manner. Despite numerous attempts at refining the procedure, dolly rollovers resulting in mostly lateral rollovers—such as specified in FMVSS 208—have not been improved upon for repeatability, and therefore constitute the bulk of the existing tests.

Of course, field rollovers occur in every conceivable way apart from being launched sideways from an inclined dolly. A large number of them occur on soil, in which furrowing is the tripping mechanism. Others occur where the vehicle encounters a rigid object such as a curb. Thus, even though soil and curb trip tests are relatively few in number, they are of considerable interest to the reconstructionist.

From the point of view of the testing laboratory, the speed at which the vehicle is launched (usually from a dolly) is the test speed, and is reported as such. In a soil trip, however, it would be a mistake to treat this as the roll initiation speed, simply because the roll is not initiated until the furrowing has taken place, the soil displaced, the leading tires slowed, and the vehicle rolled through some angle. By then, of course, the vehicle has bled off much—if not all—of its speed.

Valuable insights into these issues have been revealed by the rollover tests by Cooperrider et al.[8,16] at Failure Analysis Associates (now Exponent). Some effort was made to systematically define the beginning of the roll, specifically, a roll angle of 52°, or a roll rate of 100°/sec. However, these metrics were somewhat inconclusive, aside from the fact that the reconstructionist does not have the benefit of test instrumentation in making such a determination. Instead, the ends of the furrows or the presence of a curb provide stark evidence of where the rollover began.

Of interest to the reconstructionist is how much deceleration the vehicle experienced while furrowing occurred. Of course, the answer must depend on the soil, how deep the furrows were, and other specific conditions. That said, the vehicle velocity–time histories for the Cooperrider tests that actually produced rollovers show distinctly different slopes between furrowing and rollover. Thus it seems reasonable to mark the beginning of the rollover, and the rollover initiation speed, at the point where the slope change occurs. Doing this results in data points for the dirt-tripped rollovers in Figures 15.3, 16.1, and 16.2 being plotted at speeds considerably lower than the test speeds reported by Cooperrider et al. In fact, four of the soil-tripped vehicles had so little energy left at the end of furrowing that they failed to roll at all.

As to the deceleration during furrowing, the 1998 paper reports about 1¼ Gs for the two vehicles that rolled. For reconstruction purposes, it seems that the most appropriate procedure is to establish key vehicle positions at the beginning and end (at least) of the furrowing phase, along with estimated roll angles, and apply a suitable deceleration to the vehicle while it is furrowing. That deceleration must be estimated based on how the actual furrows compare to those in the Cooperrider and other soil trips, as well as a careful analysis of the available data. For the actual rollover trajectory after trip, an overall drag factor between 0.38 and 0.50 should provide suitable upper and lower bounds on the speed at roll initiation. In view of the low-speed data points in Figures 15.3, 16.1, and 16.2, it is advisable to account for the estimated CG heights in the analysis if the roll distances are not large.

References

1. Rizer, A.L., Obergefell, L.A., and Kaleps, I., *Simulation of Vehicle Dynamics During Rollover*, Report PB90-258096, U.S. Department of Transportation, National Highway Traffic Safety Administration, May 1989.
2. Renfroe, D.A., Partain, J., and Lafferty, J., Modeling of vehicle rollover and evaluation of occupant injury potential using MADYMO, SAE Paper 980021, SAE International, 1998.
3. Day, T.D. and Garvey, J.T., Applications and limitations of 3-dimensional vehicle rollover simulation, SAE Paper 2000-01-0852, SAE International, 2000.
4. Orlowski, K.R., Moffatt, E.A., Bundorf, R.T., and Holcomb, M.P., Reconstruction of rollover collisions, SAE Paper 890857, SAE International, 1989.
5. Martinez, J.E. and Schlueter, R.J., A primer on the reconstruction and presentation of rollover accidents, SAE Paper 960647, SAE International, 1996.
6. Jones, I.S. and Wilson, L.A., Techniques for the reconstruction of rollover accidents involving sport utility vehicles, light trucks and minivans, SAE Paper 2000-01-0851, SAE International, 2000.
7. Meyer, S.E., Davis, M., Chng, D., and Herbst, B., Accident reconstruction of rollovers—A methodology, SAE Paper 2000-01-0853, SAE International, 2000.
8. Cooperrider, N.K., Hammoud, S.A., and Colwell, J., Characteristics of soil-tripped rollovers, SAE Paper 980022, SAE International, 1998.
9. Warner, C.Y., Smith, G.C., James, M.B., and Germane, G.J., Friction applications in accident reconstruction, SAE Paper 830612, SAE International, 1983.
10. Keifer, O.F., Richardson, W.C., Layson, P.D., Reckamp, B.C., and Heilmann, T.C., Vehicle linear and rotational acceleration, velocity, and displacement during staged rollover collisions, SAE Paper 2007-01-0732, SAE International, 2007.
11. Orlowski, K.F., Bundorf, R.T., and Moffatt, E.A., Rollover crash tests—The influence of roof strength on injury mechanics, SAE Paper 851734, SAE International, 1985.

12. Bready, J.E., May, A.A., and Allsop, D.L., Physical evidence analysis and roll velocity effects in rollover accident reconstruction, SAE Paper 2001-01-1284, SAE International, 2001.
13. Carter, J.W., Habberstad, J.L., and Croteau, J., A comparison of the controlled rollover impact system (CRIS) to the J2114 rollover dolly, SAE Paper 2002-01-0694, SAE International, 2002.
14. Cooper, E.R., Moffatt, E.A., Curzon, A.M., Smyth, B.J., and Orlowski, K.F., Repeatable dynamic rollover test procedure with controlled roof impact, SAE Paper 2001-01-0476, SAE International, 2001.
15. Jordan, A. and Bish, J., Repeatability testing of a dynamic rollover test fixture, *Proceedings, 19th International Technical Conference on the Enhanced Safety of Vehicles*, U.S. Department of Transportation, National Highway Traffic Safety Administration, 2005.
16. Cooperrider, N.K., Thomas, T.M., and Hammoud, S.A., Testing and analysis of vehicle rollover behavior, SAE Paper 900366, SAE International, 1990.

17

Vehicle Structure Crash Mechanics

Introduction

Vehicles are made of engineering materials. Therefore, when they collide, one might expect to find the same applicable principles as those developed in a course on Strength of Materials, or Mechanics of Materials, to be applicable. Loads are applied to a structure, and displacements ensue. However, in a Strength of Materials course, certain restrictions are applied, which render most of the analytical techniques and results inapplicable to vehicle crashes. These restrictions, or assumptions, are categorized in Table 17.1.

As a result of these complexities, a broader-brush approach is often taken, just as we saw for rollover analysis in Chapter 16. Of course, detailed finite element models of a crashing vehicle are available, but they are necessarily complex and require considerable computational resources. They are not commonly used for reconstruction purposes, and are well beyond the scope of this book.

A number of important insights can be gained from simple models, however, and those will be the instructional tools employed in this chapter.

Load Paths

As encountered in a Strength of Materials course, a statically indeterminate (or redundant) structure is one that has more than the minimum number of structural elements required to resist an applied load. By enforcing continuity conditions (e.g., requiring equal displacements where two or more parts of the structure come together), one finds that the loads are distributed in the structure according to the stiffness of its elements. The stiffer parts carry more of the load, whereas a wet noodle would carry none.

A vehicle is a far more complex structure than those analyzed in a Strength of Materials course, but the notion of a redundant structure is still useful. Here, the structural elements are known as load paths, and the idea that a stiffer load path carries more of the load is still valid. When a vehicle

TABLE 17.1

Comparison of Limitations and Assumptions

Strength of Materials	Vehicle Crashes
Linear materials. Hooke's law pertains. Stress is proportional to strain, and the constant of proportionality is Young's Modulus, or the modulus of elasticity.	*Non-linear materials.* The stress is not proportional to strain; Hooke's Law does not apply.
Elastic materials. This means that the stress–strain relationship is the same regardless of whether strains are increasing or decreasing. The stress–strain curve could be non-linear, but when the stresses are removed, the strains disappear also.	*Inelastic (plastic) materials.* Loading and unloading occur along different paths, which may result in residual strains when the stresses are removed.
Small displacements. Some might say infinitesimal displacements, but the important concept is that the effects of the loads are not influenced by the structural displacements. A straight beam may sag under load, and its displacements may be calculated, but the displacements are so small that the beam is still essentially straight. The moment of the applied load(s) is assumed to be the same in the deflected beam as in the originally straight beam. The beam does not get turned into a hair pin.	*Large displacements.* The whole geometry of the structure is changed. For example, the bending moment due to a force is changed, because the moment arm is altered by the structural displacement.
Linear force-displacement relations. Because force is proportional to stress, and stress is proportional to strain, and strain is proportional to displacement, the displacement varies linearly with the force.	*Non-linear force-displacement relations.* The relations in loading and unloading are different, and both are nonlinear.
Simple structural shapes. Constant cross sections, such as prismatic bars, wide flange beams, and circular cross sections, are commonly encountered. Variable cross sections are sometimes treated, but the analysis is considerably complicated. Element shapes are usually straight, although angle bents and circular arcs are sometimes encountered.	*Complex shapes.* A load path is almost never straight (except, perhaps, on heavy trucks), and almost always has a variable cross-section not made from standard structural shapes. It almost always has access holes for assembly, water and paint drainage, etc.
Quasi-static loading. No inertia ensues from the loading mechanism, or from the structural movement. Loads are treated as static, which means that the change from zero load (and deflection) to full load (and deflection) is assumed to occur over a very long time. Strain rates (in./in./sec) are essentially zero.	*Dynamic loading.* The crash occurs almost within the blink of an eye, so inertial effects are important, and certainly must be accounted for. Strain rates are so high that they can only be replicated by impact testers, not conventional testing machines.

is involved in a frontal crash, the impact forces at the front of the vehicle are transmitted aft through a series of load paths that may be arranged in series or in parallel. As the loads travel through the load paths they are distributed into the vehicle body, rather like a circulatory system, until they are entirely dissipated by the time they reach the opposite (unloaded) end of the body.

Load–Deflection Curves

A commonly used concept of analyzing a vehicle crash is to treat it as a series of masses (representing the inertia of the vehicle) connected by a collection of massless load paths. Such a concept is known as a lumped-mass, or lumped-parameter, model. The application of such a model to vehicle crashes was presented in 1970 by Kamal.[1]

Traditionally, in a lumped-mass model, a detailed stress analysis of the structural elements in the load path was generally not performed, because of the complexities cited above. Instead, it was often cheaper, easier, and perhaps more accurate to subject the actual load path to a crush test, which was generally quasi-static. The relationship between applied load and resulting deflection (or perhaps between applied deflection and resisting load) was a curve, which often looked like that shown in Figure 17.1.

In keeping with the notion of an applied deflection and resisting force, it is customary practice to plot the deflection on the abscissa (horizontal axis) and treat it as the independent variable, since it is monotonically increasing. The force, on the other hand, can dip because of buckling or rupture and then rise again as the material packs together, and so is plotted on the ordinate (vertical axis). It is generally the case, by the way, that in a crashing structure, the load path loads and deflection are taken to be positive in compression, which has to be kept in mind when building a mathematical model of the structure.

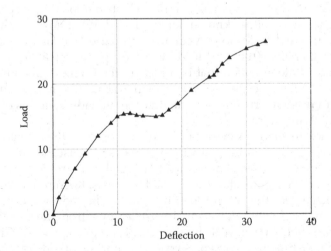

FIGURE 17.1
Typical load path load–deflection curve.

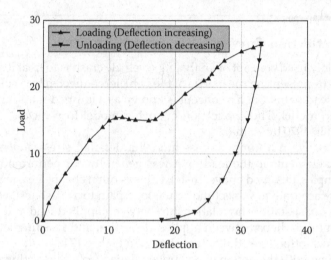

FIGURE 17.2
Typical load–unload cycle.

Now what happens the load is removed (the load path is unloaded)? In general, the load and deflection do not retreat along the original path, but descend along a new one, as shown in Figure 17.2.

Here, we see that the deflection has not retreated all the way to zero, even though the load has. The load path exhibits permanent deformation, and is said to have residual crush.

In a crash, of course, the deflection at which the loading stops and the unloading starts is neither known nor controlled. What if that path reversal starts at a different deflection? As shown in Figure 17.3, the loading would continue farther along the same loading curve; upon reversal, the unloading would proceed along a path that is similar to the original unloading curve (but slightly different, of course, since it begins at a different load and deflection). Again residual crush is present after the unloading, and is of a different amount than before.

Going back to the load cycle of Figure 17.2, what if there was a second loading cycle, such that the structure was re-loaded past the previous peak deflection? We would see behavior something like Figure 17.4. As in Figure 17.2, the first cycle is indicated by the solid triangles. Keeping the coordinate system unchanged, the second loading cycle, indicated by the hollow triangles, would begin with the residual crush left over from the first cycle. The structure would then behave as linearly elastic up to the load and deflection from which the first unloading occurred. This portion of the load–deflection curve is thus linear; the structure can load and unload along that line without incurring any additional residual crush, as long as the previous peak deflection is not exceeded.

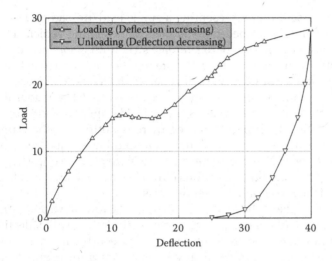

FIGURE 17.3
Unloading from a different deflection.

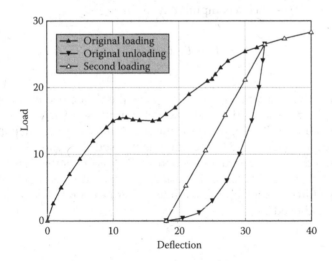

FIGURE 17.4
First cycle, followed by second loading.

Once that peak has been exceeded, however, the load and deflection continue along their original path. If, then, a second unloading occurs, it would look something like the unloading path shown in Figure 17.3.

These concepts are at the heart of the repeated impact methodology introduced by Warner et al.[2] in 1986. Repeated impacts will be discussed further in Chapter 21.

It should be said that representing a single load/unload cycle should suffice for most simple lumped-mass models of a crashing vehicle. However, if the model is complex enough to simulate structural resonance, or "ringing," then the ability to simulate an indefinite number of cycles may be needed.

If a quasi-static test were used to generate the load–deflection curve, how then are (dynamic) strain rate effects accounted for? The reader may recall that in the study dynamic systems, a viscous damper was often used in parallel with a (linear) spring. However, there is nothing in the crumpling of a vehicle structure in a crash to suggest a fluid being squeezed through an orifice. Since the highest velocities in a crash are at or near the beginning, it is readily appreciated that a viscous damping model would create excessive and unrealistic forces at the onset of a crash.

If one could subject the load paths to dynamic tests at various strain rates, then one could (theoretically) generate a different load–deflection curve for each strain rate, and then interpolate between them when running the model. Indeed, this computationally intensive (and inelegant) procedure is akin to that used in some analyses. A far simpler procedure, more applicable to lumped-mass models, is to compute a force multiplier that is proportional to the crush rate in the load path at the particular time. First introduced by Kamal,[1] the form of such a computation could be

$$F_d = F_s \left[1 + k \frac{CR}{880} \right] \tag{17.1}$$

where CR is the crush rate in the load path (the difference in the velocities of the masses at either end of the load path) in in./sec, k is a scalar factor—perhaps equal to unity—applied for the particular load path depending on the material it is made of, 880 is 50 mph expressed in in./sec, F_s is the static force obtained from the load–deflection curve, and F_d is the dynamic force. It is seen that this formulation is designed to produce perhaps a doubling of the static force at a 50 mph crush rate. Other crush rate factors have been suggested and used.[3]

Energy Absorption

Energy is the ability to do work, so when kinetic energy is converted into crush energy during a crash, work is done on the structure by overcoming the resistive force in the structure and causing it to crush. Work is force times distance. Thus the work done, and the energy absorbed, by incrementing the crush in a load path is equal to the force times the crush increment. In Figure 17.5, this is the rectangular area of the little slice under the

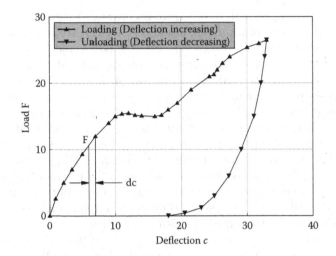

FIGURE 17.5
Differential energy absorbed.

load–deflection curve. If the crush increment dc is taken to be infinitesimal, we can then write

$$d(CE) = F(c)\,dc \qquad (17.2)$$

where $F(c)$ simply indicates that the force F is a function of the crush c. When this expression is integrated between the beginning and final crush, we see that the energy absorbed is simply the area under the load–deflection curve between those two points:

$$CE_f - CE_i = \Delta CE = \int_{c_i}^{c_f} F(c)\,dc \qquad (17.3)$$

If the process starts at zero crush and ends at the peak crush, the energy absorbed is the area under the entire curve up to peak deformation.

What about the next part of the curve, when the load path is unloading? Here, deflection is monotonically decreasing, so the crush increments are all negative. Thus the work done is also negative, which means that energy recovery is going on. The recovered energy is equal to the area under the unloading portion of the load–deflection curve. Of the total energy absorbed during the loading phase, some is recovered during unloading; the net energy absorption—or dissipation—is the area enclosed by the load–deflection curve between the initial and final conditions of zero force.

It can be appreciated from Figure 17.5 that if the deflection is to be brought back to zero, some force reversal (tensile forces) will be required. If we crush a soft drink can flat, we will have to pull on the ends to return it to its original length. If we are not quite that ambitious, we can see in Figure 17.4 that after some "shakedown" period, cyclic loading will entail a load–deflection curve that forms some sort of a loop—a hysteresis loop—that may entail sign reversals in deflection as well as load. Hysteresis is beyond the scope of this discussion, except to say that each and every loop entails the dissipation of some energy. The amount of energy dissipated depends on the amount of difference between the loading and unloading portions of the hysteresis loop.

Since energy is dissipated in both hysteresis loops and viscous dampers, we should not be surprised to encounter the concept of hysteresis damping. It also goes by the name of structural damping. Structural damping poses more modeling difficulties than does viscous damping, but it can be appreciated that it entails some analysis (and perhaps simplification) of the hysteresis loop that derives from the load–deflection curve.

Another damping mechanism is Coulomb, or friction, damping. As the name implies, the mechanism involves energy dissipation through the friction between two elements that slide against one another. While the friction force may have a constant magnitude, it changes direction during each cycle, because it always opposes the relative velocity. This renders the mechanical system nonlinear, and will not be discussed further here. Suffice it to say that while friction is very important in the discussion of tire behavior and post-crash vehicle run-outs, it is generally not to be found in the consideration of crush energy.

Restitution

In general usage, "restitution" means restoring something to its original state. In physics, full restitution occurs when the relative exit velocity, at the point of contact and normal to the contact surface between two objects, is equal (and opposite) to the relative impact velocity at that point. No restitution occurs when there is no relative exit velocity. As usual, the physics definition is more complicated. One might say that, in a sense, full restitution means that all of the original velocity has been restored.

We recognize that even in the ideal world of smooth, frictionless spheres and such, the colliding bodies deform during the impact, and that a non-zero relative exit velocity is generated when the original shapes of the collision partners are restored, at least in part. So there must be some connection between the load–deflection curve and restitution. But what is it?

We start with the simplest example: a vehicle rolling freely into a fixed, rigid barrier. Its kinetic energy at the instant of impact is

$$KE_i = \frac{1}{2}mV_i^2 \tag{17.4}$$

As the structure crushes, energy is absorbed, and the vehicle speed is reduced. Finally (in the blink of an eye), the vehicle arrives at a state of zero velocity, and hence zero kinetic energy. Crush is at its peak. All of the kinetic energy has been converted into crush energy, which, as we have seen, is the area under the loading portion of the load–deflection curve for all of the structure that participates in the crush. Mathematically, we can write

$$KE_i = CE_{absorbed} = Area_{loading} \qquad (17.5)$$

If the structure is to have some recovery, it continues to exert force on the barrier (in fact, the force may be at peak levels also) and starts to push the car away from the barrier face. In accordance with Newton's Second Law, the car starts to accelerate away from the barrier. This motion leads to a decrease in the compressive crush; the crush rate has gone from positive to negative. Depending on the shape of the unloading portion of the load–deflection curve, the compressive force exerted by the barrier decreases. The crush energy also decreases, in equality with the increasing kinetic energy.

Eventually (in another blink of an eye), the force decreases to zero. Beyond that point there is no further increase in the vehicle's kinetic energy, and hence no further decrease in its crush energy either. Instead of further decreasing the crush, the vehicle's motion serves to open a gap between it and the barrier. We say the vehicle has rebounded.

At this point, the vehicle possesses a final kinetic energy equal to the recovered crush energy, which is the area under the unloading portion of the load–deflection curve. Mathematically, we have

$$KE_f = CE_{recovered} = Area_{unloading} \qquad (17.6)$$

The final, or rebound, velocity is

$$V_f = -\sqrt{\frac{2KE_f}{m}} \qquad (17.7)$$

where the minus sign indicates that motion is in the opposite direction. Using Equations 17.4 through 17.7 in the physics definition for the coefficient of restitution ε, we have

$$\varepsilon = -\frac{V_f}{V_i} = \sqrt{\frac{Area_{unloading}}{Area_{loading}}} \qquad (17.8)$$

It is important to note that since Equations 17.5 and 17.6 apply only to a particular vehicle into a rigid, fixed barrier, Equation 17.8, as derived, is valid only for that condition as well. However, the equation can be extended to a collision with deformable fixed barrier by recognizing that the vehicle's structure would be placed in series with the deformable barrier face. Suppose these two structural elements are considered as one, with a combined load–deflection characteristic (i.e., necessarily different from that of the vehicle structure alone). Equation 17.8 is still valid as long as the loading and unloading areas refer to the combined load–deflection characteristic. In all likelihood, the resulting restitution coefficient will be different from that of the vehicle structure alone being forced to absorb all the energy.

The logic can be extended to the impact of two deformable structures (vehicles). It just has to be remembered if a restitution coefficient is to be estimated, the colliding structures have to be analyzed like two springs in series combined into a single load spring, and treated as a single load path having a (combined) load–deflection curve. The bottom line is that the restitution coefficient pertains to the specific collision, and the specific structures participating in it. It does *not* pertain to either individual structure. A particular vehicle will experience a different restitution coefficient for every different structure with which it collides.

The velocity change into a fixed, rigid barrier is given by

$$\Delta V = V_f - V_i = -V_i(1 + \varepsilon) \tag{17.9}$$

where the minus sign means that the vehicle velocity decreases. An important fact of life is that in a barrier collision, the ΔV is not equal to the impact speed, but is in fact augmented by the presence of restitution. The basic source of restitution is an unloading curve that does not drop vertically to zero, but has area under it.

We will see in Equation 19.12 that Equation 17.9 can be extended to uniaxial vehicle-to-vehicle collisions if ΔV is replaced by the combined magnitude of the two ΔVs and V_i is replaced by the closing velocity.

As might be expected from the way the restitution coefficient depends on the collision, not just the vehicle structure, there is a considerable amount of scatter in observed restitution coefficients. In barrier impacts, values have been observed from about 0.3–0.7 for ΔVs below 5 mph,[4] and 0.3–0.4 for ΔVs between 8 and 27 mph.[5] Other barrier impacts with average crush values <19 in. involved restitution coefficients between about 0.13 and 0.33.[6] Vehicle-to-vehicle impacts with closing speeds between 2 and 13 mph have produced restitution coefficients of 0.22–0.62,[7] and those with closing speeds between 7 and 20 mph entailed restitution coefficients of 0.07–0.62.[5] Restitution coefficients observed in other vehicle-to-vehicle front-to-rear tests at closing speeds from 3 to 15 mph ranged from 0.28 to 0.52.[8]

These data show that, in general, restitution coefficients trend downward as the closing speed increases. Since bumpers are designed to resist damage (and thus be resilient) at low speeds, this is not surprising. ΔV is more influenced by the restitution coefficient at low speeds and thus assumes more importance in low-speed accident reconstructions. Hence, emphasis is laid on low speeds in the literature dealing with restitution coefficients.

At higher speeds, permanent damage is predominant, with a lower fraction of the crush energy being recovered upon rebound. In many of the severe crashes that dominate the NHTSA crash database, the restitution coefficients can be calculated by examining the test reports or the accelerometer data, using the methods of Chapters 11 through 13, subject to the considerations of accelerometer locations discussed in Chapters 1, 13, and 21. In general, barrier crashes at 30–35 mph result in restitution coefficients of about 0.1–0.2.

Structural Dynamics

Plenty has already been said about the complexity of the structural dynamics involved in vehicle crashes. However, important insights can be obtained from analyzing the rigid barrier impact of a simple one-mass model with a single load path having a linear load–deflection curve. Such a model may not be valid for detailed studies or for reconstruction, but it will demonstrate the basic techniques used, and give some idea of the order of magnitude of the various parameters.

One might think of replacing the curve of Figure 17.1 with a straight line passing through the origin, such that in a given crash, the force, deflection, and crush energy are replicated. However, the line to be drawn has only one parameter to adjust, so it will not be possible to match all three criteria. That said, the gross crash behavior of many vehicle structures has been found to be explainable by a linear model to a surprising degree. Such models lie at the very foundation of crush energy assessment, as will be discussed at length in Chapter 21.

Plowing ahead with a linear structure, the model then looks a lot like the usual spring–mass system, or harmonic oscillator, as seen in Figure 17.6, except that the spring is a compression-only device. The single mass represents the vehicle, its occupants and its cargo. The load–deflection "curve" represents all of the crushable structure, and is just a straight line having a slope—or stiffness—of k, which is in consistent units (e.g., lb/ft). The mass moves in the positive or negative X-direction without friction (no tire forces). At time zero (initial contact), its velocity is V_i in the positive X-direction and its displacement is zero. Since the barrier does not move, the crush of the structure is equal to the displacement X (again in consistent units—ft, for example) of the mass. The time derivative (i.e., the rate of change with respect to time) of X is the velocity, denoted by $\dot{X}$, measured in consistent units (e.g., ft/sec). The time derivative of the velocity is the acceleration, denoted by $\ddot{X}$,

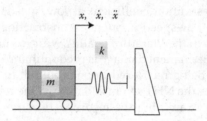

FIGURE 17.6
(See color insert.) Simple barrier crash model.

and measured in consistent units (e.g., ft/sec²). The acceleration of gravity is 32.2 ft/sec² in the example units.

Since this simple model has only one degree of freedom (one mass having motion described completely by one variable), there is only one differential equation governing that motion:

$$m\ddot{X} + kX = 0 \tag{17.10}$$

The solution to this differential equation is obtained directly, and is

$$X = \frac{V_i}{\omega}\sin\omega t \tag{17.11}$$

$$\dot{X} = V_i\cos\omega t \tag{17.12}$$

$$\ddot{X} = -V_i\omega\sin\omega t \tag{17.13}$$

where the quantity ω is given by

$$\omega = \sqrt{\frac{k}{m}} = \sqrt{\frac{gk}{W}} \tag{17.14}$$

and has units sec⁻¹, or rad/sec, like the roll rates discussed in Chapter 16. It is sometimes called the "circular frequency."

Equations 17.10 through 17.13 hold whenever the structure is in contact with the barrier. When the structure disengages from the barrier, the acceleration has returned to zero. After that, the acceleration remains at zero, and the equations no longer apply. The time at which this happens is the total duration of the impact T, which is found from Equation 17.13 by observing when the sine function returns to zero (when its argument ωt is π). Thus,

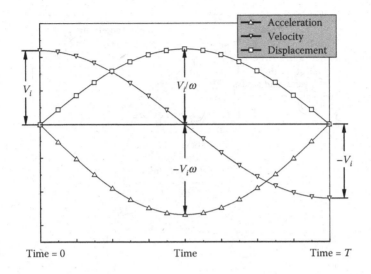

FIGURE 17.7
Barrier crash time response.

$$\omega = \sqrt{\frac{k}{m}} = \sqrt{\frac{gk}{W}} \qquad (17.15)$$

We see that the crash duration is entirely independent of the initial velocity V_i, and is a function of the vehicle mass and stiffness only.

When the solution is plotted as a function of time, we get graphs of the acceleration, velocity, and displacement, as shown in Figure 17.7. The most useful of these curves for analytical purposes is the velocity–time plot, often called the V–t curve for short. Its utility comes from the fact that at any time t, the slope of the V–t curve is the acceleration, and for any interval of time between t_1 and t_2, the area under that portion of the curve is equal to the amount of displacement that occurs in the time interval (see Figure 17.8). Thus, all three quantities are contained in the V–t curve. As we have seen in actual crashes, the plot of velocity is much smoother than that of acceleration, which is the quantity actually measured and therefore always plotted, whereas the V–t curve requires integration, which is not always carried out by the reporting agency.

It should not escape notice that with this crash model, the load–deflection curve is linearly elastic, and the velocity goes from V_i and the beginning to $-V_i$ at the end. Thus, ΔV is $-2V_i$. Since the loading and unloading curves are the same, the areas under them are also the same. We could have anticipated from Equation 17.8 that the coefficient of restitution would be unity, as is the case in a perfectly elastic collision. Equation 17.9 would have warned us of a doubling of the ΔV.

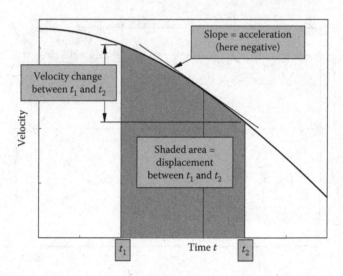

FIGURE 17.8
Uses of the velocity–time curve.

It might be tempting to argue a longer duration would be beneficial for occupant protection (which is beyond the scope of this book). Consider, however, that occupants cannot be rigidly coupled to the vehicle. As a result, the occupants continue to move in the direction of travel, while the vehicle is starting its change in velocity. Thus, a "velocity debt" is built up between the occupant and the vehicle, and that debt will have to be paid off when the occupant contacts the restraint system or compartment interior. For unrestrained occupants, this can be after the vehicle has reached maximum deformation. Thus, it is entirely possible for the vehicle to be in rebound when the second collision occurs. This rebound can actually add to the occupant's ΔV.

Vehicles are sometimes criticized because the crush is deemed to be excessive and thus makes them "look bad" after a crash. They may indeed "look bad," but acting to make the structure bounce back more—and recover more of its original shape—is going to tend to increase the ΔV, and possibly raise the injury exposure.

On the other extreme from the purely elastic crash is the perfectly plastic collision. In that case, the loading portion of the load–deflection curve can be linear (and in fact identical to the one we just analyzed), but the unloading curve is different. The instant the vehicle reaches zero velocity in a barrier crash, the structure does not try to recover and does not push on the barrier after maximum crush is reached. Instead, the load–deflection curve drops vertically to zero load. The vehicle stays stopped, nestled up against the barrier face.

In a perfectly plastic barrier impact, the crash is over as soon as the velocity reaches zero. In Equation 17.12, the cosine does just that when its argument ωt reaches $\pi/2$. Thus, the duration of the crash of this kind of structure is

$$T = \frac{\pi}{2\omega} = \frac{\pi}{2}\sqrt{\frac{m}{k}} = \frac{\pi}{2}\sqrt{\frac{W}{gk}} \tag{17.16}$$

which is half the duration of a purely elastic collision. Thus, a graph of the differential equation solution would entail only the left half of Figure 17.7; the acceleration would go vertically to zero from its minimum value of $-V_i\omega$ and stay that way, and the displacement would remain constant forever at its peak of V_i/ω.

For structures having linear loading portions of their load–deflection curves, be they perfectly elastic, perfectly plastic, or something in between, the peak crush, acceleration, and force can be obtained from Equations 17.11 and 17.13 as

$$\text{Max crush} = V_i/\omega = V_i\sqrt{\frac{m}{k}} = V_i\sqrt{\frac{W}{gk}} \tag{17.17}$$

$$\text{Min acceleration} = -V_i\omega = -V_i\sqrt{\frac{k}{m}} = V_i\sqrt{\frac{gk}{W}} \tag{17.18}$$

$$\text{Max force} = -mV_i\omega = V_i\sqrt{km} = V_i\sqrt{\frac{kW}{g}} \tag{17.19}$$

where W is the weight of the vehicle $= mg$.

It is interesting to note that the maximum crush is a linear function of the impact velocity. A linear relationship has also been observed to hold for the residual crush (i.e., that remaining after the crash) for many different kinds of vehicles, almost all of which have nonlinear load–deflection curves. This observation was used by Kenneth Campbell in a landmark paper[9] in 1974 to analyze the crush energy involved in crashing—a subject that is discussed further in Chapter 21.

While a linear load–deflection characteristic may be an oversimplification of the actual behavior of an automotive structure, it turns out that the overall results from such a model are, in many cases, not a bad representation of what happens.

Restitution Revisited

As discussed previously, actual restitution coefficients are somewhere between 0.0 and 1.0 because of the way the unloading portion of the load–deflection curve behaves. An analytical model of partial restitution that retains linear loading and unloading segments is shown in Figure 17.9.

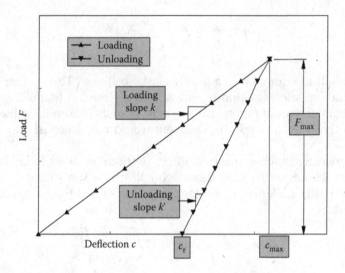

FIGURE 17.9
Simplified load–deflection curve for partial restitution.

To formulate an analytical solution that includes unloading, the differential equation has to be broken up into two parts: loading and unloading. The loading portion will have the same solution, but will be valid only up to F_{max}. For unloading, a second solution will have to be obtained, such that it matches up with the loading segment at F_{max}. This is facilitated by resetting the clock, so to speak, at F_{max}, then shifting the time base after the solution is obtained.

We are not going to that trouble here, however. Rather, we will consider the crash from the view point of an energy balance between the initial contact, maximum deformation, and separation from the barrier. That was the principle behind Equation 17.8. For this situation, we see that the area under the loading portion of the load–deflection curve is $\text{Area}_{loading} = F_{max}^2/2k$. Similarly, the area under the unloading portion is $\text{Area}_{unloading} = F_{max}^2/2k'$. Thus, the restitution coefficient can be directly obtained from Equation 17.8 as

$$\varepsilon = \sqrt{\frac{F_{max}^2/2k'}{F_{max}^2/2k}} = \sqrt{\frac{k}{k'}} \qquad (17.20)$$

Since the unloading stiffness k' is generally much greater than the loading stiffness k, the restitution coefficient is less than unity. The higher the unloading stiffness is, the lower the restitution will be, and the less the rebound.

The maximum crush that occurs during the impact, C_{max}, is called the dynamic crush. Since there is no such thing as a perfectly plastic crash, the dynamic crush will be larger than the residual crush c_r measured after the crash. If the dynamic crush were integrated twice, the dynamic crush could be plotted against the displacement. One might think that c_r could be read off such a plot, but it is almost never the case that the final integrated deflection matches the measured residual crush. For one thing, actual force–deflection curves do not display the sharp corner drawn in Figure 17.9; often the return to zero force is subtle and hard to read off a plot. That makes it hard to distinguish between structural restoration and vehicle rigid-body motions. For another thing, it appears that many structures continue to recover after the instrumentation record ends.

Equation 17.20 suggests that if a restitution coefficient of 1/10 (10%) is observed, the unloading stiffness must be 100 times that of the loading stiffness. Moreover, Equation 17.8 shows us that the recovered energy must be only 1/100 (1%) of the loading energy. A restitution coefficient of 1/5 (20%) implies a recovered energy fraction of 1/25 (4%). So even though restitution can (and does) contribute significantly to ΔV, its influence on crush energy can safely be ignored in a general discussion of the topic. This effect is explored in more detail in Chapter 19.

Small Car Barrier Crashes

Having in hand a simple model and solution for crash dynamics, it is an interesting exercise to apply values typical of various kinds of automobiles, and see what the general results are. As an example, consider a small car weighing 2000 lb. A typical value of stiffness for the front structure of such an automobile is 2000 lb/in., or 24,000 lb/ft. Without restitution and with this stiffness and weight, the crash duration is 0.080 s, or 80 msec, independent of the impact velocity. In a barrier impact of 30 mph, we see from Equation 17.17 that the predicted maximum crush is 2.24 ft, or 26.9 in. The predicted maximum force applied to the crushing structure is 53,700 lb from Equation 17.19, and therefore the maximum deceleration is 26.7 G. It should be kept in mind that actual crashes produce much more jagged acceleration–time histories whereas the model produces a smooth sine wave, so the actual peak acceleration in a crash test report will probably be much higher, and is subject to the filtering employed, as discussed in Chapter 13. In any case, the model does a good job of predicting displacements, velocities, and average values of force and acceleration.

Large Car Barrier Crashes

Repeating the above analysis for a large car produces somewhat different results. A typical value of weight for a large car is 4000 lb, and a typical value of stiffness is 2600 lb/in., or 31,200 lb/ft. The predicted time to bring the vehicle to a stop against the barrier face is 0.099 sec, or 99 msec. Predicted maximum crush at this point (dynamic crush) is 33.3 in. in a 30 mph crash. Maximum force is 86,600 lb, which produces a maximum deceleration of 21.7 G.

Any restitution will cause the car to rebound away from the barrier. If the restitution coefficient is 0.1, the unloading slope is 100 times that of the loading stiffness, or 260,000 lb/in. The time to go down the slope from the maximum load to zero load is 1/10 the time it took to go up the loading slope, or about 0.010 s, or 10 msec, which extends the duration of the entire event from 99 to 109 msec. Because the unloading stiffness is 100 times that of the loading stiffness, the crush is reduced by 0.33 in., such that the residual crush measured when the crash is over 33.0 in. The vehicle moves away from the barrier at 1/10 the impact speed, so the ΔV in this 30 mph barrier crash is 33 mph.

Small Car/Large Car Comparisons

A comparison of the barrier crash experience in a large car and a small car points out some interesting differences. The dynamic crush in the large car is 33.3 in., whereas in a small car it is 26.9 in. This difference reflects, perhaps more than anything else, the physical size of the vehicles and the fact that the large car has more room in the front to crush and be used in absorbing energy. Of course, the force level in the large car is significantly higher than in the small car. However, the small car experiences higher peak decelerations. Again, this reflects the fact that it simply has to be stopped in a shorter distance. This shorter distance causes the event duration to be shorter for the small car.

Narrow Fixed Object Collisions

So far, we have looked at aligned frontal collisions into a barrier. In such impacts, the entire front structure of a vehicle is loaded, and loaded uniformly. This gives the crushing structure the best opportunity to generate forces and absorb energy (and in some cases to produce a high level of compartment deceleration). In many accidents, however, the impact is not "square on," is not centered on the front of the vehicle, and does not involve the entire front structure. Examples of such collisions might be impacts into poles, bridge abutments, walls at an angle, or other narrow objects. In such impacts, only a portion of the structure is exercised, while other portions may be only slightly damaged, if at all. In such cases, the entire brunt of energy absorption is carried by only a part of the structure, and quite different behavior ensues.

A simple means of modeling such impacts is to reduce the structural stiffness used in the analytical models to account for the limited structural involvement. Suppose, for example, that only half of the front structure is involved. One might argue that only half of the force is going to be produced for a given value of crush. Hence, one is justified in reducing the effective structural stiffness by 50%. In our example of a large car, the stiffness k would be reduced to 15,600 lb/ft. This reduced stiffness value causes the crush phase of the crash event to be extended to 140 msec. The application of the other equations developed previously shows that a maximum crush of 47.1 in. results (though only for a portion of the structure). The maximum structural deceleration is 15.3 G. As is so often the case in narrow fixed object collisions, the pulse has gotten much longer.

If we were only looking at compartment crash pulses, we might conclude that such impacts carry less injury potential than aligned frontal barrier crashes to, since we would find that the interior contact velocity is much reduced in these cases. However, the mechanism of the entire crash tends to be changed, and occupants may now be threatened by intrusion that might not have been present in the aligned frontal crash. The fact that we now have 47 in. of crush as compared to 33 in. is indicative of this threat. Moreover, all of the loads are being carried by only a part of the structure, which means that individual load paths can be much more severely loaded in this kind of impact. Such statements apply, for example, to door latches, and the threat of a door coming open increases the risk of ejection. There is also an increase in the risk of partial ejection and contact with objects outside the vehicle. Finally, there may be an increased risk of lower leg injuries, lower limb entrapment, and contact with structure that has now moved into the stroking space (interior room to move) needed by the occupant. It is rarely the case, though, that intrusion becomes so great as to cause a physical crushing of the occupant.

Vehicle-to-Vehicle Collisions

When one vehicle strikes another, both vehicles move and both vehicles usually crush. This simple observation tells us that the models developed previously for barrier collisions will probably not be applicable to collisions with moving, deformable crash partners. Instead, it is necessary to rethink the analysis from scratch.

As with the model of Figure 17.6, we will assume linear load–deflection relations for both vehicles; the analysis will take us up to the points of maximum crush (and force and acceleration), but will not address the subsequent structural unloading. The new model is shown in Figure 17.10.

The subscript 1 refers to vehicle 1, and 2 refers to vehicle 2. The vehicles have different masses and stiffnesses. Each vehicle moves only in the

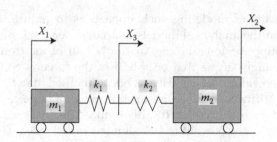

FIGURE 17.10
(**See color insert.**) Vehicle-to-vehicle crash model.

(positive or negative) X direction. A coordinate X_3 is introduced as the interface between the vehicles, so that the crush in each structure can be tracked individually. The crush in vehicle 1's structure is $(X_1 - X_3)$, positive in compression, and the crush in vehicle 2's structure, again positive in compression, is $(X_3 - X_2)$. We recognize that the two structures are in series; the load F is the same in each (because of Newton's Third Law), and the crush of the two structures combined is $(X_1 - X_2)$. Thus, to simplify the analysis, we can replace the two individual structures with a single combined structure having a stiffness k_T. To calculate this stiffness, we enforce the equality of forces

$$F = k_T (X_1 - X_2) = k_1 (X_1 - X_3) = k_2 (X_3 - X_2) \qquad (17.21)$$

Solving the right-hand Equation 17.21 for X_3 yields

$$X_3 = \frac{k_1 X_1 + k_2 X_2}{k_1 + k_2} \qquad (17.22)$$

Substitution into the load–deflection equation for vehicle 1, we find that

$$F = k_1 [X_1 - X_3] = k_1 \left[X_1 - \frac{k_1 X_1 + k_2 X_2}{k_1 + k_2} \right]$$

$$= \frac{k_1 k_2}{k_1 + k_2} (X_1 - X_2) \qquad (17.23)$$

Equating like terms of Equation 17.23 with the right-hand Equation 17.21, we find the desired stiffness k_T of the combined structure to be

$$k_T = \frac{k_1 k_2}{k_1 + k_2} \qquad (17.24)$$

This is the familiar formula for two springs in series.

Since the two masses move separately along the X-axis, the model has two degrees of freedom. Consequently, the motion of the two-mass system is governed by two differential equations, as follows:

$$m_1\ddot{X}_1 = -F = -k_T(X_1 - X_2) \tag{17.25}$$

$$m_2\ddot{X}_2 = +F = k_T(X_1 - X_2) \tag{17.26}$$

where the tire forces are assumed negligible, and where the signs on the compressive force F are a recognition that F opposes the motion of Mass 1, but augments the motion of Mass 2. The solution of these two simultaneous linear differential equations, in their present form, can entail some messy algebra, so it is useful to note that the right-hand sides of these equations suggest the variable substitution for the combined crush X_1–X_2:

$$\xi = X_1 - X_2 \tag{17.27}$$

Differentiating Equation 17.27 results in

$$\dot{\xi} = \dot{X}_1 - \dot{X}_2 \tag{17.28}$$

$$\ddot{\xi} = \ddot{X}_1 - \ddot{X}_2 \tag{17.29}$$

Equation 17.29 suggests that it can be used as a variable substitution if the two differential equations are subtracted. Therefore, dividing Equations 17.25 and 17.26 by the appropriate masses produces

$$\ddot{X}_1 = -\frac{k_T}{m_1}\xi \tag{17.30}$$

$$\ddot{X}_2 = \frac{k_T}{m_2}\xi \tag{17.31}$$

Now subtracting the two equations yields

$$\ddot{X}_1 - \ddot{X}_2 = \ddot{\xi} = -k_T\left[\frac{1}{m_1} + \frac{1}{m_2}\right]\xi \tag{17.32a}$$

or

$$\ddot{\xi} + k_T\left[\frac{m_1 + m_2}{m_1 m_2}\right]\xi = 0 \tag{17.32b}$$

This is an equation that has a well-known solution. The equation is homogeneous because there are no external forces acting on the two-vehicle system. The initial conditions for the solution of this equation are

At time $t = 0$, Relative deformation in the combined structure $= \xi = 0$ (17.32)

At time $t = 0$, the relative deformation rate in
the combined structure $= \dot{\xi} = V_{cl}$ (17.33)

where V_{cl} is the closing rate $(V_1–V_2)$ between the vehicles at impact. The solution to this initial-value problem is

$$\xi = \frac{V_{cl}}{\omega} \sin \omega t$$

$$\dot{\xi} = V_{cl} \cos \omega t$$ (17.34)

$$\ddot{\xi} = -V_{cl}\omega \sin \omega t$$

where

$$\omega^2 = k_T \left[\frac{m_1 + m_2}{m_1 m_2} \right]$$ (17.35)

Substitution for the combined stiffness k_T yields the expression for the circular frequency ω:

$$\omega = \sqrt{\frac{k_1 k_2}{k_1 + k_2} \frac{m_1 + m_2}{m_1 m_2}} = \sqrt{\frac{g k_1 k_2 (W_1 + W_2)}{(k_1 + k_2) W_1 W_2}}$$ (17.36)

This is the complete analog for the solution of the barrier crash problem in Equations 17.11 through 17.14. However, it is important to note that this ω is a function of the stiffnesses and the masses of both vehicles, and that is a significant difference. It relates directly to the duration of the crush phase of the event, just as it does in barrier crashes. The circular frequency is unsubscripted because it applies to both vehicles. Physically, this means that they both come to rest in unison, with respect to the system center of mass.

This solution is expressed only in the *relative*—not absolute—motions of the vehicles. So what happened to the absolute motion parameters X_1 and X_2? The mathematics suggested that the solution be formulated in terms of the relative displacements $(X_1 - X_2)$ and relative velocities. There is a second relationship that comes from adding the differential Equations 17.25 and 17.26, instead of subtracting them to obtain Equation 17.32. Summing these two equations leads to

$$m_1 \ddot{X}_1 + m_2 \ddot{X}_2 = 0 \qquad (17.37)$$

or

$$\ddot{X}_2 = -\frac{m_1}{m_2} \ddot{X}_1 = -\frac{W_1}{W_2} \ddot{X}_1 \qquad (17.38)$$

which is to say that the accelerations are related by the mass ratio of the vehicles. This is a consequence of Newton's Third Law, which was enforced when we assigned the same force F to both structures. Knowing $\ddot{\xi}$ from Equations 17.34 and using the definition in Equation 17.29 with Equation 17.38, we can solve for the two accelerations.

As for the individual displacements and velocities, we cannot find them from the initial-value problem because the collision is not tied to Mother Earth. After all, both vehicles could roll on frictionless wheels. All that is required is that Newton's Laws be enforced in an inertial reference frame (that moves without acceleration, either with zero displacement or constant linear velocity). Whereas the analysis of barrier or other fixed object crashes yields solutions for absolute variables because the objects are tied to Mother Earth (by definition), in a vehicle-to-vehicle collision it matters not whether the crash occurred in Paris, Texas or Paris, France. All that matters is what goes on within the two-vehicle system.

In that regard, once we know ξ, we know the crush force from $F = k_T \xi = k_1 c_1 = k_2 c_2$, where c_i is the crush in load path i. Thus,

$$c_1 = \frac{k_T}{k_1} \xi = \frac{k_2}{k_1 + k_2} \xi$$

$$(17.39)$$

$$c_2 = \frac{k_T}{k_2} \xi = \frac{k_1}{k_1 + k_2} \xi$$

Differentiating these equations, the crush rates can be found as

$$\dot{c}_1 = \frac{k_2}{k_1 + k_2} \dot{\xi}$$

$$(17.40)$$

$$\dot{c}_2 = \frac{k_1}{k_1 + k_2} \dot{\xi}$$

where $\dot{\xi}$ is the (instantaneous) relative velocity between the vehicles. We see that the crush and crush rate in one vehicle are proportional to the other vehicle's stiffness. The relative stiffnesses apportion the relative velocity into the two crush rates. We will see in Chapter 19 how the two masses apportion the initial closing velocity into the vehicle ΔVs in an analogous manner.

Because the movements X_1 and X_2 were both assumed positive in the X direction, the solution is completely symmetric in the subscripts 1 and 2. We should, however, be able to extract the barrier collision from these equations as a special case, by letting m_2 and k_2 go to infinity. Indeed, the formula for ω reduces to the same form we had previously. Since the barrier does not move, V_2 is zero, and the closing velocity V_{cl} is equal to V_1. Thus, the equations for displacement, velocity, and acceleration are identical.

On the contrary, if both vehicles have identical masses and identical stiffnesses, and equal and opposite velocities, we would expect the interface between the vehicles to be stationary before, during, and after the collision. In other words, the two car collision would be like two barrier crashes head-to-head. Indeed, if one substitutes $k_2 = k_1 = k$ and $m_2 = m_1 = m$, one finds that Equation 17.36, the expression for ω, again reduces to that used for the barrier collision. Since the fixed, rigid barrier does not move (by definition), $V_2 = 0$, $x_2 = 0$, $\xi = X_1$, and $V_{cl} = V_1$. This renders Equations 17.34 identical to Equations 17.11 through 17.13. Therefore, our physical intuition is correct: the predicted results are the same.

In fact, our solution applies for any uniaxial collision between two bodies, whether they are fixed or moving, rigid or (linearly) deformable. The collision could be front-to-front or front-to-rear. The solution could also apply to front-to-side impacts, except that in the side-struck vehicle there is no rigid separation between the deformable side structure and the compartment. Thus, side impacts are particularly complex and require a special analysis that is beyond the scope of this book. An illuminating discussion is offered in the 1984 paper by Strother et al.[10]

Large Car Hits Small Car

Now, we find out what happens when we set up a crash between the large car and the small car we have previously modeled. We configure the crash so that at the point when the small car starts to unload (at the end of the crush phase), it has accumulated the same ΔV as in the barrier crash (−30 mph, or −44 ft/sec). Call the small car Vehicle 1. Since the large car is twice as massive, its ΔV is half as much at that point: 15 mph, or 22 ft/sec. This implies a closing velocity at impact of 66 ft/sec.

Now, we tie the crash to Mother Earth. We set up the crash so as to minimize the post-crash run-outs; that is, balance the momenta of the two vehicles so that they are both at rest at the end of the crush phase (as the vehicles were in the barrier crashes). This was done in DOT Test 5683, for which we have looked at some results in Chapters 1 and 13. In our "test," Vehicle 1, the small car, is required to enter the crash at 44 ft/sec; the large car is impacted at half the speed, or 22 ft/sec.

The difference between the vehicle-to-vehicle crash and separate barrier impacts can be found in the circular frequency ω. Applying Equation 17.36, we find a value of ω of 18.1/sec, as compared to 19.7/sec for the impact in which the small car hit the barrier. From Equation 17.16, we find that the duration for both vehicles to reach maximum force, maximum acceleration, and maximum crush is 87 msec, as compared to 80 msec for the small car barrier impact. We may see that this value of 87 msec is some kind of average of the values seen in barrier collisions.

At maximum total crush, $\omega t = \pi/2$, and the sine function in Equations 17.34 has a value of unity. There the maximum total crush is given by

$$\xi_{max} = \frac{V_{cl}}{\omega} = 3.65 \text{ ft} \tag{17.41}$$

From Equation 17.39, the crush in Vehicle 1 is found to be 2.06 ft, or 24.7 in., which is about 8% less than in the barrier collision. On the other hand, the large car experiences a maximum crush of 19.0 in., whereas in a 15 mph barrier impact it is 16.7 in., from Equation 17.17. We conclude from this that the crash experienced by each vehicle is not directly related to a barrier crash at the same ΔV; rather, the structural characteristics of the two vehicles act together to redistribute the crush as compared to a barrier test. In this instance, the stiffer small car has crushed less, and forced the large car structure to accept a larger share of the crush.

The remaining crash parameters can be ascertained from the relationships developed previously. In particular, the small car experiences a peak acceleration of 24.7 G, which is less than the value of 26.7 G in the barrier crash. Similarly, the maximum force experienced by the small car is 49,460 lb compared to 53,720 lb in the barrier test. For these particular vehicles, the small car has been able to use some of the crush of the large car to its advantage. For the large car, the maximum crush is 14.6 in., the peak deceleration is 12.4 G, and the maximum force is 49,460 lb (as one would expect).

Once the force is known, one can calculate the energy absorbed during the crush phase of the crash from the following:

$$CE = \frac{F^2}{2k} \tag{17.42}$$

As discussed in connection with Equation 17.20, a 10% restitution coefficient will result in only a 1% energy recovery, so Equation 17.42 provides a good first approximation of the net crush energy absorbed. The application of this equation for crush energy stored during the crush phase shows that the small car absorbs 60,100 ft-lb of energy in a barrier crash. In a car-to-car crash at the same ΔV, however, the crush energy absorbed is 51,000 ft-lb. To emphasize the point, the ΔVs for the small car were set up to be equal in the barrier crash and in the car-to-car collision. However, the crush energies are decidedly not equal. Furthermore, the duration of the crush event is not equal, being 0.080 s in the barrier crash and 0.087 s in the car-to-car crash.

Barrier Equivalence

As has been seen in our example crashes, a barrier impact cannot be said to be "equivalent" to a vehicle-to-vehicle collision, because of the additional role that structural stiffnesses play in the latter. The crush energies are different, the amount of crush is different, and the duration of the event is different. Of course, we could adjust the speed of the barrier test so that the crush energies were equal. This would be some sort of "energy equivalent speed" for a barrier test. In fact, that very term is defined in Chapter 21. In such a test, however, the velocity changes would not be equal, and one would be hard pressed to argue that this collision is equivalent to the vehicle-to-vehicle impact either. In any case, the duration of the crush phase of the event is independent of speed, so no matter what speed is chosen, the crush phase duration will in general always be different for the vehicle-to-vehicle crash than the barrier.

The reason for the inherent difference between the two crash types is the effect of the other vehicle being hit. One can have true equivalence between the two crash test modes only if the entire velocity time histories are the same. For this to occur, the ΔVs must be the same, and the circular frequencies must also be equal. In this regard, we must note the dependence of ω on the stiffness and mass of both vehicles.

From this, one can conclude that there is no such thing as "barrier equivalence." This statement can be made even without considerations of such further complicating factors as underride/override, the effect of vehicle architecture, and so on.

Nevertheless, the auto safety literature is replete with references to "barrier equivalent velocity." It is a term bandied about with carelessness, meaning different things to different people. Sometimes it means equal ΔV, sometimes it means equal crush energy without restitution, sometimes it includes restitution, and so on. What is one to do? If someone insists on using the term, every effort should be made to elicit a precise definition. Meanwhile, the term should be banished from one's own vocabulary.

Load–Deflection Curves from Crash Tests

Now that we have looked at a load–deflection curve for an idealized (linear) model, what does an actual curve look like? In Chapters 11 and 13, we discussed the nature and use of various acceleration data channels for DOT Test 5404, a 35.0 mph barrier impact of a 2005 Toyota Corolla. In Chapter 14, we discussed the analysis of load cell barrier data for that test, and presented a force–deflection curve for the front structure. Obtained by summing the 36 load cell data traces and filtering the results to CFC 60, it is presented here as a solid line in Figure 17.11.

Obviously, one would get a different curve—one much more jagged—if the load cell data were filtered to a higher channel class. There are still ups and downs at CFC 60. However, a straight line be fitted through the peaks and valleys, and the results would look generically similar to the idealization of Figure 17.9.

Note that to generate this curve, we must use a barrier crash, not a vehicle-to-vehicle crash such as DOT 5683 mentioned above. This is because in a vehicle-to-vehicle crash test, we have no way of knowing where the interface lies between the vehicles—the counterpart of X_3 in our analysis—is located. Therefore, we would not have a way of determining how the total crush is apportioned between the vehicles.

However, there is more than one way to skin a cat—or obtain a force–deflection curve from a barrier test. Another way is to multiply the acceleration in Gs by the test weight of the vehicle, using a minus sign to reflect

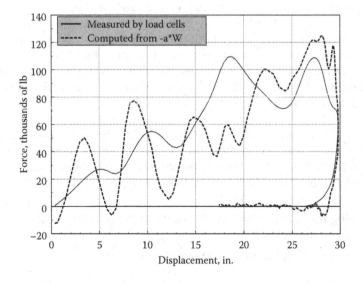

FIGURE 17.11
Load–deflection curve for test DOT 5404.

forces being positive in compression. The result is shown in the dashed line of Figure 17.11. How can the barrier forces be tensile (negative) in places? And why are the curves different from each other?

When encountering a load–deflection plot, the presence of noticeable negative forces is a good clue that the plot was derived from acceleration ($F = ma$) rather than barrier force measurements. Recall that in Chapter 13, the difference in accelerations observed in various parts of the vehicle was discussed. The presence of various masses, such as the engine and the occupants, and their effects on measured acceleration, was also discussed. Those and perhaps other effects render the use of accelerations to generate a load–deflection curve questionable at best, although it is not uncommon to encounter such plots. Preferred practice is to use barrier force, or not attempt a load–deflection plot at all. Sometimes, one finds a plot of acceleration vs. displacement.

Measures of Crash Severity

What is a severe crash? A moderately severe crash? Severity could be measured in a number of ways, such as the amount of repair cost, the disruption of traffic, and so on. However, everything pales in comparison with threat to life and limb. Therefore, almost all discussions on crash severity focus on the exposure of a vehicle's occupants to injury or fatality. Of all the experts analyzing a crash, the reconstructionist is the one who is most often relied upon for an assessment of crash severity.

For a very long time, the velocity change ΔV has been widely used as a severity metric. A 1976 paper by Kahane et al.[11] presented plans for the National Crash Severity Study (NCSS). In that paper, the very question of what severity metric to use was discussed. Even at that time, ΔV was "one of the more frequently used measures of collision severity." Other variables considered were: pulse width, peak acceleration, average acceleration, onset rate, and "pulse shape factor." ΔV was selected because it

a. "Is relatively easy and inexpensive to obtain through reconstruction of accident events and therefore can be used on a large numbers of accidents. The estimates can be obtained from vehicle damage as well as from the complete data collected at the accident site.

b. Is a meaningful quantitative characteristic of the acceleration trace during the collision phase. ΔV is a vector thus having a direction as well as a magnitude.

c. Has a meaning that can be understood by users."

The authors did note that ΔV could "be applied uniformly to collisions of relatively short duration—say less than approximately 200 milliseconds."

Thus, we have seen ΔV as the ubiquitous measure of the exposure to serious injury or fatality in frontal and rear crashes. In more recent times however, advanced restraint technology and much improved belt usage rates in the United States have significantly lowered fatality rates (and probably serious injury rates too). Perhaps not coincidentally, the Insurance Institute for Highway Safety (IIHS) and others have placed greater emphasis on structural performance—especially intrusion—as a measure of injury threat. Thus vehicles have become more intrusion resistant and, according to some researchers, stiffer as well. But intrusion is a measure of structural performance, not crash severity.

In side impacts, the story is somewhat different. When it is a door that is struck, it is not reasonable to expect a 100 lb door, connected with hinges and a latch, to keep a 4000 lb striking vehicle out of the compartment. The same logic holds for fixed object crashes. Intrusion will occur, and a near side occupant will be affected by it. In fact, the interior collision between the occupant and the intruding door will not occur at the ΔV because it will happen before the vehicle reaches its ΔV. Instead, the collision will occur at something closer to the closing velocity. It can thus be argued that for a side impact with a near-side occupant, the closing velocity is probably a better measure of crash severity. Restraint systems, such as side air bags and side curtain air bags, tend to be more effective at reducing the injury than are side structure improvements.

In rollovers, we have already seen that ΔV is not the measure of crash severity. Rather, one can cite roll initiation speed, roll distance, and number of rolls as more appropriate metrics, all of which should be addressed by the reconstructionist. Some have fingered roof deformation as a measure of injury and fatality exposure; others have argued that roof deformation is a consequence of severity, not a cause of injury. Without getting into the debate, suffice it to say that as in other crash modes, roof crush is a measure of structural performance, not crash severity. Indeed, the requirements for structural stiffness of the roof have been raised, and greenhouse stiffness has increased.

References

1. Kamal, M.M., Analysis and simulation of vehicle to barrier impact, SAE Paper 700414, SAE International, 1970.
2. Warner, C.Y., Allsop, D.L., and Germane, G.J., A repeated-crash test technique for assessment of structural impact behavior, SAE Paper 860208, SAE International, 1986.
3. Radwan, R.A. and Hollowell, W.T., System identification of vehicle structures in crash loading environments, SAE Paper 900415, SAE International, 1990.
4. King, D.J., Siegmund, G.P., and Bailey, M.N., Automobile bumper behavior in low-speed impacts, SAE Paper 930211, SAE International, 1993.

5. Tanner, C.B., Wiechel, J.F., Bixel, R.A., Cheng, P.H., Guenther, D.A., and Cassidy, M.P., Coefficients of restitution for low and moderate speed impacts with non-standard impact configurations, SAE Paper 2001-01-0891, SAE International, 2001.
6. Lawrence, J., Rix, R., Ho, A., King, D.J., and D'Addario, P., Front and rear car crush coefficients for energy calculations, SAE Paper 2010-01-0069, SAE International, 2010.
7. Cipriani, A.L., Bayan, F.P., Woodhouse, M.L., Cornetto, A.D., Dalton, A.P., Tanner, C.B., Timbario, T.A., and Deyerl, E.S., Low speed collinear impact severity: A comparison between full scale testing and analytical prediction tools with restitution analysis, SAE Paper 2002-01-0540, SAE International, 2002.
8. Rich, A., Wright, B., and Fish-Rich, M.L., E.D.R. delta-V reliability and restitution values for six low and moderate speed collinear central crash tests, *Accident Reconstruction Journal*, 20(6), Victor Craig, November/December 2010.
9. Campbell, K.L., Energy basis for collision severity, SAE Paper 740565, SAE International, 1974.
10. Strother, C.E., Smith, G.C., James, M.B., and Warner, C.Y., Injury and intrusion in side impacts and rollovers, SAE Paper 840403, SAE International, 1984.
11. Kahane, C.J., Smith, R.A., and Tharpe, K.J., The National Crash Severity Study, *Proceedings, 6th Internal Technical Conference on Experimental Safety Vehicles*, U.S. Department of Transportation, National Highway Traffic Safety Administration, Washington, DC, 1976.

18

Impact Mechanics

Crash Phase Duration

Typically, the crash phase is thought of as having a distinct duration: from the time the undeformed vehicles first touch, to the time when they cease to be in contact with one another. The former is a little easier to define and detect than the latter, which can be appreciated by analyzing crash test data. (Even with contact switches in the crush zones, it may be difficult to tell just when the vehicles separate. We can generally expect various locations in the contact surfaces to separate at various points in time, particularly if the vehicles exit the crash with some yaw velocity relative to one another.) Watching high-speed video coverage may not be that edifying. Often the crash test analyst must be content with defining a certain resultant acceleration below which a vehicle is deemed to be acted on by tire forces only, and not contact forces from the collision partner. After that time, the analyst deems the vehicles to be separated.

Generally, the duration does not depend very much on the speed of the impact. An understanding of the reason why may be gained by considering a simple harmonic oscillator (a lumped mass attached to a weightless spring that has a linear force–deflection characteristic), in which the system comes into contact with a rigid, infinitely massive barrier, spring end first, at some defined speed. (One might argue, with some correctness, that a linear spring is not a very good model of a vehicle structure, but on the other hand such an assumption by Campbell[1] in 1974 forms the basis of the vast majority of crush energy analysis methods used today.) When one solves the differential equation of motion for this system, one finds that the crash duration (when the displacement, the force, and the acceleration return to zero) is the time for a sine wave to reach half a cycle. This leads to a simple expression for the duration:

$$t_d = \pi\sqrt{\frac{m}{k}}$$

(18.1)

where m is the vehicle mass and k is its stiffness (force per unit crush), and where the impact speed is entirely absent from the equation. (Of course, that velocity may come into the picture in an indirect way if the structure is nonlinear—if the stiffness varies with the crush.) Anyway, for a structure with a defined stiffness, we can see the inverse variation of duration with stiffness. This is a reminder that in making judgments about duration, we must keep in mind that the way the structures interact affects their combined stiffness. One may find a crash phase duration as short as 100 ms in a square-on full frontal barrier crash, and over twice that long if not all the structure engages the barrier, which has the effect of reducing the overall structural stiffness. Similarly, in a vehicle-to-vehicle collision, the duration is also increased because the two structures are smashed together in series (end-to-end), and the stiffness of two structures in series is less than that of one of them into a barrier. The duration is extended even more if two structures are only partially engaged with each other. As long as the structural deformation mode is similar to that in which its stiffness was measured in the first place, speed does not affect the impact duration.

Degrees of Freedom

In general, the position of a solid body (i.e., a body that maintains its form) can be completely described by six parameters: three coordinates by which to specify the location of the center of mass, and three angles (roll, pitch, and yaw) to describe its orientation about the center of mass. We say that the body has six degrees of freedom. Generally, Cartesian coordinates (X, Y, and Z) are used to locate the center of mass. The process of time-wise differentiation produces six components of velocity and acceleration (three translational and three rotational). If one views a vehicle as a system of sprung and unsprung masses (such as the body and the wheels), the sprung mass still has six degrees of freedom, even if the vehicle is operating on a planar surface.

For the study of coplanar collision mechanics, such refinements are not considered. Each vehicle is considered to be a single entity in which all its motion is describable by the position and orientation of its center of mass relative to the horizontal X- and Y-axis. Elevation changes (Z) during the collision are ignored, as are pitch and roll. Therefore, at the time of initial contact each vehicle possesses three velocity quantities: the yaw rate and the translational velocity (having X and Y components, or equivalently, magnitude and direction). Thus, in a two-vehicle crash, there are a total of six unknown impact quantities: three for each vehicle.

Because the center of mass is above the plane in which the tire forces act, pitch and roll moments cause a redistribution of vertical forces at the wheels, even if pitch and roll angles are ignored. These "weight transfer" effects could be incorporated into the longitudinal and lateral force calculations. It is worth noting, however, that the most widely used simulation programs SMAC[2] and EDSMAC[3] do not incorporate such effects.

Mass, Moment of Inertia, Impulse, and Momentum

Momentum is defined as the product of mass and velocity, as follows:

$$\vec{L} = m\vec{V} \tag{18.2}$$

where the over arrows indicate that both $\vec{L}$ and $\vec{V}$ are vector quantities. Mass is a scalar quantity and is independent of the (Newtonian) coordinate system in use. Impulse is the time integral of force; thus,

$$\vec{p} = \int \vec{F}dt \tag{18.3}$$

Mass is the measure of material in a particle or body, and is the constant of proportionality that connects force and acceleration, as discussed in Chapter 1. It is the resistance to acceleration. When particles are connected together to form a body, rotation of that body, and its resistance to rotational acceleration, are subject to Newton's Laws. The resistance to rotational acceleration is termed a body's mass moment of inertia. It plays the same role that mass does in linear motions. The mass moment of inertia is the constant of proportionality between torque and acceleration. Of course, torque is defined as being in the direction of a particular axis, and so is angular acceleration. Thus, mass moment of inertia is defined for a particular axis. For a three-dimensional vehicle, there are three axes to consider: roll (about the x-axis), pitch (about the y-axis), and yaw (about the z-axis). Thus there are three mass moments of inertia: roll, pitch, and yaw. As with any irregular-shaped body, these generally have three different magnitudes. Usually, the roll moment of inertia is considerably smaller than the pitch and yaw moments of inertia. If a location in the body is measured by its distance r from an axis through the center of mass, then the mass moment of inertia about that axis is calculated by the volume integral

$$I = \iiint \rho^2 dm \tag{18.4}$$

or

$$I = mk^2 \tag{18.5}$$

where k is the "radius of gyration," understood to apply only to that particular axis. A study conducted by this author showed that as a rule of thumb, for many vehicles, the yaw radius of gyration is approximately 49% of the wheelbase. That same study also showed that treating a vehicle as a rectangular prism (even though it has a nonuniform density) allows a good approximation of the yaw moment of inertia to be obtained as

$$I \approx \frac{0.92}{12} m \left[(OAL)^2 + (OAW)^2 \right]$$

(18.6)

where *OAL* is the overall length, *OAW* is the overall width, *m* is the mass, *I* is the yaw moment of inertia, and all four quantities are expressed in consistent units (e.g., slugs and feet). Of course, moments of inertia have been measured for many vehicles and although there may be measurement issues, the measurements can probably be used in preference to the above estimate.

General Principles of Impulse–Momentum-Based Impact Mechanics

By "impulse–momentum based," we mean the application of Newton's Second Law, and only that principle, to the motion of vehicles involved in a crash. (Energy conservation is not utilized to obtain a solution.) As we discussed in Chapter 1, Newton's Second Law (change in momentum equals impulse) takes the form

$$d(m\vec{V}) = \vec{F}dt$$

(18.7)

where $\vec{F}$ is the sum of the external forces acting on the system. The impulse $\vec{F}dt$ is a vector, which is seen to be equal to (and therefore collinear with) the change in momentum.

Different formulations of impulse–momentum-based impact mechanics may be obtained, depending on how one defines a system. Consider two ice-resurfacing Zamboni machines crashing on an ice rink. If a single Zamboni is defined as a system, the system is clearly receiving an impulse from the adverse Zamboni. The ice also provides a reaction force, which may be ignored because it acts perpendicular to the crash impulse. It contributes nothing to the crash other than to resist the acceleration of gravity and to maintain the Zamboni's Z (not for Zamboni) coordinate on the ice. The Zamboni is also receiving an impulse from the friction associated with the reaction forces, which may also be ignored because the ice renders it very small compared to the collision impulse. The collision impulse is the main player in this event and must be calculated.

However, if one takes both Zambonis together as a system, we know from Newton's Third Law that the inter-vehicle forces (and therefore the impulses) are equal and opposite. They are internal to the system and cancel each other out. The only impulse external to the system, and contributing to the crash, comes from the friction forces. Again, these may be taken as second-order, because they are so much smaller than the collision forces. This same assumption is made in most vehicle crashes, even though they do not occur on ice (usually). For low-speed crashes, however, the collision forces may be low enough to be comparable to the tire forces, in which case the neglect of tire forces must be reconsidered. In some low-speed accident reconstructions, a time-forward simulation is used to avoid having to make what is, for those cases, an assumption that could prove to be embarrassing.

At any given time, the velocities of the two vehicles are $\vec{V}_1$ and $\vec{V}_2$. The two-mass system is equivalent to a single mass $m_1 + m_2$ that acts as if all the mass were located at the system center of mass. This location has a velocity $\vec{V}_{scm}$, the velocity of the system center of mass. The system momentum is

$$\vec{L} = \vec{L}_1 + \vec{L}_2 = \vec{L}_{scm} = m_1 v \vec{V}_1 + m_2 v \vec{V}_2 = (m_1 + m_2)\vec{V}_{scm} \qquad (18.8)$$

from which

$$\vec{V}_{scm} = \frac{m_1 \vec{V}_1 + m_2 \vec{V}_2}{m_1 + m_2} \qquad (18.9)$$

Note that nothing was said, in the derivation of Equation 18.9, about a collision. With or without a collision, Newton's First Law for this two-vehicle system says that as long as tire forces are ignored, there is no exterior impulse, and there is no change in the velocity vector of the center of mass—neither in its magnitude nor in its direction. The velocity of the system center of mass can be calculated at any point in time, whether the vehicles are at impact, in the crash phase, or at separation, or even if there has been no crash at all.

Suppose there is a crash. A reference frame attached to the system center of mass moves at a constant velocity. Therefore, an observer riding along on that reference frame will see the individual vehicles come to rest in that reference frame at the time the crash phase ends. (Of course, later on they will come to rest with respect to Mother Earth once the run-out phase is complete. However, by then impulses will have been delivered to the vehicles through the tires, and the system center of mass will have also stopped moving.)

Eccentric Collisions and Effective Mass

A "concentric" impact is one in which the line of action of the impulse passes directly through the center of mass of the struck body. If it does not, the moment arm of the impulse may be obtained by drawing a perpendicular line from its line of action to the center of mass, and measuring the length of that line. If the length is nonzero, the impact is "eccentric." In baseball, an eccentric impact to the ball could result in a foul tip, a chopper into the ground, or a pop fly; a concentric impact might be called "getting all the wood on the ball," and might result in a home run. The difference is not so much due to the speed of the bat or the speed of the ball at impact, but the eccentricity of the impact. Obviously, a single collision between two vehicles could be concentric for one and eccentric for the other, because of their irregular shapes.

The idea behind "getting all the wood on the ball" is analogous to the concept of effective mass. Obviously, the mass of the baseball bat does not change, but how much of its mass is brought to bear on the ball depends very much on the eccentricity of the hit. The spin of the ball as it comes off the bat is also affected.

Consider, then, an impulse that is delivered at distance h from the center of mass. Because of the eccentricity, less of the body is brought to bear in resisting the impulse, and an impulse torque is applied to the body. It is as if the body had an "effective mass," smaller than the total mass, that acted like a particle in receiving a linear momentum change from the impulse. At the same time, the body resisted the impulsive moment the same as if there were a "companion mass" acting like a particle struck eccentrically. See Figure 18.1.

The effective mass and the companion mass act like a lopsided barbell with all the mass concentrated in two places, and connected via a massless

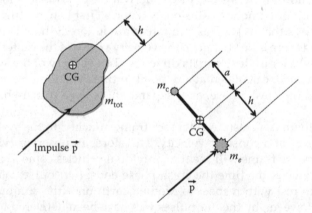

FIGURE 18.1
Effective mass concept.

link. The effective mass is struck concentrically and the companion mass is not struck at all. The two point mass system, being rigid, will respond to the impulse the same as the original body if:

   a. The total mass is the same.
   b. The location of the center of mass is the same.
   c. The total moment of inertia is the same.

The object of the calculations is to determine the distance a, the effective mass m_e, and the companion mass m_c. These three conditions are sufficient to find the three unknowns. Expressing the conditions mathematically:

   a. $m_{tot} = m_e + m_c$ (18.10)
   b. $m_c a = m_e h$ (18.11)
   c. $I_{tot} = m_{tot} k^2 = m_c a^2 + m_e h^2$ (18.12)

where k is the body's radius of gyration about an axis perpendicular to the plane defined by the line of action of the impulse and a line from the effective mass to the center of mass, and h is the eccentricity of the impact. From Equation 18.10, we have

$$m_c = m_{tot}\left(1 - \frac{m_e}{m_{tot}}\right) \tag{18.13}$$

and from Equations 18.11 and 18.13 we find that

$$a = h\left[\frac{m_e / m_{tot}}{1 - (m_e / m_{tot})}\right] \tag{18.14}$$

If we define the effective mass ratio γ as

$$\gamma = \frac{m_e}{m_{tot}} \tag{18.15}$$

then

$$m_c = m_{tot}(1 - \gamma) \tag{18.16}$$

and

$$a = h\left(\frac{\gamma}{1 - \gamma}\right) \tag{18.17}$$

Substituting into Equation 18.12, we obtain

$$\gamma = \frac{k^2}{h^2 + k^2} \tag{18.18}$$

The location a of the companion mass is then

$$a = \frac{k^2}{h} \tag{18.19}$$

It is important to note that the all-important effective mass ratio γ is not determined strictly on the basis of body geometry or other body properties. It depends on the nature of the impact; specifically, where the line of action of the impulse (the "impulse center") was and what the impulse direction was. The reconstructionist will not find it labeled on the vehicle, unless someone is trying to play a practical joke. This necessity of defining an impulse center is a feature possessed by all impulse–momentum methods that define a single vehicle as a system, and thus attempt to deal with eccentric collisions. The reconstructionist must make a judgment, and that judgment may be hard to make, because the impulse is not delivered at a single point, but rather distributed over an area. One commonly used method for specifying the impulse center is to use the centroid of the damage area. Another method is to realize that the forces are not uniformly distributed over the damage area, and to take the nonuniformity into account by using the centroid of the crush profile. The careful reconstructionist will realize, however, that the profile itself changes continuously as the vehicles crush together, and that the centers of impulse for the two vehicles must actually be at the same physical point in space.

A second issue is the direction of the impulse. Some procedures, such as Crash 3[4], address this question by requiring the user to specify the Principal Direction of Force (PDOF) as an input to the reconstruction. This may be technically consistent with the very meaning of PDOF, but since the direction of the impulse vector must match the direction of the velocity change vector, this is rather like requiring the reconstructionist to specify an important part of the answer before he or she begins.

Again, it must be realized that the impulse is delivered via crush forces that are distributed over an area, not concentrated at a single point. Of course, the line of action of distributed forces is located at a specific point, by definition. However, the tendency of a vehicle to rotate because of the impulse eccentricity will be resisted by any change that would have to occur to the crush profile to maintain a contact surface between the vehicles. Thus, the effects of impact eccentricity (i.e., vehicle yaw motions) can be somewhat muted, or shared by the two vehicles.

Consider, for example, a side impact in which the front of one vehicle hits the side of another. The striking vehicle might appear to be in a concentric collision, since the impact forces are distributed across its front. However, it is seen to rotate. Why? If the struck vehicle is moving forward at impact, it will impose friction forces lateral to the front of the striking vehicle. Thus, the resultant contact force will change angles and perhaps develop some eccentricity relative to the striking vehicle center of mass. The friction forces will be augmented by crush forces if the struck vehicle deforms inward so that the striking vehicle "snags" on hard points such as the struck vehicle door pillars or suspension, changing the force distribution and resultant force direction further. Also, as the struck vehicle rotates, it changes the force distribution on the striking vehicle. So the striking vehicle rotates, even though it might, at first glance, appear to be in a concentric collision.

Indeed, vehicle yaw rotation is a hallmark of eccentric collisions. Therefore, if the run-out analysis indicates a significant yaw rate at separation, the reconstructionist does well to ask where the rotation came from, and some accounting for eccentricity should probably occur during the crash analysis.

The best way to establish a line of action of the impulse (and impulse center) to be used in analysis is to create a scale drawing of the two vehicles at full engagement, and draw a single line common to both vehicles through the chosen impulse center, at an appropriate angle. After the analysis is done, and the direction of the ΔV vectors is obtained, that direction should be compared to the direction of the assumed impulse line. The angle of the line, and corresponding eccentricity, should be adjusted until acceptable agreement with the direction of the ΔV vectors is achieved.

Using Particle Mass Analysis for Eccentric Collisions

The preceding discussion provides some insights of how particle collision analysis may be utilized for eccentric impacts, since mass distribution and yaw moment of inertia are accounted for by the calculation of effective mass. One simply calculates the effective mass for each of the vehicles and uses the results in a particle-type analysis. The yaw moment of inertia does not have to be adjusted, since one of the predicates of the effective mass concept was the equivalence of yaw moments of inertia. However, the results of using the effective mass for analysis have to be interpreted in terms of what happens at the center of mass.

Since the companion mass has no impulse delivered to it during the impact, its motion is (initially) unaffected. The motion of the body (vehicle) at separation can be partitioned into its motion at impact, plus its motion due to the crash during the crush phase. The companion mass will not be affected by the crash, which means that the instantaneous center of crash motions is located at the companion mass. The ΔV at the companion mass is zero. Because the center of mass is fixed relative to the rest of the body, the ΔV at the center of mass is given by

$$\frac{\Delta V_{com}}{a} = \frac{\Delta V_{me}}{a+h} \qquad (18.20)$$

where ΔV_{me} is the ΔV of the effective mass, as calculated from the physics of the impact, and ΔV_{com} is the ΔV of the center of mass. So

$$\Delta V_{com} = \frac{a}{a+h}\Delta V_{me} = \frac{k^2/h}{(k^2/h)+h}\Delta V_{me} = \gamma\Delta V_{me} \qquad (18.21)$$

These thoughts illustrate that for a rotating vehicle, the ΔV that is experienced at a particular location (the occupant's seat, or the center of mass, for two examples) depends on the location in question. It can be less or more than that of the vehicle as a whole.

Momentum Conservation Using Each Body as a System

Taking each vehicle as a separate system, we apply Newton's Second Law (change in momentum equals impulse) separately to each system (vehicle), as follows:

$$m\Delta\vec{V} = \vec{p} \qquad (18.22)$$

$$I\Delta\omega = h\vec{p} \qquad (18.23)$$

where I is the vehicle yaw moment of inertia, h is the moment arm of the impulse $\vec{p}$ about the center of mass, and the over arrows reflect the fact that velocity and force are both vectors, with a component along each arbitrarily chosen (but unaccelerated) orthogonal axis. Since we are limiting ourselves to motions in the plane, the over arrows indicate that Equation 18.22 has two components, and Equation 18.23, which deals with angular momentum, has only one component, normal to the plane. Each vehicle therefore has three degrees of freedom.

With three scalar equations and three degrees of freedom for each vehicle, one might think we could then solve for each vehicle's three unknown velocity entities (two translational velocity components, and one yaw rate). However, in the above equations, we have introduced two more unknowns: the two components of $\bar{p}$ (assuming that we are justified in treating h as a known quantity). So with three equations and five unknowns, the system of equations cannot be solved, unless we are willing to make some assumptions about $\bar{p}$ and h. However, if we combine the two three-equation sets into a single set of six equations, we do not double the unknowns from 5 to 10. This is because Newton's Third Law tells us that the impulse $\bar{p}$ is equal and opposite on the two vehicles. Therefore, the combined equation set leaves us with six unknown velocity quantities (three for each vehicle), plus two unknown impulse quantities (the two components of $\bar{p}$), for a total of eight. However, we still have only six equations. We need two more.

The Planar Impact Mechanics Approach

One approach to this conundrum has been planar impact mechanics (PIM), by Ray Brach.[5] He defines (or rather, expects the reconstructionist to define) a vertical planar contact surface over which the contact forces are transmitted between the vehicles. At a particular point on the surface, he calculates the relative velocity between the vehicles, normal to the plane. The relative velocity at separation is related to its counterpart at impact, through the restitution coefficient, which is treated as a known quantity (even though, in this author's opinion, it is not, since it is highly variable, as we saw in Chapter 17). This is a seventh equation. For an eighth equation, he introduces the impulse ratio coefficient, also considered as a known, that is, the ratio of the tangential to normal impulse components. These last two equations act as constraint equations on the six unknown velocities. The system of eight independent equations is linear, so the existence and uniqueness theorems from linear algebra apply, along with the solution methods.

In what must surely be a prodigious feat of algebra, Brach solves these equations in closed form. The solution requires the assumption of values for the restitution coefficient and the impulse ratio coefficient. This seems problematic. The restitution coefficient is a well-known (but often misunderstood) concept, but the coefficient is specific to the particular collision. For example, crashing Car A into a rigid barrier may produce one restitution coefficient, Car B into a barrier produces a different restitution coefficient, crashing Car A into Car B results in yet a third restitution coefficient, and repeating the car-to-car crash probably produces a fourth restitution coefficient, particularly if the crash occurs at a different speed. Brach presents the means that allow one to infer the third value from the first two, but even barrier crashes produce highly variable restitution coefficients from nominally identical vehicles. As we saw in Chapter 17, a negative correlation between impact speed and restitution coefficient is often seen.

We will see that the calculated closing velocity for uniaxial impacts is remarkably insensitive to variations in the restitution coefficient, and that the ΔV calculation is less insensitive (Chapter 19). This author has not explored the effects on calculated impact speeds in coplanar collisions, since the restitution coefficient is not assumed *a priori* in the analysis.

The impulse ratio coefficient is even more problematic. Brach concedes that it requires a good deal of care and judgment to select the impulse ratio coefficient, and that one needs to ensure that the selection does not result in the violation of the principles of thermodynamics. The impulse ratio coefficient is not a familiar quantity to many engineers and is probably unique to Brach's formulation. An approach that could avoid specifying these quantities *a priori* would appear advantageous.

The Collision Safety Engineering Approach

Another approach was taken by Collision Safety Engineering (CSE), first by Woolley with his IMPAC program,[6,7] and then by Smith with a program called PLASMO.[8] Again, the CSE approach is an impulse–momentum look at vehicle-to-vehicle collisions, but is forward looking instead of backward looking. In other words, it is an initial-value problem in which the starting conditions are specified and the ending conditions are computed, as opposed to Brach's approach. Therefore, the calculations must be iterated until the ending conditions match those obtained from the run-out analysis.

In the CSE approach, linear momentum conservation is applied to the two-vehicle system as a whole, which produces two equations. Angular momentum conservation is applied to each vehicle separately, which produces two more equations. These last two equations require the specification by the user of an impulse center, which is "the point in the crush zone with a moment arm such that on the average the cross product of that moment arm with the impulse produces the correct rotational impulse ... [which seems like a circular definition]. After a 'solution' is obtained for a specified impulse center, it is appropriate to make additional runs to test the sensitivity of the calculation to the center of impulse."[6]

Since IMPAC and PLASMO seek the six unknown velocity quantities, two more equations are needed. These are constraint equations obtained by requiring that, for the two vehicles, the velocity vector components at the impulse center (as distinct from either center of mass) be the same following the exchange of momentum. Physically, this implies a fully plastic (zero restitution) collision.

For collisions in which the vehicles can slip relative to one another (e.g., sideswipes), these programs allow for a change in the constraint equations. Here, a slip plane through the impulse center is specified at some angle. Along the slip plane, the slip speed is specified as a fraction of the approach velocity at the center of impulse. Normal to the slip plane, no relative velocity is allowed. Again, this implies zero restitution.

The six-equation system is linear. It is solved in IMPAC by iteration using Newton's method, and in PLASMO by matrix operations. Smith points out that the velocity constraints at the impulse center are applied at separation, not at impact. Therefore, the coefficient matrix at impact is singular (it has two rows of zeros); its inverse does not exist. Consequently, the equations cannot be inverted (as was the case with the fully plastic uniaxial crashes analyzed with momentum conservation). This means that the approach must remain an initial-value problem.

As with Brach's approach, the CSE approach requires the specification of a property (the impulse center), on which the results depend, but that is not of primary interest to parties requesting the reconstruction. The CSE approach is also similar in that it requires the user to specify a coefficient of slip and a coefficient of restitution. Again, these coefficients are not so easily measured or documented and they are of lesser interest to those requesting the reconstruction, but not necessarily of lesser importance. This is not an ideal situation.

Methods Utilizing the Conservation of Energy

In the discussion of coplanar impact mechanics so far, nothing has been said about another fundamental principle of physics: conservation of energy. This principle holds that (in a nonrelativistic world, anyway) energy can neither be created nor destroyed; it merely changes form. One will find that in a crash, the total kinetic energy possessed by the two vehicles at impact has somehow been reduced by the time they exit the crash. The obvious destination for this lost kinetic energy is crush energy: the energy required to deform the vehicle structures. This loss will be reflected in the results from an impulse–momentum calculation. However, the principle could be enforced (and would provide an additional equation) if the crush energy could be calculated from a knowledge of vehicle crush and structural behavior (e.g., as discovered from crash testing), and the result set equal to the lost kinetic energy. Additionally, the assessment of crush energy will facilitate a damage comparison between the accident vehicles and those subjected to crash tests. This comparison is a desirable (some might say necessary) check on the validity of the results. This approach will be discussed in Chapters 19, 21, 23, 24, 25, and 26.

References

1. Campbell, K., Energy basis for collision severity, SAE Paper 740565, SAE International, 1974.
2. McHenry, R.R., Jones, I.S., and Lynch, J.P., *Mathematical Reconstruction of Highway Accidents: Scene Measurement and Data Processing System*, US Department of Transportation, Washington, DC, 1974.
3. *EDSMAC: Simulation Model of Automobile Collisions*, Version 4, Fifth Edition, Engineering Dynamics Corporation, 2006.
4. *Crash3 User's Guide and Technical Manual*, DOT HS 805732, National Highway Traffic Safety Administration, 1982.
5. Brach, R.M. and Brach, R.M., Vehicle accident analysis and reconstruction methods, SAE International, 2005.
6. Woolley, R.L., The 'IMPAC' computer program for accident reconstruction, SAE Paper 850254, SAE International, 1985.
7. Woolley, R.L., The 'IMPAC' program for collision analysis, SAE Paper 870046, SAE International, 1987.
8. Smith, G.C., Conservation of momentum analysis of two-dimensional colliding bodies, with or without trailers, SAE Paper 940566, SAE International, 1994.

19

Uniaxial Collisions

Introduction

Uniaxial (or, equivalently, co-linear) collisions could be viewed as an almost trivial case (but an important one, nevertheless). However, the general approaches to a solution for uniaxial collisions, and the nature of the results, are typical of a broad range of impacts. The uniaxial case is therefore a good point of departure for the application of Newtonian physics to the reconstruction of accidents. For uniaxial collisions, all velocity vectors are parallel by definition, and Equation 18.7 reduces to a single component, which is a scalar equation.

Conservation of Momentum

In reconstructing vehicle crashes, Equation 18.7 is integrated over the duration of the crush phase. This results in a computation of the total impulse delivered over that phase, and a calculation of the momentum change that has occurred during the interval, as seen in Equation 18.22. The velocities at impact are V_1 and V_2 for vehicles 1 and 2, respectively, and the velocities at separation are V'_1 and V'_2. When both vehicles are taken together as a single system, there is no external impulse delivered to the system during the crash phase as long as tire/roadway friction forces can be ignored. In that event, Equation 18.22 then becomes

$$m_1V_1 + m_2V_2 = m_1V'_1 + m_2V'_2 \qquad (19.1)$$

Note that nothing has been said (or assumed) about a collision having occurred. The only thing for sure is that momentum has been conserved. It follows that Equation 19.1 applies as well when there is no collision (or when there is no transfer of momentum between the vehicles). It is also the case that the crush phase does not have to be infinitesimally short. Finite time intervals (and finite configuration changes in the vehicles) are admissible, as long as the external impulse delivered to the system during that time interval is negligible.

For consistency, in 'any derivation, the velocities should all be assumed positive in the same direction (whereupon negative signs will consistently indicate velocities in the opposite direction). Thus, one can easily visualize a uniaxial collision with positive velocity directions as a front-to-rear crash in which a faster vehicle overtakes a slower-moving vehicle.

In Equation 19.1, we have four velocities, two of which may be considered as the unknowns. However, we have exhausted the equations available to us from momentum conservation. An additional equation is needed.

It is usually the case that practitioners of impulse–momentum mechanics obtain the additional equation by introducing the concept of restitution, and defining the coefficient of restitution as follows:

$$\text{Closing Velocity} = V_{cl} = V_1 - V_2 \tag{19.2}$$

$$\text{Separation Velocity} = V_{sep} = V_2' - V_1' \tag{19.3}$$

$$\text{Restitution Coefficient} = \varepsilon = \frac{V_{sep}}{V_{cl}} \tag{19.4}$$

where Vehicle 1 is assumed to be moving faster than Vehicle 2 at impact, and slower than Vehicle 2 at separation. (Minus signs will indicate results contrary to the assumptions.) Note that the coefficient of restitution is defined for the two vehicles in this particular crash mode at this particular closing velocity. It will be different from that seen in a rigid barrier test, that seen between two other vehicles, or even that seen between the same two vehicles in the same crash mode at a different closing velocity. However, if one is still willing to take ε as a known quantity (and this is less problematical for uniaxial collisions than for other modes), then the equations can be solved—subject to a condition that has some interesting implications.

The restitution coefficient ε will vary between 0 and 1. As can be seen in Equation 19.4, if $\varepsilon = 0$, the separation velocity V_{sep} is zero. For uniaxial collisions, both vehicles move along the same line, which is assumed normal to the contact surface between them (otherwise the velocity vectors would probably not be parallel), and a zero separation velocity means that there is no post-impact separation of the vehicles from this plane. In other words, they are "stuck together" post-impact. This is a perfectly plastic collision. (In more general impact cases, the condition of zero separation velocity normal to the contact surface would not imply that the vehicles are "stuck together," since there could still be post-impact relative motion along the contact surface (sliding), and the motions at the centers of mass could be different than those at the contact surface.)

If $\varepsilon = 1$, the separation velocity V_{sep} is equal to the closing velocity V_{cl}. This is known as a perfectly elastic collision.

Notice that we have not yet identified which two velocities are known, and which two are unknown. If we wish to obtain a "backwards" solution in which the impact velocities are the unknowns, then Equations 19.2 through 19.4 are going to have to be manipulated to obtain an expression for one of the impact velocities (V_1 or V_2), and then the expression will have to be substituted into Equation 19.1. For example,

$$V_2 = V_1 - V_{cl} = V_1 - \frac{V_{sep}}{\varepsilon} = V_1 - \frac{V_2' - V_1'}{\varepsilon} \quad \varepsilon \neq 0 \tag{19.5}$$

As noted, Equation 19.5 is not valid for $\varepsilon = 0$, because ε is in the denominator. An equation that results from solving for V_1 or V_2 will also have ε in the denominator, either implicitly or explicitly. In this situation, a completely general "backwards" solution is precluded because perfectly plastic collisions will cause Equation 19.5 to blow up. This is understandable physically. If we observe two vehicles during post-impact run-out that are stuck together, an indefinitely large number of possibilities exist for their speeds at impact.

A "time-forward" solution is possible, however, because ε can be kept out of the denominator. If we manipulate Equations 19.2 through 19.4 to obtain an expression for V'_2, and substitute it into Equation 19.1, we obtain

$$V_1' = \frac{(m_1 - m_2\varepsilon)V_1 + m_2(1 + \varepsilon)V_2}{m_1 + m_2} \tag{19.6}$$

Similarly,

$$V_2' = \frac{(m_2 - m_1\varepsilon)V_2 + m_1(1 + \varepsilon)V_1}{m_1 + m_2} \tag{19.7}$$

The equations are symmetric in the vehicle subscripts, as we might have expected.

Now let us define the velocity change as the velocity at separation, minus the velocity at impact:

$$\Delta V_1 = V_1' - V_1 \tag{19.8}$$

$$\Delta V_2 = V_2' - V_2 \tag{19.9}$$

Again, these are vector equations that reduce to scalar expressions for the uniaxial case under consideration. Substitution of Equations 19.8 and 19.9 into Equations 19.6 and 19.7 results in

$$\Delta V_1 = -\frac{m_2}{m_1 + m_2}(1 + \varepsilon)V_{cl} \tag{19.10}$$

$$\Delta V_2 = \frac{m_1}{m_1 + m_2}(1 + \varepsilon)V_{cl} \tag{19.11}$$

Note that the ΔVs always have opposite signs (but they are not, in general, equal). The negative sign in Equation 19.11 indicates that ΔV_1 will turn out to be in the opposite direction from the other velocity quantities; the sign of the numeric result will indicate whether that is actually the case. The difference between the ΔVs will be

$$\Delta V_2 - \Delta V_1 = (1 + \varepsilon)V_{cl} \tag{19.12}$$

This means that the difference between the ΔVs is dependent only on the restitution coefficient and the closing velocity. The difference in ΔVs is allocated to each vehicle according to the ratio of the other vehicle's mass to the total mass, as seen in Equations 19.10 and 19.11. For a perfectly plastic collision, $\varepsilon = 0$, and the ΔV difference equals the closing velocity; for a perfectly elastic collision, $\varepsilon = 1$, and the ΔV difference is double the closing velocity. The ΔVs are similarly affected by ε.

Energy comes in many forms, and anything that happens in the physical world requires that energy be expended. Without a source of energy, nothing happens. A stationary, burning car involves the expenditure of chemical energy stored in vehicle fuel and/or other energy sources. In vehicle collisions, the energy source is the motion of the vehicles. The energy of motion, which is called kinetic energy, signified by the symbol KE, and defined as

$$KE = \frac{1}{2}mV^2 \tag{19.13}$$

which is why speed plays such an important role. Therefore, once we have the two velocities pre- and post-impact, it is possible to calculate the kinetic energy loss resulting from the crash. We can write

$$\Delta KE = KE' - KE = \frac{1}{2}m_1V_1^2 + \frac{1}{2}m_2V_2^2 - \frac{1}{2}m_1(V_1')^2 - \frac{1}{2}m_2(V_2')^2$$

$$= \frac{1}{2}m_1[V_1^2 - (V_1')^2] + \frac{1}{2}m_2[V_2^2 - (V_2')^2]$$

$$= \frac{1}{2}m_1(V_1 - V_1')(V_1 + V_1') + \frac{1}{2}m_2(V_2 - V_2')(V_2 + V_2') \tag{19.14}$$

Substituting Equations 19.8 through 19.11 results in

$$\Delta KE = \frac{m_1 m_2}{2(m_1 + m_2)}(1 + \varepsilon)V_{cl}(2V_{cl} + \Delta V_1 - \Delta V_2) \tag{19.15}$$

But it is seen from Equation 19.12 that

$$\Delta V_1 - \Delta V_2 = -(1 + \varepsilon)V_{cl} \tag{19.16}$$

Therefore,

$$\Delta KE = \frac{1}{2}\frac{m_1 m_2}{m_1 + m_2}(1 - \varepsilon^2)(V_{cl})^2 \tag{19.17}$$

It is seen that the kinetic energy loss is proportional to the square of the closing velocity, and depends on the restitution coefficient. In fact, for a perfectly elastic collision, $\varepsilon = 1$, and no kinetic energy is lost. For a perfectly plastic collision, $\varepsilon = 0$, and some (but not all) of the kinetic energy is lost. Solving the closing velocity, we have

$$V_{cl} = \sqrt{\frac{2}{1 - \varepsilon^2}\frac{m_1 + m_2}{m_1 m_2}\Delta KE} \tag{19.18}$$

Conservation of Energy

One of the great conservation laws discussed in Chapter 1 is the conservation of energy. We can see, from Equation 19.18, that in general there is a loss of kinetic energy in a vehicle collision. Where did the energy go? In physics, "heat" and maybe "light" are the answers that show up when rounding up the usual suspects. The same can be said for vehicle crashes, but heat and light are not the major players. The main destination of dissipated kinetic energy is, in fact, crush energy, which is obvious from looking at the vehicle(s) post-crash. The Law of Energy Conservation can be expressed simply as

$$\Delta KE = CE_{tot} \tag{19.19}$$

where CE_{tot} is the total crush energy among the various vehicles involved. Substitution into Equation 19.18 results in

$$V_{cl} = \sqrt{\frac{2}{1 - \varepsilon^2}\frac{m_1 + m_2}{m_1 m_2}CE_{tot}} \tag{19.20}$$

This simple equation has important implications for all manners of reconstructions—not just uniaxial crashes. First of all, the closing velocity is directly related but not proportional to the total crush energy. This is a classic "damage only" solution, which is devoid of scene information except for restitution coefficient (which is, of course, related to the vehicle exit velocities). A "damage-only" reconstruction may be appealing in a case where the scene information is missing or nonexistent. However, crush energy about only one vehicle will not suffice; one needs to know *all* crush energies to effect a damage-only solution. This is an important fact to keep in mind when planning how to reconstruct an accident.

Second, a reconstruction solution may be checked for reasonableness by looking at the reconstructed closing velocity. If that quantity is out of whack, the likely culprit is the calculation of one or both of the vehicle crush energies. It is another way the mathematics whispers in the ear of the reconstructionist as to where the truth resides.

These observations have been derived from the equations for uniaxial collisions; comparable equations for two- or three-dimensional crashes do not show such obvious connections between crush energy and closing velocity. However, years of experience have shown the observations to be valid, nevertheless.

In terms of vehicle weights, Equation 19.20 may be written as

$$V_{cl} = \sqrt{\frac{2CE_{tot}(m_1 + m_2)}{m_1 m_2 (1 - \varepsilon^2)}} = \sqrt{\frac{2gCE_{tot}(W_1 + W_2)}{W_1 W_2 (1 - \varepsilon^2)}} \qquad (19.21)$$

This equation is as notable for what it does *not* contain as for what it does. It contains no vehicle velocity, at either impact or at separation, for either vehicle. An analysis of the vehicle damage will not yield the traveling speed of either vehicle—only the relative velocity between them at impact. It is as if the damage-only analysis is disconnected from Mother Earth (which, come to think of it, is as it should be, since ground forces are considered negligible in this analysis). Also, it is noteworthy that the equation contains the restitution coefficient, which is generally not known to very good accuracy. However, ε only appears as a squared quantity subtracted from unity, which renders the results extraordinarily insensitive to errors in ε. See Table 19.1.

For example, a 50% change in restitution coefficient leads to a change of <1% in the calculation of closing velocity.

Earlier work also allows us to obtain expressions for the vehicle velocity changes. Substitution of Equation 19.21 into Equations 19.10 and 19.11 yields

$$\Delta V_1 = \sqrt{\frac{2gCE_{tot}(1 + \varepsilon)}{W_1(1 + r_1)(1 - \varepsilon)}} \qquad (19.22)$$

$$\Delta V_2 = \sqrt{\frac{2gCE_{tot}(1 + \varepsilon)}{W_2(1 + r_2)(1 - \varepsilon)}} \qquad (19.23)$$

TABLE 19.1

Sensitivity of Closing Velocity Calculation to Restitution Coefficient

Restitution Coefficient ε	% Change in ε	$R_{cl}(\varepsilon)^a$	% Change in $R_{cl}(\varepsilon)$
0.10	—	1.005	—
0.15	50	1.011	0.64
0.20	100	1.021	1.55
0.25	150	1.033	2.76
0.30	200	1.048	4.30

a $R_{cl}(\varepsilon) = \sqrt{1/(1 - \varepsilon^2)}$

TABLE 19.2

Sensitivity of Velocity Change Calculation to Restitution Coefficient

Restitution Coefficient ε	% Change in ε	$R_{\Delta V}(\varepsilon)^a$	% Change in $R_{\Delta V}(\varepsilon)$
0.10	—	1.106	—
0.15	50	1.163	5.21
0.20	100	1.225	10.78
0.25	150	1.291	16.77
0.30	200	1.363	23.27

a $R_{\Delta V}(\varepsilon) = \sqrt{(1 + \varepsilon)/(1 - \varepsilon)}$

where the mass ratios are given by

$$r_1 = \frac{m_2}{m_1} = \frac{W_2}{W_1} \tag{19.24}$$

$$r_2 = \frac{m_1}{m_2} = \frac{W_1}{W_2} \tag{19.25}$$

Note the linkage between crush energy and velocity change. The velocity changes are fairly insensitive to errors in restitution coefficient, but not as insensitive as the closing velocity. See Table 19.2.

For example, a change of 50% in the restitution coefficient leads to a change in the ΔV of just over 5%.

20

Momentum Conservation for Central Collisions

In this chapter, we consider a somewhat more general case, in which two bodies experience momentum conservation while colliding in a plane. Rotational effects are not considered, which means the impact is central (contact forces pass through the centers of mass), that rotational motions are not significant, or that rotational inertia is insignificant (i.e., the bodies are particles). Each body thus has two degrees of freedom, instead of one in uni-axial collisions, or three if rotational effects are considered. Since the two bodies are taken together as a system, contact forces do not enter the picture. If forces external to the system are negligible, momentum is conserved during the crash phase. This can be expressed mathematically as

$$L_X = L_X' \tag{20.1}$$

$$L_Y = L_Y' \tag{20.2}$$

or

$$m_1 V_{X1} + m_2 V_{X2} = m_1 V_{X1}' + m_2 V_{X2}' \tag{20.3}$$

$$m_1 V_{Y1} + m_2 V_{Y2} = m_1 V_{Y1}' + m_2 V_{Y2}' \tag{20.4}$$

Again, the primed velocities may be considered as known, being determined during the analysis of the run-out phase. The velocities without primes are the unknowns. We see that we have four unknowns, with only two equations available from momentum conservation considerations. Again, there is a gap to be filled.

Instead of generating additional equations by introducing constants such as the coefficient of restitution, we keep the number of equations at two and reduce the number of unknowns. This is done by assuming that the directions of the impact velocity vectors are known, perhaps through scene evidence or other knowledge of the accident sequence. The magnitudes of the impact velocity vectors are V_1 and V_2, which are now the unknowns. The directions of the impact vectors, or approach angles, are θ_1 and θ_2, and are considered to be known quantities. To implement this revised point of view, we rewrite Equations 20.3 and 20.4 as

$$m_1 V_1 \cos(\theta_1) + m_2 V_2 \cos(\theta_2) = m_1 V'_{X1} + m_2 V'_{X2} \tag{20.5}$$

$$m_1 V_1 \sin(\theta_1) + m_2 V_2 \sin(\theta_2) = m_1 V'_{Y1} + m_2 V'_{Y2} \tag{20.6}$$

This system of two equations and two unknowns can now be solved. Note that since we have not introduced the restitution coefficient (or other constants that might be zero), this system can be solved in a backwards direction. In other words, the impact conditions can be expressed in terms of the separation conditions. Solution of the equations, utilizing the trigonometric identities for angle differences, results in

$$V_1 = \frac{[V'_{X1} + (m_2/m_1)V'_{X2}]\sin(\theta_2) - [V'_{Y1} + (m_2/m_1)V'_{Y2}]\cos(\theta_2)}{\sin(\theta_2 - \theta_1)}$$

$$\sin(\theta_1 - \theta_2) \neq 0 \tag{20.7}$$

$$V_2 = \frac{[V'_{X2} + (m_1/m_2)V'_{X1}]\sin(\theta_1) - [V'_{Y2} + (m_1/m_2)V'_{Y1}]\cos(\theta_1)}{\sin(\theta_1 - \theta_2)}$$

$$\sin(\theta_1 - \theta_2) \neq 0 \tag{20.8}$$

Alternatively, one can choose to express the separation conditions in terms of the departure angles θ'_1 and θ'_2. In that case, momentum conservation is expressed as

$$m_1 V_1 \cos(\theta_1) + m_2 V_2 \cos(\theta_2) = m_1 V'_1 \cos(\theta'_1) + m_2 V'_2 \cos(\theta'_2) \tag{20.9}$$

$$m_1 V_1 \cos(\theta_1) + m_2 V_2 \cos(\theta_2) = m_1 V'_1 \cos(\theta'_1) + m_2 V'_2 \cos(\theta'_2) \tag{20.10}$$

Solution of these equations, and again utilizing the trigonometric identities for angle differences, results in

$$V_1 = \frac{V'_1 \sin(\theta_2 - \theta'_1) + (m_2/m_1)V'_2 \sin(\theta_2 - \theta'_2)}{\sin(\theta_2 - \theta_1)} \qquad \sin(\theta_1 - \theta_2) \neq 0 \tag{20.11}$$

$$V_2 = \frac{V'_2 \sin(\theta_1 - \theta'_2) + (m_1/m_2)V'_1 \sin(\theta_1 - \theta'_1)}{\sin(\theta_1 - \theta_2)} \qquad \sin(\theta_1 - \theta_2) \neq 0 \tag{20.12}$$

Either way, the denominator is the same. It is important to note that for the degenerate condition of a uniaxial condition, $\sin(\theta_1 - \theta_2) = 0$, and the equations blow up. As this condition is approached, small changes in the approach angles θ_1 and θ_2 can cause large changes in the results. At some point, these equations have to be replaced by Equations 20.6 and 20.7 for the uniaxial condition.

Mathematically, the uniaxial collision is a singularity that arises from a trivial $(0 = 0)$ equation for the enforcement of two-dimensional momentum conservation in the direction perpendicular to the crash.

This is clearly an undesirable, if not ridiculous, situation. One always likes to have the degenerate condition arise from the general condition as a special case. Crash3 suffers from the same problem. According to the *Crash3 User's Guide*,[1] "The user should note that transition from the oblique (i.e., conservation of linear momentum) to the axial solution form is made in Crash3 when the initial velocity vectors are 10 degrees from parallel. The transition may sometimes produce abrupt changes in speed results when the heading angles were changed by only one degree" (p. 3-11). It would be better to develop a formulation without this behavior, as will be done in Chapter 23.

The Law of Energy Conservation is not utilized in this analysis. However, the dissipation of kinetic energy can still be calculated, once all the velocities are known:

$$\Delta KE = \frac{1}{2}m_1(V_1^2 - V'^2_1) - \frac{1}{2}m_2(V_2^2 - V'^2_2) \tag{20.13}$$

These equations, Equations 20.11 and 20.12 in particular, may seem to be of limited application, but in fact they are widely used. They have the great advantage of being simple and easily programmed into a spreadsheet. The sensitivity of the calculations to being close to a uniaxial crash can be easily tested, and the equations for uniaxial crashes used instead if indeed the occasion warrants.

It is also the case that the equations apply to central collisions only. However, eccentric collisions can be handled through the use of effective masses, as discussed in Chapter 18.

Reference

1. *Crash3 User's Guide and Technical Manual*, DOT HS 805732, National Highway Traffic Safety Administration, Washington, DC, April 1982.

21

Assessing the Crush Energy

Introduction

If one is to apply the principle of energy conservation, calculating the crush energy dissipated by each vehicle in crush is not just important, but absolutely necessary. To contemplate a pile of tangled-up sheet metal that was formerly a vehicle, and take on the task of computing how much energy was required to effect the change, is a daunting task. However, the seminal paper by Campbell[1] in 1974 considered a number of vehicle crash tests conducted by General Motors, made some observations about how the vehicle deformation varied with test speed. From those observations, he developed an analytical model that is the basis of most crush energy assessment methods in use today. Because his observations implied that the residual crush varied linearly with some kind of force applied to the structure, Campbell's approach can be called a constant-stiffness model, in the same sense that a linear spring is a constant-stiffness device.

Constant-Stiffness Models

Campbell observed crash-tested vehicles post-impact, for which the residual crush had been measured. (Residual crush, sometimes called static crush, is the crush that remains after the crash. This is distinct from dynamic crush, which would vary dynamically during a crash. Generally, the peak dynamic crush is greater than the residual crush, due to structural recovery that occurs when the crash is over.) He noticed that the average residual crush (i.e., spatial average over the crush surface) bore a roughly linear relationship with the speed of a barrier crash test.

It was well known that crush energy came from the conversion of kinetic energy, which varies as the square of velocity. Therefore, as we observed in Equation 19.21, there is a linear relation between the closing speed V_{cl} (test speed, in Campbell's case) and the square root of the crush energy (CE). It follows, then, that Campbell's observation can be expressed as

$$\bar{c} = d_0 + d_1\sqrt{CE} \tag{21.1}$$

where $\bar{c}$ is the average crush, given by

$$\bar{c} = \frac{1}{L}\int_0^L c(x)dx \tag{21.2}$$

Here, c is the residual crush, or the displacement of a point from its initial position on the undamaged vehicle to its final position on the damaged vehicle. The location on the undamaged vehicle where the measurement is made is described by x; c is a function of x. L is the crush width. The coefficients d_0 and d_1 are characteristics of the portion of the vehicle structure being crushed.

To obtain crush energy from residual crush, we somehow have to develop a force–deflection characteristic. For guidance in this regard, we know that the classical linear spring has the force–deflection characteristic

$$F = k\chi \tag{21.3}$$

where F is the applied force, taken positive in compression, χ is the (dynamic) crush, also positive in compression, and k is the slope of the force–deflection characteristic; that is, the spring stiffness. The area under the force–deflection characteristic, from zero crush to maximum crush χ, is the crush energy CE, given by

$$CE = \frac{1}{2}k\chi^2 \tag{21.4}$$

(In the classical linear spring, the restitution coefficient would be unity, all the crush energy would be recovered during rebound, and the residual crush would be zero.) We notice that

$$\chi = \sqrt{\frac{2}{k}}\sqrt{CE} \tag{21.5}$$

which is somewhat analogous to Equation 21.1. We expect, therefore, that a linear relationship between residual crush and force, analogous to Equation 21.3, will lead to a linear relationship between residual crush c and the square root of the crush energy, as Campbell observed.

However, vehicles are not classical linearly elastic springs, and residual crush is not dynamic crush, so for crashed vehicles we construct a linear, but somewhat more generalized, relationship between "force" per unit width and residual crush as follows:

$$F = A + Bc \tag{21.6}$$

"Force" is in quotation marks because at the time residual crush is being measured, there is no external force being applied to the vehicle. However, F is an indicator of the maximum level of force, per unit width, that had been applied during the crush phase. B may be thought of as the structural stiffness (the resistance to crush per unit of crush, per unit width). A may be thought of as the initial resistance to crush per unit width, wherein force is generated even though no residual crush ensues (as with a no-damage bumper, for example). We keep in mind that the residual crush is a function of measurement location x.

In a strip of structure of differential width dx, the energy absorbed is the area under the force–deflection curve, from zero to the residual crush. Integrating Equation 21.6 with respect to c results in

$$d(CE) = \left(Ac + \frac{1}{2}Bc^2 + G \right)dx \tag{21.7}$$

where G may be thought of as a constant of integration. To obtain the crush energy for all of the deformed structure, we must integrate across the crush width, as follows:

$$CE = \int_0^L \frac{d(CE)}{dx}dx = \int_0^L \left(Ac + \frac{1}{2}Bc^2 + G \right)dx \tag{21.8}$$

At this point, a key assumption is made: that the structural parameters A, B, and G are all constant in the range of integration; that is, along the entire width of the crush profile. (If there are "hard" points that invalidate the assumption, the integration can be broken up into separate regions within which A, B, and G are constants.) It is also the case that A, B, and G are defined per unit width, meaning that the structural characteristics are assumed not to vary as a function of height. If that assumption is not valid (as in override crashes, for example), it will be necessary to subdivide the structure vertically, as shown by Struble.[2]

Equation 21.8 becomes

$$CE = A\int_0^L c(x)dx + \frac{1}{2}B\int_0^L c^2(x)dx + GL \tag{21.9}$$

Substituting Equation 21.2 results in

$$CE = A\overline{c}L + \frac{1}{2}B\int_0^L c^2(x)dx + GL \tag{21.10}$$

In all derivations presented by others, c is taken as a constant and set equal to the average crush $\overline{c}$. This will lead to negligible error in square-on frontal or rear barrier crashes, but as the crush profile deviates from uniformity (as in offset impacts, for example), the error will increase. To be correct about it, we define a dimensionless shape function $f(x)$ such that

$$c(x) = \overline{c}f(x) \tag{21.11}$$

For a uniform crush profile, $f(x)$ is identically equal to unity. Substitution into Equation 21.10 yields

$$CE = A\overline{c}L + \frac{1}{2}B\overline{c}^2\int_0^L f^2(x)dx + GL = \left[A\overline{c} + \frac{1}{2}B\overline{c}^2\left(\frac{1}{L}\right)\int_0^L f^2(x)dx + G\right]L \tag{21.12}$$

We now introduce the form factor β as

$$\beta = \frac{1}{L}\int_0^L f^2(x)dx \tag{21.13}$$

Since $f(x)$ is dimensionless, β will be as well. How to calculate β for nonuniform crush profiles will be discussed later.

Substitution into Equation 21.12 leads to

$$CE = \left[A\overline{c} + \frac{1}{2}B\beta\overline{c}^2 + G\right]L \tag{21.14}$$

Now, we are ready to relate Campbell's observation, Equation 21.1, to the calculation of crush energy as given by Equation 21.14. Substituting the former into the latter results in

$$CE = \left[A(d_0 + d_1\sqrt{CE}) + \frac{1}{2}B\beta(d_0 + d_1\sqrt{CE})^2 + G\right]L$$

$$= \left[\left(Ad_0 + \frac{1}{2}B\beta d_0^2 + G\right) + (Ad_1 + B\beta d_0 d_1)\sqrt{CE} + \left(\frac{1}{2}B\beta d_1^2\right)CE\right]L \tag{21.15}$$

Equating coefficients of $(CE)^N$ to balance the two sides of Equation 21.15, we have

$$\frac{1}{2}B\beta d_1^2 L = 1 \tag{21.16}$$

from which

$$d_1^2 = \frac{2}{B\beta L} \tag{21.17}$$

$$Ad_1 + B\beta d_0 d_1 = 0 \tag{21.18}$$

from which

$$d_0 = -\frac{A}{B\beta} \tag{21.19}$$

Finally,

$$Ad_0 + \frac{1}{2}B\beta d_0^2 + G = 0 \tag{21.20}$$

from which

$$G = -d_0\left[A + \frac{1}{2}B\beta d_0\right] \tag{21.21}$$

Substituting Equation 21.19 into the above yields

$$G = -\left(-\frac{A}{B\beta}\right)\left[A + \frac{1}{2}B\beta\left(-\frac{A}{B\beta}\right)\right] = \frac{A^2}{2B\beta} \tag{21.22}$$

This important result shows that G is not an independent constant, but is a function of A and B. Substituting Equation 21.22 into Equation 21.14, we obtain

$$CE = \left[A\bar{c} + \frac{1}{2}B\beta\bar{c}^2 + \frac{A^2}{2B\beta}\right]L \tag{21.23}$$

This form of the result shows that Campbell's formulation is a two-parameter model; that is, it depends on two parameters A and B. It is a quadratic in $\bar{c}$, and the square can be completed as follows:

$$CE = \frac{1}{2B\beta}\left[2AB\beta\bar{c} + (B\beta)^2\bar{c}^2 + A^2\right]L = \frac{L}{2B\beta}\left(A + B\beta\bar{c}\right)^2 \qquad (21.24)$$

To the best of this author's knowledge, the first to show the completion of the square (but without the form factor) was Ron Woolley.

The total "force" implied in the crush profile is obtained by integrating the "force" per unit width in Equation 21.6 over the length of the profile. Doing so results in

$$F_{tot} = \int_0^L F(x)dx = \int_0^L [A + Bc(x)]dx = A\int_0^L dx + B\int_0^L c(x)dx \qquad (21.25)$$

We recognize the second integral as the area under the crush profile, which is equal to the product of the average crush and the profile length. See Equation 21.2. Consequently, Equation 21.25 becomes

$$F_{tot} = AL + B\bar{c}L = (A + B\bar{c})L \qquad (21.26)$$

Sample Form Factor Calculation: Half-Sine Wave Crush Profile

To illustrate the calculation of form factor, consider an idealized crush profile that has the shape of a half-sine wave, as follows:

$$c(x) = C_{MAX}\sin\left(\frac{\pi x}{L}\right) \quad 0 \le x < L \qquad (21.27)$$

We see that this smooth profile has zero values at the ends, and a peak crush of C_{MAX} at the middle, at $x = L/2$. The average crush is given by

$$\bar{c} = \frac{1}{L}\int_0^L c(x)dx = \frac{1}{L}\int_0^L C_{MAX}\sin\left(\frac{\pi x}{L}\right)dx = \frac{2}{\pi}C_{MAX} \qquad (21.28)$$

Therefore, the dimensionless shape function is

$$f(x) = \frac{1}{c}c(x) = \frac{\pi}{2C_{MAX}}C_{MAX}\sin\left(\frac{\pi x}{L}\right) = \frac{\pi}{2}\sin\left(\frac{\pi x}{L}\right) \qquad (21.29)$$

and the form factor β is given by

$$\beta = \frac{1}{L}\int_0^L f^2(x)dx = \frac{1}{L}\int_0^L \left[\frac{\pi}{2}\sin\left(\frac{\pi x}{L}\right)\right]^2 dx = \frac{\pi^2}{4L}\int_0^L \sin^2\left(\frac{\pi x}{L}\right)dx \qquad (21.30)$$

To integrate this expression, we use the trigonometric identity

$$\sin^2\alpha = \frac{1}{2}(1 - \cos 2\alpha) \qquad (21.31)$$

We then obtain

$$\beta = \frac{\pi^2}{4L}\int_0^L \frac{1}{2}\left[1 - \cos\left(\frac{2\pi x}{L}\right)\right]dx = \frac{\pi^2}{8} = 1.2337 \qquad (21.32)$$

Sample Form Factor Calculation: Half-Sine Wave Squared Crush Profile

A narrow-object impact may generate even more contour in the crush profile than was analyzed above. A half-sine wave squared profile (or haversine, as it is often called) has such a characteristic, as can be seen in Figure 21.1. This crush profile has the functional form

$$c(x) = C_{MAX}\sin^2\left(\frac{\pi x}{L}\right) \qquad (21.33)$$

The average crush is given by

$$\bar{c} = \frac{1}{L}\int_0^L C_{MAX}\sin^2\left(\frac{\pi x}{L}\right)dx = \frac{C_{MAX}}{2L}\int_0^L \left[1 - \cos\left(\frac{2\pi x}{L}\right)\right]dx = \frac{C_{MAX}}{2} \qquad (21.34)$$

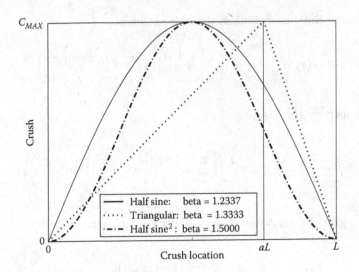

FIGURE 21.1
Three crush shapes and their form factors.

The dimensionless shape function is

$$f(x) = \frac{1}{c}c(x) = \frac{2}{C_{MAX}}\left[C_{MAX}\sin^2\left(\frac{\pi x}{L}\right)\right] = 2\sin^2\left(\frac{\pi x}{L}\right) \qquad (21.35)$$

Thus, the form factor β is

$$\beta = \frac{1}{L}\int_0^L \left[2\sin^2\left(\frac{\pi x}{L}\right)\right]^2 dx = \frac{4}{L}\int_0^L \left\{\frac{1}{2}\left[1 - \cos\left(\frac{2\pi x}{L}\right)\right]\right\}^2 dx$$

$$= \frac{1}{L}\int_0^L \left[1 - 2\cos\left(\frac{2\pi x}{L}\right) + \cos^2\left(\frac{2\pi x}{L}\right)\right]dx \qquad (21.36)$$

To integrate the third term in this expression, recall that

$$\cos^2\alpha = \frac{1}{2}(1 + \cos 2\alpha) \qquad (21.37)$$

Using this trigonometric identity, we obtain

$$\beta = \frac{1}{2L}\int_0^L \left[3 - 4\cos\left(\frac{2\pi x}{L}\right) + \cos\left(\frac{4\pi x}{L}\right)\right]dx = \frac{3}{2} = 1.5000 \qquad (21.38)$$

We see that the more contoured the crush profile, the higher is its form factor.

Form Factors for Piecewise-Linear Crush Profiles

It may be a useful academic exercise to calculate form factors for smooth crush profiles that can be represented by a single mathematical equation, but what we really need is a completely general formulation that can represent any kind of crush profile. So we turn to the analysis of a piecewise linear crush profile, in which straight line segments are drawn between discretely spaced crush measurements, such that a reasonable representation of any actual profile is achieved. The number of crush measurements is N, and the segments between them number $N - 1$. The ith segment of such a profile appears as shown in Figure 21.2.

The ith segment lies between x_i and x_{i+1}, and the segment length is

$$\Delta x_i = x_{i+1} - x_i \tag{21.39}$$

The average crush in that segment is

$$\bar{c}_i = \frac{C_i + C_{i+1}}{2} \tag{21.40}$$

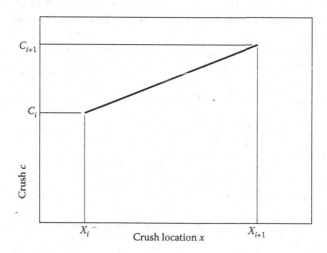

FIGURE 21.2
Segment of a piecewise linear crush profile.

and the area under the crush profile in that segment is

$$(Area)_i = \bar{c}_i \Delta x_i \tag{21.41}$$

The area under the entire profile is

$$Area = \sum_{i=1}^{N-1}(Area)_i = \sum_{i=1}^{N-1}\bar{c}_i \Delta x_i \tag{21.42}$$

The average crush over the entire profile is

$$\bar{c} = \frac{Area}{L} = \frac{1}{L}\sum_{i=1}^{N-1}\bar{c}_i \Delta x_i \tag{21.43}$$

where L is the length of the entire crush profile, given by

$$L = \sum_{i=1}^{N-1} \Delta x_i \tag{21.44}$$

Note that when Equation 21.43 is expanded, the interior crush measurements appear twice, whereas the end points only appear once. Therefore, the average crush $\bar{c}$ as computed from Equation 21.43 is *not* a numeric average. It is a weighted average, such that when the weighted average is multiplied by the total crush length, the total crush area is obtained. Note also that there has been no assumption that the segment lengths Δx_i are equal, unlike the case in Crash3.

However, if the segment lengths are all equal to Δx, as in the NASS protocol and in NHTSA crash tests, then

$$L = (N-1)\Delta x \tag{21.45}$$

and Equation 21.43 reduces to

$$\bar{c} = \frac{1}{2(N-1)}\sum_{i=1}^{N-1}(C_i + C_{i+1}) \tag{21.46}$$

For three equally spaced crush measurements, Equation 21.46 becomes

$$\bar{c} = \frac{1}{4}(C_1 + 2C_2 + C_3) \tag{21.47}$$

For six equally spaced crush measurements, we have

$$\bar{c} = \frac{1}{10}(C_1 + 2C_2 + 2C_3 + 2C_4 + 2C_5 + C_6) \tag{21.48}$$

From Figure 21.2, we see that the crush is a piecewise linear function of x, and at any point in the ith segment, the crush is given by

$$c = c(x) = C_i + \frac{C_{i+1} - C_i}{\Delta x_i}(x - x_i), \quad x_i \le x < x_{i+1}, \quad i = 1, 2, \ldots, N - 1 \tag{21.49}$$

The dimensionless shape function for the ith segment is

$$f(x) = \frac{c(x)}{\bar{c}} = \frac{1}{\bar{c}}\left[C_i + (x - x_i)\frac{C_{i+1} - C_i}{\Delta x_i}\right] \tag{21.50}$$

Over the entire crush profile, the form factor β is given by

$$\beta = \frac{1}{L}\int_0^L f^2(x)dx = \frac{1}{L}\sum_{i=1}^{N-1}\int_{x_i}^{x_{i+1}}\frac{1}{\bar{c}^2}\left[C_i + (x - x_i)\frac{C_{i+1} - C_i}{\Delta x_i}\right]^2 dx \tag{21.51}$$

The integration is made simpler with the variable substitution

$$\xi = x - x_i \tag{21.52}$$

Then, we have

$$\beta = \frac{1}{L\bar{c}^2}\sum_{i=1}^{N-1}\int_0^{\Delta x_i}\left[C_i + \frac{C_{i+1} - C_i}{\Delta x_i}\xi\right]^2 d\xi \tag{21.53}$$

or

$$\beta = \frac{1}{3L\bar{c}^2}\sum_{i=1}^{N-1}\left[C_i^2 + C_iC_{i+1} + C_{i+1}^2\right]\Delta x_i \tag{21.54}$$

As before, note that no assumption has been made about uniform crush intervals. However, in dealing with NHTSA crash test data and NASS accident investigations, uniform crush intervals Δx pertain. Using that fact, and substitution of Equation 21.45 into the above, results in

$$\beta = \frac{1}{3(N-1)\bar{c}^2} \sum_{i=1}^{N-1} \left[C_i^2 + C_i C_{i+1} + C_{i+1}^2 \right] \tag{21.55}$$

For two equal segments ($N = 3$), the above reduces to

$$\beta = \frac{1}{6\bar{c}^2} \left[C_1^2 + C_1 C_2 + 2C_2^2 + C_2 C_3 + C_3^2 \right] \tag{21.56}$$

For five equal segments ($N = 6$), we have

$$\beta = \frac{1}{15\bar{c}^2} \left[\begin{array}{c} C_1^2 + C_1 C_2 + 2C_2^2 + C_2 C_3 + 2C_3^2 + C_3 C_4 \\ +2C_4^2 + C_4 C_5 + 2C_5^2 + C_5 C_6 + C_6^2 \end{array} \right] \tag{21.57}$$

Sample Form Factor Calculation: Triangular Crush Profile

Perhaps the simplest type of piecewise linear crush is the triangular profile shown in Figure 21.1, where the maximum crush C_{MAX} is located at $x = aL$, and where a is a number between zero and unity. The average crush for this profile is

$$\bar{c} = \frac{C_{MAX}}{2} \tag{21.58}$$

independent of a. Substitution into Equation 21.54 yields

$$\beta = \frac{1}{3L(C_{MAX}/2)^2} \left\{ \left[0 + 0 + (C_{MAX})^2 \right](aL) + \left[(C_{MAX})^2 + 0 + 0 \right](1-a)L \right\}$$

$$= \frac{4L(C_{MAX})^2}{3L(C_{MAX})^2} [a + 1 - a] = \frac{4}{3} = 1.3333 \tag{21.59}$$

again independently of *a*. Even though this crush profile is pointed, it has less overall contour, as can be seen in Figure 21.1, and therefore a lower form factor, compared to the half-sine-squared profile. (Upon encountering such a profile, one might be suspicious that the representation of the crush profile is overly crude.)

Constant-Stiffness Crash Plots

Now that we have explored the nature of the form factor, we can return to the main agenda of assessing crush energy and characterizing vehicle structures. If we multiply Equation 21.24 by $2/L$ and take the square root of both sides, we obtain

$$\sqrt{\frac{2(CE)}{L}} = \frac{1}{\sqrt{B\beta}}[A + B\beta\bar{c}] = \frac{1}{\sqrt{\beta}}\left[\frac{A}{\sqrt{B}} + \sqrt{B}\beta\bar{c}\right] \tag{21.60}$$

Multiplying through by $\sqrt{\beta}$, we obtain

$$\sqrt{\frac{2\beta(CE)}{L}} = \sqrt{B}(\beta\bar{c}) + \frac{A}{\sqrt{B}} \tag{21.61}$$

If the left-hand side of the equation is plotted on the ordinate (vertical y axis) and $(\beta\bar{c})$ is plotted on the abscissa (horizontal x axis), we recognize the above as the equation of a straight line: $y = mx + b$, where the slope of the line is m, and the intercept is b. A graph of this equation will have the following properties:

$$Ordinate = \sqrt{\frac{2\beta(CE)}{L}} = ECF \tag{21.62}$$

$$Abscissa = \beta\bar{c} \tag{21.63}$$

$$Slope = \sqrt{B} \tag{21.64}$$

$$Intercept = \frac{A}{\sqrt{B}} \tag{21.65}$$

Such a graph is known as a "crash plot." Suppose that a particular vehicle structure is crashed in a particular crash mode, at various speeds such that the average crush varies from test to test. Suppose also that each test is represented by a single data point, with the ordinate (vertical coordinate) and abscissa (horizontal coordinate) of each such point calculated as above. Campbell's observation regarding test speed and residual crush holds that the points will lie along a straight line, as long as the structure maintains a linear relationship between force and residual crush, as postulated in Equation 21.6. The slope and intercept of the crash plot will yield the structural characteristics A and B. This process is known as characterizing the structure.

The left-hand side of Equation 21.61, or right hand side of Equation 21.62, has also been called the energy of crush factor, or ECF.[3]

If the straight line can be constructed through the various data points, and its equation written, then the structural parameters A and B can be found from Equations 21.64 and 21.65 as follows:

$$A = (Slope)x(Intercept) \tag{21.66}$$

and

$$B = (Slope)^2 \tag{21.67}$$

Example Constant-Stiffness Crash Plot

To learn how a crash plot is actually constructed, we will characterize the front structure of a 2000 Pontiac Grand Am. Our first task is to gather all the frontal crash tests we can find that could be used for this purpose. We start by considering all vehicle makes and models that are structurally similar (ideally, identical) to the vehicle in question. We consult the Vehicle Year & Model Interchange List (Sisters & Clones List)[4] for a 2000 Pontiac Grand Am, and find that the production run for this vehicle was 1999–2005, and that it was interchangeable with the Oldsmobile Alero during those years.

Then we go to the NHTSA crash test database and look for frontal crash tests of that range of vehicles. Vehicle-to-vehicle tests are not particularly useful, since we have to know the crush energy absorbed by just the frontal structure of the vehicle in question, and any test in which two or more crushable structures are involved immediately raises questions about how the crush energy was divided up among the crash partners. Frontal fixed barrier tests are the easiest to work with, since the test vehicle has all the kinetic energy to begin with, and winds up with all of the crush energy being dissipated in its own structure. We find two such tests: DOT Test 2967 of a 1999 Pontiac Grand Am SE at 35.1 mph, and DOT Test 3617 of a 2001 Pontiac Grand AM SE at 34.7 mph. We download these two test reports.

Ideally, we would like to have tests over a range of crash severities, so having only two tests at about 35 mph is not optimal. However, it is unusual to find frontal barrier crash at speeds other than about 30 and 35 mph.

A potential difficulty arises with "data clustering," as pointed out by Strother et al.[5] The purpose of a crash plot is to characterize the structure using a regression line, so that vehicles in crashes that are necessarily different from the barrier crash tests can have their crush energies evaluated. Clustered data introduce and accentuate trend line uncertainties, whereas distributed data make for a more reliable characterization. To reduce the uncertainties, it is common practice to make the trend line reflect a "damage onset" of about 5 mph; that is, measurable crush damage starts to accumulate at barrier crash speeds above the damage onset of about 5 mph.[6] It is this author's practice to implement a onset speed of about 5 mph by calculating what the data point would be if each vehicle, tested at some other speed, were also tested at about 5 mph. A fictitious "test" is then created by way of a rough average of the various 5-mph data points. A single such "test" is enough to eliminate data clustering uncertainties, while reflecting a damage onset barrier test speed of about 5 mph. The pertinent test data are shown in Figure 21.3.

The calculations are constructed so as to be able to handle moving-barrier impacts, to be discussed later, so the barrier weights are specified very large in order to simulate a fixed barrier. Restitution coefficients are determined from the velocity–time traces from the actual tests. Since crush was documented in both tests with six evenly spaced measurements, Equations 21.48 and 21.57 were used to calculate the average crush and form factor, respectively. For Test 3617, for example, its average crush was 16.26 in., and its form factor was 1.01936. The form factor is very close to unity, as expected, because a barrier crash produces a very flat crush profile.

Since the structural parameters A and B are reported in lb/in. and lb/in.^{2}, respectively, the crush is measured in in., and the weights are in lb, the crush energy was converted from foot-lb to in.-lb for the crash plot. Consequently, the ordinate, or vertical axis (the energy of crush factor) is calculated using a multiplier of 12. For Test 3617, the crash plot abscissa, or horizontal coordinate, is the product of the form factor and the average crush, or 16.57, as shown in Figure 21.3.

Since the vehicle was crushed across its entire width in this test, for computation purposes, the crush width was taken as its overall width. It has not been this author's practice to use reported crush widths for severe full-width frontal and rear barrier tests because the entire width is known to have been involved in the crush, whether the testing agency measured all of it or not; the overall vehicle width is a much more reliable measure of

DOT#	Speed mph	Crush Energy ft-lb	O/A Width in.	Test Weight lb	Barrier Weight+ lb	Restit. Coeff.	Crush Values, in.						Cavg in.	Form Factor* —	Multi-plier	Crash Plot		Ord. @ 5 mph
							C_1 in.	C_2 in.	C_3 in.	C_4 in.	C_5 in.	C_6 in.				Abscissa	Ord.	
2967	35.11	145,473	70.4	3567	1.000E+09	0.097	17.4	19.6	19.8	19.8	19.6	14.8	19.01	1.00412	6.3563	19.09	223.15	31.781
3617	34.73	137,916	70.4	3488	1.000E+09	0.137	10.2	15.1	18.1	18.2	18.1	13.2	16.26	1.01936	6.3027	16.57	218.92	31.513

Sample Calculations

C_1	C_2	C_3	C_4	C_5	C_6	Cavg	FF
17.0	19.5	16.1				18.03	1.002440
17.9	19.0	19.3	19.5	18.7	17.8	18.87	1.000677

FIGURE 21.3
Pertinent data for crash plot.

the actual crush width. The effect is to reduce the structural parameters slightly from the standard NHTSA practice because the crush is spread over a greater numerical width. However, the effect is canceled out because accident vehicles are treated the same way. There is complete consistency in the treatment of both test vehicles and vehicles involved in field accidents.

The crash plot ordinate is calculated using Equation 21.62. For Test 3617, the ordinate is 218.92, as is also shown in Figure 21.3. The data points for the crash plot are in the third and second columns from the right. Ordinates corresponding to crush onset at 5 mph were also calculated, and are in the last column.

The actual crash plot is shown in Figure 21.4.

The line is a linear regression on three points: the two test data points, and the fictitious "test" at 5 mph. One can readily see that a regression line through only the two actual data points would have produced a serious data clustering issue: the results would have been accurate for barrier crashes in the 35 mph range, of course, but other crash conditions would have been fraught with uncertainties.

The slope of the regression line is 10.49 $lb^{1/2}$/in., and the intercept is 33.27 $lb^{1/2}$. From Equation 21.67, the B parameter is the square of the slope, or 110.1 $lb/in.^2$. From Equation 21.66, the A parameter is the product of the slope and the intercept, or 348.0 lb/in. To the extent that the regression line passes near the points for the 35 mph tests, reconstructing those impacts using the measured crush values and the derived A and B values will produce a good approximation of the barrier test speeds. If the regression line goes directly through the data point for a certain test, its speed will be reconstructed with complete accuracy, given the actual crush

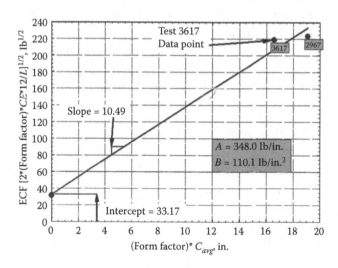

FIGURE 21.4
Example crash plot.

measurements. Other crashes producing distributed crush profiles and comparable average crushes can have accurate crush energy assessments. However, there are no actual data points for average crush values below 16 in. Therefore, there can be no expectation that crashes with crash plot abscissa values well below 16 in. (low-speed impacts, say) will be subject to accurate crush energy assessment.

It is always the case that crush energy assessments will be the most accurate when using structural parameters that are based on crashes that exercise the structure in a similar way, and that are of a similar severity.

Constant-Stiffness Crash Plots for Uniaxial Impacts by Rigid Moving Barriers

Suppose we have a uniaxial crash in which one of the collision partners is rigid. This is the situation with a crash test involving a rigid moving barrier, whether it hits a vehicle in the front, the rear, or the side. Then all of the crush energy CE_{tot} in Equation 19.21 is assignable to the tested vehicle, as simply CE. Squaring both sides, we can write

$$CE = \frac{V_d^2 W_B W_V (1 - \varepsilon^2)}{2g(W_B + W_V)} \tag{21.68}$$

where the subscripts B and V refer to the barrier and to the tested vehicle, respectively. Dividing numerator and denominator by W_B yields

$$CE = W_V \left[\frac{1}{1 + W_V / W_B} \right] \frac{V_d^2 (1 - \varepsilon^2)}{2g} \tag{21.69}$$

We now define the vehicle's effective weight W_{EFF} as

$$W_{EFF} = W_V \left[\frac{1}{1 + W_V / W_B} \right] \tag{21.70}$$

and Equation 21.69 becomes

$$CE = \frac{W_{EFF} V_d^2 (1 - \varepsilon^2)}{2g} \tag{21.71}$$

Note that as the barrier becomes infinitely massive,

$$\lim_{W_B \to \infty} W_{EFF} = \lim_{W_B \to \infty} W_V \left[\frac{1}{1 + W_V/W_B} \right] = W_V \tag{21.72}$$

which implies that Equation 21.71 is valid for crashes into an infinitely massive (fixed) barrier as well as a moving one. Now, we are in a position to substitute Equation 21.71 into Equation 21.61. The result is

$$\sqrt{\frac{W_{EFF}\beta(1 - \varepsilon^2)}{gL}}(V_{cl}) = \sqrt{B}(\beta\bar{c}) + \frac{A}{\sqrt{B}} \tag{21.73}$$

This important result shows that for uniaxial impacts into barriers of either finite or infinite weight, Campbell's original observation that the crush is proportional to the test speed is indeed valid if the structure can be characterized by a constant-stiffness model. Equation 21.73 also provides an alternative construction for crash plots, in which the ordinate from Equation 21.62 can be replaced with the left-hand side of Equation 21.73.

Quantities in all these equations are assumed to be expressed in terms of consistent units. It is often the case, however, that common usage is anything but consistent. For example, global (scene) coordinates, such as position, velocity, and energy, are often in feet (or meters) or miles per hour (or kilometers per hour), whereas vehicle quantities such as crush are often in inches (or millimeters). To complicate matters further, speed in earlier U.S. crash tests was usually reported in miles per hour. The units of A and B are usually in lb/in. and lb/in.^{2}, respectively. If the crush profile is in inches and the crush energy CE is in ft-lb, then Equation 21.61 becomes

$$ECF = \sqrt{\frac{24\beta(CE)}{L}} = \sqrt{B}(\beta\bar{c}) + \frac{A}{\sqrt{B}} \tag{21.74}$$

On the other hand, if we keep Equation 21.73 in an in.-lb-sec unit system but wish to enter the closing velocity in mph, and use a gravitational acceleration of 32.2 ft/sec^2, then these quantities must be converted to in./sec and in./sec^2, respectively. Equation 21.73 then becomes

$$\sqrt{\frac{1}{32.2(12)}} \sqrt{\frac{W_{EFF}\beta(1 - \varepsilon^2)}{L}}[12(1.4667)V_{cl}] = \sqrt{B}(\beta\bar{c}) + \frac{A}{\sqrt{B}} \tag{21.75}$$

or

$$ECF = 0.8954\sqrt{\frac{W_{EFF}\beta(1 - \varepsilon^2)}{L}}(V_{cl}) = \sqrt{B}(\beta\bar{c}) + \frac{A}{\sqrt{B}} \tag{21.76}$$

The left-hand equality has been incorporated into the spreadsheet formulas that calculate the crash plot ordinate, or energy of crush factor ECF, for each crash test in the crash plot.

Segment-by-Segment Analysis of Accident Vehicle Crush Profiles

Form factors come into play when it is desired to analyze crash test data for the purpose of structural characterization. The reason is that the characterization is accomplished through a crash plot, the abscissa and ordinate of which involve the calculation of parameters that pertain to the entire crush profile: average crush and form factor. Typically, the crash test crush profiles are piecewise linear with uniform measurement intervals, in which case the formulas for the average crush $\bar{c}$, Equation 21.46, and for the form factor β, Equation 21.55, pertain.

Once A and B have been found, however, the calculation of crush energy for an accident vehicle can be done using the average crush $\bar{c}$ and the form factor β to characterize, if you will, the crush profile. However, the accident vehicle crush is usually measured with a piecewise linear profile, often using irregular measurement intervals. It is advantageous, therefore, to develop a means of analyzing the profile segment-by-segment—calculating the contribution in each segment to the crush energy and the force, and summing the results at the end. This is made theoretically possible because the integral in Equation 21.10 can be subdivided into segments, just like the profile itself. Moreover, the nonintegral terms in Equation 21.10 can be expressed in integrals, which can also be broken up into segments.

In a similar fashion, the intensity of the force, which can and does vary along the profile, is expressed appropriately as a density function (per unit length), so the total collision force can also be expressed as an integral, and subdivided into segments, just like the crush energy calculation is. Therefore, it is not necessary to fit a mathematical curve to the crush profile. Such curve-fitting was presented above only for the purpose of illustrating the concept, and showing how the average crush and form factor were influenced by the curve shape.

That said, Equation 21.10 and its predecessor 21.7 both involve the structural constant G. It was shown in Equation 21.22 that G is a function of A and B, but the relationship depends on the form factor β, which is calculated for the crush profile as a whole. Therefore, for segment-by-segment calculations to work, all the crush profile segments for the accident vehicle have to be processed, and the overall form factor calculated, before the segment-by-segment calculations can be completed. If β were always unity, of course, this little detour would not be necessary, but then if all field accidents were square-on flat barrier crashes, probably a trained monkey could figure out what the speed was. From Equation 21.54, we can write

$$\beta = \frac{1}{3L\bar{c}^2} \sum_{i=1}^{N-1} (Coeff)_i \tag{21.77}$$

where $(Coeff)_i$, the contribution to the calculation of β from each segment, is given by

$$(Coeff)_i = \left(C_i^2 + C_i C_{i+1} + C_{i+1}^2 \right) \Delta x_i \tag{21.78}$$

Here, the intent is to calculate Δx_i, $(Area)_i$ using Equation 21.41, and $(Coeff)_i$ for each segment; then at the end calculate L, $\bar{c}$, and β, by summing up the contribution to those various parameters from all the segments. Using Equation 21.8, we can proceed to calculate the contributions to the force and the crush energy as follows:

$$CE = \int_0^L d(CE) = \sum_{i=1}^{N-1} (CE)_i \tag{21.79}$$

where

$$(CE)_i = \int_{x_i}^{x_{i+1}} \left(Ac + \frac{1}{2} Bc^2 + G \right) dx \tag{21.80}$$

and where G has already been calculated for the profile in question, using Equation 21.22, which requires that the form factor β has been calculated as well. Using the expression for the crush variation in the ith segment, Equation 21.49, along with the variable transformation introduced in Equation 21.52, we can write

$$(CE)_i = A \int_0^{\Delta x_i} \left[C_i + \frac{C_{i+1} - C_i}{\Delta x_i} \xi \right] d\xi + \frac{1}{2} B \int_0^{\Delta x_i} \left[C_i + \frac{C_{i+1} - C_i}{\Delta x_i} \xi \right]^2 d\xi + G \int_0^{\Delta x_i} d\xi \tag{21.81}$$

Recalling the expression for form factor β, Equation 21.53, the second integral in Equation 21.81 has already been evaluated. Performing the other integrations leads to

$$(CE)_i = A\overline{c}_i\Delta x_i + \frac{1}{2}B\left[C_i^2 + C_iC_{i+1} + C_{i+1}^2\right]\frac{\Delta x_i}{3} + G\Delta x_i$$

$$= \left(A\overline{c}_i + G\right)\Delta x_i + \frac{B}{6}(Coeff)_i \tag{21.82}$$

where $(Coeff)_i$ has been defined above in Equation 21.78. Using Equation 21.25, the total force required to produce the profile crush is

$$F_{tot} = \int_0^L F dx = \sum_{i=1}^{N-1} F_i \tag{21.83}$$

where

$$F_i = \int_{x_i}^{x_{i+1}} (A + Bc) dx \tag{21.84}$$

Substituting the function for a piecewise linear crush profile, Equation 21.49, and using the variable transformation of Equation 21.52, results in

$$F_i = A\int_0^{\Delta x_i} d\xi + B\int_0^{\Delta x_i}\left[C_i + \frac{C_{i+1} - C_i}{\Delta x_i}\xi\right]d\xi \tag{21.85}$$

These integrations were performed in deriving Equation 21.81. Using them again, plus the definition of segment average crush $\overline{c}_i$ in Equation 21.40, yields

$$F_i = \left(A + B\overline{c}_i\right)\Delta x_i \tag{21.86}$$

Now the crush energy and the force contributed by the various segments can be summed, which is what we set out to be able to do.

Constant-Stiffness Crash Plots for Repeated Impacts

The conceptual basis underlying repeated-impact testing is that vehicles are made primarily of metal parts that are stressed beyond the elastic limit in a crash, and that once a structure has been loaded and unloaded, upon reloading, its force–deflection relationship approximately retraces the previous unloading slope until the original force–deflection curve is encountered. Past that point, the structure resumes following that original curve.

For repeatedly impacted vehicles, this concept leads to the hypothesis that the relationship between residual crush and absorbed energy is the same whether the absorbed energy comes from a single crash or a series of crashes in which the total absorbed energy is the same. Of course, the strain rate in a single test is different from (and usually greater than) the strain rates in a series of tests at lower speeds. Therefore, for the repeated impact method to "work," the crush energy must be unaffected by the strain rate, and by any differences in the unloading and reloading curves.

The hypothesis was originated by Warner et al.[7] in 1986. It has been put to the test for several different vehicles,[8] and crash plots have shown very good agreement between a single impact, and repeated impacts. It has become a generally accepted procedure for investigating structural behavior.

Suppose a series of R crashes is conducted. The restitution energy in test $R-1$ was re-absorbed by the structure during the loading portion of test R, as was the case for all the prior impacts. Thus, the total absorbed energy for the first $R-1$ tests is simply the sum of their individual kinetic energies at impact. During test R, some energy is recovered in restitution. From Equation 21.71, the total absorbed energy for the first R tests is then

$$CE_R = \frac{W_{EFF}}{2g} \sum_{K=1}^{R-1} (V_{cl})_K^2 + \frac{W_{EFF}}{2g} (V_{cl})_R^2 (1 - \varepsilon_R^2) \qquad (21.87)$$

This quantity can then be substituted into Equation 21.62 to obtain the desired crash plot ordinate (ECF) that pertains after R tests in the series. A new point, consisting of the R tests run to date, will be created on the crash plot for each new test that is run. In this way, a single vehicle can be used in repeated crash tests to produce an entire crash plot. One can thereby determine not only the values of A and B, but whether the constant-stiffness model applies over the entire range of crash severities under consideration.

By hypothesis, the absorbed energy after R repeated tests is the same as that after a single test S. Thus,

$$CE_S = \frac{W_{EFF}}{2g} (V_{cl})_S^2 (1 - \varepsilon_S^2) \qquad (21.88)$$

The question is: What is the speed of that single test? Equating the above expressions, we have

$$(V_{cl})_S^2 = \frac{\sum_{K=1}^{R-1} (V_{cl})_K^2 + (V_{cl})_R^2 (1 - \varepsilon_R^2)}{1 - \varepsilon_S^2} \qquad (21.89)$$

If one expects the restitution coefficient in the single test to be the same as in test R, then the above reduces to

$$(V_{cl})_S^2 = \frac{1}{1 - \varepsilon_R^2} \sum_{K=1}^{R-1} (V_{cl})_K^2 + (V_{cl})_R^2 \tag{21.90}$$

Constant Stiffness with Force Saturation

If there are enough points on a crash plot spread sufficiently well over a range of crash severities, one can observe whether a straight line through all the points represents the best fit to the data. It is sometimes the case that linearity exists over a portion of the data, but there is a noticeable deviation at the higher severities. The data may rise above linearity, due to stacking up of hard elements in the engine compartment, or they may fall below linearity, due to structural separations or large-scale buckling. In the latter case, a better fit to the data may be achieved by assuming that at some point, the force in the structure becomes saturated; that is, with increasing crush beyond a certain point, the force no longer rises, but remains at some plateau. This concept was introduced by Strother et al.[5] in 1986. The crush at which force saturation occurs is called the saturation crush. The force–deflection characteristic giving rise to this behavior is shown in Figure 21.5.

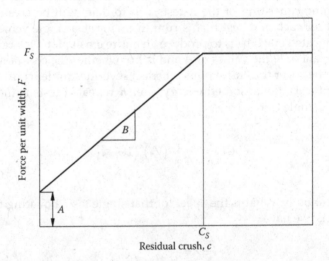

FIGURE 21.5
Constant-stiffness model with force saturation.

We see that in contrast to the conventional constant-stiffness model with parameters A and B, this model has been truncated at the saturation crush C_S, after which point the force F remains constant at the saturation force F_S. It is assumed that A and B have been determined by a regression fit to crash test data for which, at a minimum, the average crush values are less than C_S. As we shall see, in the underlying tests it is even better to have all the crush measurements, not just the average crush, be less than C_S.

For crashes at less than the saturation crush, Equation 21.24 suffices to find the crush energy. At higher severities, some of the crush values may exceed the saturation crush, at which location the force is F_S, and Equation 21.24 no longer applies. At other points on the same crush profile, the crush may be less than C_S, and Equation 21.24 is still valid. Needless to say, this complicates the crush energy assessment.

Before getting into that subject, however, it is useful to see what force saturation looks like in a constant-stiffness crash plot. We avoid the complications mentioned above by considering an example in which the crush is a uniform value C_U, for which case the form factor β is unity. For crush values below the saturation crush, Equation 21.61 for the regression line in the crash plot becomes

$$\sqrt{\frac{2CE}{L}} = \sqrt{B}C_U + \frac{A}{\sqrt{B}} \qquad C_U \le C_S \tag{21.91}$$

This straight line terminates at the saturation crush C_S. At that point, the saturation force F_S may be obtained by applying Equation 21.6, as follows:

$$F_S = A + BC_S \tag{21.92}$$

The energy of crush factor (ECF) may be obtained by applying Equation 21.24, as follows:

$$\frac{CE}{L} = \frac{1}{2B}[A + BC_S]^2 = \frac{F_S^2}{2B} \tag{21.93}$$

Therefore,

$$\sqrt{\frac{2CE}{L}} = \frac{F_S}{\sqrt{B}} \tag{21.94}$$

To calculate the crush energy beyond the saturation crush, to the crush energy expressed in Equation 21.93, we add the product of the saturation force, multiplied by the additional crush past crush saturation. In equation form,

$$\frac{CE}{L} = \frac{F_s^2}{2B} + F_s(C_u - C_s) \qquad C_u > C_s \qquad (21.95)$$

Therefore, beyond the saturation crush, we have

$$\sqrt{\frac{2CE}{L}} = \sqrt{\frac{F_s^2}{B} + 2F_s(C_u - C_s)} \qquad C_u > C_s \qquad (21.96)$$

We see that beyond the saturation crush, the energy of crush factor is no longer linear with C_u, but varies according to a square root function.

To get a feel for what the crash plot would look like when force saturation exists, consider a structure for which $A = 200$ lb/in., $B = 100$ lb/in.^{2}, and $C_s = 20$ in. The resulting crash plot is shown in Figure 21.6.

Notice that there is no sudden "kink" in the plot when force saturation is reached, even in this special case where all the structure saturates all at once. The deviation from linearity is gradual. Looking at a series of crash test data points, it may not be all that obvious where force saturation occurs. This may be turned to advantage, in that one may be able to adjust the saturation crush C_s so as to get the best fit for the data points beyond force saturation. Note also that in postulating the existence of force saturation, we have created a three-parameter model, as distinct from the usual two-parameter constant stiffness model. More information on two- and three-parameter models is contained in a 1999 paper by Welsh.[9]

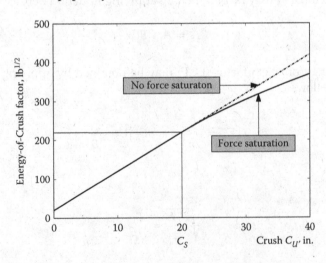

FIGURE 21.6
Example crash plot with uniform crush and force saturation.

Constant Stiffness Model with Force Saturation, Using Piecewise Linear Crush Profiles

In practice, crush profiles are not uniform, but variable, and represented with piecewise linear functions. In profiles where there is force saturation, generally it occurs in some, but not all, of the segments. For segments in which saturation crush has not been exceeded, the analysis can proceed as usual with a constant stiffness model. The crush energy contribution from the segment is given in Equation 21.82, repeated here:

$$(CE)_i = (A\bar{c}_i + G)\Delta x_i + \frac{B}{6}(Coeff)_i \qquad \bar{c}_i \leq C_S \qquad (21.97)$$

where $(Coeff)_i$ is given by Equation 21.78, also repeated here

$$(Coeff)_i = \left(C_i^2 + C_iC_{i+1} + C_{i+1}^2\right)\Delta x_i \qquad C_i, C_{i+1} \leq C_S \qquad (21.98)$$

The force contribution for the segment is given by Equation 21.86, repeated here:

$$F_i = (A + B\bar{c}_i)\Delta x_i \qquad \bar{c}_i \leq C_S \qquad (21.99)$$

In segments where there is a transition into force saturation, additional crush data points can be generated (by interpolation) where $c(x) = C_S$, so that any given segment is either entirely above or below C_S. Where the entire segment is precisely at crush saturation, $C_i = C_{i+1} = C_S$, $(Coeff)_i = 3C_S^2\Delta x_i$, and

$$(CE)_i = (A\bar{c}_i + G)\Delta x_i + \frac{B}{6}(3C_S^2\Delta x_i) = \left(AC_S + \frac{1}{2}BC_S^2 + G\right)\Delta x_i \qquad (21.100)$$

which is what we would expect from Equation 21.14, with $\beta = 1$. The force per unit length in the segment has reached the saturation force F_S, given by

$$F_S = A + BC_S \qquad (21.101)$$

See Equation 21.6. Now the only remaining task is to analyze the segments in which saturation crush has been exceeded.

This was the main motivation for performing the analysis segment-by-segment. In a given segment, the crush can be evaluated as to whether the saturation crush has been exceeded. This is a logical branch point; the crush energy and force are computed in the manner we have discussed if there is no crush saturation, and computed another way if there is. Our task now is to figure out what that other way is: how to do the analysis if crush saturation is present.

In any segment in crush saturation, the structure would have reached the crush energy and force specified in Equations 21.100 and 21.101, and passed beyond that point in the force–deflection curve. Beyond saturation, the force remains constant at the saturation force, so

$$F_i = F_S \Delta x_i = (A + BC_S)\Delta x_i \qquad c > C_S \qquad (21.102)$$

Referring to Figure 21.5 and Equation 21.7, the differential crush energy is

$$d(CE) = AC_S + \frac{1}{2}BC_S^2 + G + F_S(c - C_S) \qquad c > C_S \qquad (21.103)$$

Again, the crush profile in the segment is linear, and is given by Equation 21.49. Substitution of that equation into the above, and integrating over the segment length gives

$$CE_i = \int_{x_i}^{x_{i+1}} \left[AC_S + \frac{1}{2}BC_S^2 + G + F_S(c - C_S) \right] dx \qquad c > C_S \qquad (21.104)$$

This can be rewritten as

$$CE_i = \left[AC_S + \frac{1}{2}BC_S^2 + G - F_SC_S \right] \int_0^{\Delta x_i} d\xi + F_S \int_0^{\Delta x_i} \left[C_i + \frac{C_{i+1} - C_i}{\Delta x_i}\xi \right] d\xi$$

$$C_i, C_{i+1} > C_S \qquad (21.105)$$

after having used (again) the variable transformation in Equation 21.52. These integrals have been seen before, in the development of Equation 21.82. The result of the integration is

$$CE_i = \left[AC_S + \frac{1}{2}BC_S^2 + G - F_SC_S \right] \Delta x_i + F_S\bar{c}_i\Delta x_i \qquad \bar{c}_i > C_S \qquad (21.106)$$

or

$$CE_i = \left[AC_S + \frac{1}{2}BC_S^2 + G + F_S(\overline{c}_i - C_S) \right] \Delta x_i \qquad \overline{c}_i > C_S \qquad (21.107)$$

This is readily seen to match up with Equation 21.100 when the crush is exactly at saturation.

Again, Equation 21.107 only applies to the segments where both C_i and C_{i+1} exceed C_S.

The force in the segment is simply the product of the (constant) saturation force per unit width F_S, multiplied by the segment width Δx_i. We then have

$$F_i = F_S \Delta x_i \qquad c > C_S \qquad (21.108)$$

where F_S is given by Equation 21.92, and the result is identical to Equation 21.102.

In summary, for a constant-stiffness model and a piecewise linear crush profile, we can add a point (if necessary) where the crush equals the saturation crush C_S, rendering the crush throughout each segment either saturated or unsaturated. For all unsaturated segments, the crush energy and force contributions are calculated by Equations 21.97 and 21.99, respectively. For all saturated segments, the crush energy and force are calculated by Equations 21.107 and 21.108, respectively. The crush energy and force contributions are then summed over all the segments.

An example of this procedure is shown in Figure 21.7, which illustrates a constant stiffness model with force saturation, for a side impacted vehicle. The table is necessarily truncated, in that more than six columns would normally be used for a side impact crush profile.

The cells shaded with light gray are the values entered by the user. The x locations shaded in darker gray indicate additional locations that were calculated, by linear interpolation, on the basis that the crush depth at that point reached the saturation value of 4.6 in. Each segment is represented by a column of numbers which is below x_i and to the left of x_{i+1}. In that column are computed the segment length Δx_i, $(Area)_i$, and $(coeff)_i$ in order, followed by the force contribution F_i and the crush energy contribution CE_i. These last two calculations are done either of two ways, depending on whether the average crush in the segment exceeds the saturation crush C_S or not. In the seventh column of numbers, the segment results are summed. The total crush energy and force for the profile are presented additionally.

The procedure is designed to calculate both crush energy and force according to two models: constant stiffness and constant force (which is discussed below). In this case, the constant force calculations were not done because a blank cell was used for the values of F_S.

Profile: Right Side Doors

A = 248.5 lb/in.   B = 137.30 lb/in² C_s = 4.6 in. F_s = ☐ lb/in

Constant Stiffness Model
CE = 75,709 ft-lb
Force = 65,755 lb

Constant Force Model
CE = ☐ ft-lb
Force = ☐ lb

| Depth C (in.): | 0.00 | 4.60 | 21.30 | 23.60 | 20.40 | 4.60 | 0.00 |
| Location X (in.): | −4.90 | −3.84 | 0.00 | 5.80 | 28.10 | 63.6 | 73.90 |

								Sum	
delx, in.	1.06	3.84	5.80	22.30	35.47	10.33		78.80	= L
AvgC, in.	2.30	12.95	22.45	22.00	12.50	2.30			
Area, sq. in.	2.4	49.8	130.2	490.6	443.4	23.8		1140.2	= Area
coeff, in²	22	2201	8777	32437	18842	219		62497	
Const. Stiff. F, lb	597	3381	5104	19626	31219	5828		65755	
Const. Stiff. CE, ft-lb	109	3241	8934	33612	28752	1062		75709	
Const. Force F, lb									
Const. Force CE, ft-lb									

Average crush C_{avg} = 14.469 in.
Form factor β = 1.26281
Crash plot abscissa βC_{avg} = 18.27 in.
G = 178.08 lb
Saturation force $F_s = A + Bc_s$ = 880.08 lb/in.
Saturation Energy = 2773.81 lb

Sample Calculations

Point No.	i-1	-i-	i+1
Crush Depth C	−2	0	3
Location X	1	3.0	6

Sample Calculations

Point No.	0	i-1	-i-	i+1	i+2	0
Crush Depth C	0	4.5	3	2	1.5	0
Location X	50.0	5	20	30	35	50.0

☐ ==> outside edge of vehicle
(#.#) ==> negative crush depth; treat as zero
☐ Input quantities are in BOLD type

Crush Extrapolation left: $C_{i-1} = C_i - [(C_{i+1} - C_i)/(X_{i+1} - X_i)](X_i - X_{i-1})$

Crush Extrapolation right: $C_{i+2} = C_{i+1} + [(C_{i+1} - C_i)/(X_{i+1} - X_i)](X_{i+2} - X_{i+1})$

Location of Crush value C_i: $X_i = [X_{i-1}(C_{i+1} - C_i) + X_{i+1}(C_i - C_{i-1})]/(C_{i+1} - C_{i-1})$

Left X zero crossing: $X_0 = X_i - C_i/[(C_{i+1} - C_i)/(X_{i+1} - X_i)]$

Right X zero crossing: $X_N = X_i - C_i/[(C_i - C_{i-1})/(X_i - X_{i-1})]$

Force Calc begins at: −4.90 in.   Force calc ends at 73.90 in.   Weighting factor = 1.000

FIGURE 21.7
Segmented crush energy and force calculations with saturation.

Constant-Force Model

A constant-force model assumes that the structure's force–deflection characteristic, per unit width, is as shown in Figure 21.8.

This looks a lot like Figure 21.5, but there are crucial differences. In the constant-stiffness model with force saturation, the initial part of the force–deflection curve does not necessarily pass through the origin. More importantly, in a constant-stiffness model the saturation crush C_S is a third parameter and is determined from deviations from linearity in the crash plot, from physical observations of crash behavior, or from a desire to be conservative when crush measurements in the field exceed those seen in crash tests. By contrast, in a constant-force model, the saturation crush C_S, along with the saturation force F_S, are to be determined in a way analogous to A and B, and constitute the characteristics of a two-parameter model. It is assumed, therefore, that the crash-tested vehicles have already achieved force saturation (i.e., the measured crush values are greater than C_S), whereas in the constant-stiffness model the opposite assumption is made.

Referring to Figure 21.8, the energy absorbed, per unit width, by the time the residual crush at that location has reached c, is

$$d(CE) = \left[F_S c - \frac{1}{2} F_S C_S \right] dx \qquad c > C_S \tag{21.109}$$

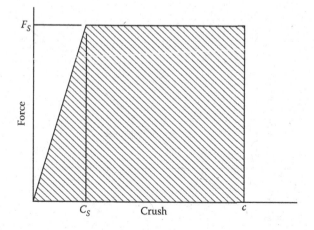

FIGURE 21.8
Constant-force model.

Integrating over the length L of the crush profile yields the total crush energy:

$$CE = F_S \int_0^L c(x)dx - \frac{1}{2}F_S C_S \int_0^L dx \qquad c > C_S \qquad (21.110)$$

assuming that C_S and F_S are constant throughout the crush length. Recalling Equation 21.2, the definition of average crush $\bar{c}$, we can write

$$\frac{CE}{L} = F_S \bar{c} - \frac{1}{2}F_S C_S \qquad \bar{c} > C_S \qquad (21.111)$$

which is again a linear relationship involving the average crush $\bar{c}$, of the form $y = mx + b$. A plot of this relationship looks like Figure 21.9, where the slope is F_S and the intercept is $-\frac{1}{2}F_S C_S$.

This plot is used in the same way, and is the counterpart of the constant-stiffness crash plot: a linear regression through the data points is used to find the parameters F_S and C_S. Of course, the analysis only applies for tests in which $\bar{c}$ is indeed greater than C_S. If $\bar{c}$ in some of the tests is less than C_S, such tests must be excluded from the regression analysis.

To apply the model to a specific crash, determine the average crush and use Equation 21.110 to compute the crush energy, as long as $c > C_S$. The total force that produced the crush is simply the product of the saturation force (per unit crush) F_S and the crush width. Thus,

$$F_{tot} = F_S L \qquad c > C_S \qquad (21.112)$$

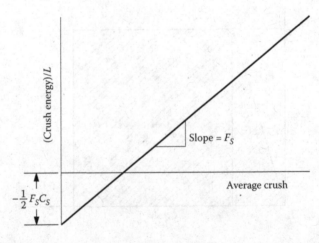

FIGURE 21.9
Crash plot for a constant-force model.

For field accidents in which $c \leq C_S$, the analysis proceeds as follows:

$$d(CE) = \left[\frac{1}{2} \frac{F_S}{C_S} c^2 \right] dx \qquad c \leq C_S \qquad (21.113)$$

Again, refer to Figure 21.8. Again assuming that F_S and C_S are constant over the crush length L, we can integrate over that length as follows:

$$CE = \frac{1}{2} \frac{F_S}{C_S} \int_0^L c^2(x)dx \qquad c \leq C_S \qquad (21.114)$$

We have seen this integral before. Recalling Equations 21.11 and 21.13, we can write

$$CE = \frac{F_S L}{2C_S} \beta \bar{c}^2 \qquad c \leq C_S \qquad (21.115)$$

or

$$CE = \frac{F_S L}{2\beta C_S} (\beta \bar{c})^2 \qquad c \leq C_S \qquad (21.116)$$

The expression is put in this last form because $\beta \bar{c}$ is the abscissa of the familiar constant-stiffness crash plots. The form factor β is calculated in the same way as before.

Constant-Force Model with Piecewise Linear Crush Profiles

As discussed earlier regarding constant-stiffness models with force saturation, as applied to piecewise linear crush profiles, a similar approach can be taken. In each segment between crush measurements, the crush is assumed to vary linearly with position. If there is a transition to force saturation in any given segment, a point can be added to the profile at $c(x) = C_S$, so that in any given segment, the crush is either entirely above or entirely below force saturation. Each segment can be analyzed individually, and the force and crush energy contributions from all of them can be added together.

In each segment, the crush profile looks like Figure 21.2, and the crush is given by Equation 21.49. If the segment is unsaturated (i.e., $c(x) < C_S$ for

$x_i \leq x < x_{i+1}$), we see from Figure 21.8 that the differential crush energy is given by

$$d(CE_i) = \frac{1}{2}\left(\frac{F_S}{C_S}\right)c^2 dx \qquad c \leq C_S \tag{21.117}$$

where the subscript i denotes that the calculation will be for the ith segment only. Integrating over the length of the segment, we have

$$CE_i = \frac{1}{2}\frac{F_S}{C_S}\int_{x=x_i}^{x=x_{i+1}} c^2(x)dx \qquad c \leq C_S \tag{21.118}$$

This may look familiar, which it should be because the force is proportional to the crush, and the crush energy is once again a function of the square of the crush. This integral was already evaluated during the development of Equation 21.53. We thus obtain

$$CE_i = \frac{F_S}{6C_S}(Coeff)_i \qquad c \leq C_S \tag{21.119}$$

where $(Coeff)_i$ is given by Equation 21.78. In a similar fashion, the force contribution from the ith segment is

$$F_i = \int_{x_i}^{x_{i+1}} \frac{F_S}{C_S}c(x)dx \qquad c \leq C_S \tag{21.120}$$

We can substitute the piecewise linear function for $c(x)$ of Equation 21.49, employ the variable transformation of Equation 21.52, and perform the indicated integration. This was all done in the development of Equation 21.82. The result is

$$F_i = \frac{F_S}{C_S}\bar{c}_i \Delta x_i \qquad c \leq C_S \tag{21.121}$$

If the entire segment is precisely at crush saturation, $C_i = C_{i+1} = C_S$, $(Coeff)_i = 3C_S^2 \Delta x_i$, and Equation 21.119 becomes

$$CE_i = \frac{F_S}{6C_S}\left(3C_S^2 \Delta x_i\right) = \frac{1}{2}F_S C_S \Delta x_1 \tag{21.122}$$

From Equation 21.121, the force in that segment is

$$F_i = F_S \Delta x_i \qquad c > C_S \qquad (21.123)$$

This is identical to Equation 21.108. However, Equation 21.108 was derived for a constant stiffness model with force saturation, so the saturation force F_S was calculated by applying the saturation crush C_S to the force–deflection relation; that is, finding the force that corresponded to the saturation crush C_S. In this case, the saturation force F_S is obtained from the crash plot, as is the saturation crush C_S.

Beyond saturation, the crush occurs at a constant force F_S, and Equation 21.123 continues to apply. Additional crush energy is dissipated at that constant force, so from the force–deflection curve of Figure 21.8, we obtain

$$CE_i = \frac{1}{2} F_S C_S \Delta x_i + F_S (c - C_S) \Delta x_i = F_S \left(c - \frac{C_S}{2} \right) \Delta x_i \qquad c > C_S \qquad (21.124)$$

In summary, for a constant-force model and a piecewise linear crush profile, we can add a point (if necessary) where the crush equals the saturation crush C_S, rendering the crush throughout any given segment either saturated or unsaturated. For all unsaturated segments, the crush energy and force contributions are calculated by Equations 21.119 and 21.121, respectively. For all saturated segments, the crush energy and force are calculated by Equations 21.124 and 21.123, respectively. The crush energy and force contributions are then summed over all the segments.

Structural Stiffness Parameters: Make or Buy?

Now we have on hand the wherewithal to calculate the structural stiffness parameters for constant-force models and constant-stiffness models using crash plots, and to use those parameters with or without force saturation. The necessary calculations have been incorporated into spreadsheets. Using the developments in this chapter, one has the ability to explain the work, starting from basic principles. The equations and the resulting numbers have complete transparency.

However, the subject does entail some subtleties and complexities. Not everyone has the time, the training, or the inclination to work through the numbers themselves. For them, a valuable service is provided by vendors such as Neptune Engineering (NEI),[10] which offers for sale structural stiffness parameter values, based on the application of a constant-stiffness model to NHTSA crash test data. At the same time, anyone relying on the information has an obligation. As stated on NEI's web site (http://www.neptuneeng. com), "It is important that the user of stiffness coefficients understands how the coefficients were determined .... The coefficients are determined using concepts and procedures presented in Society of Automotive Engineers papers 920607, 940913, 950358, 960896, 980024 & 1999-01-0105."

To follow this advice and understand how the coefficients were determined, an example was considered: Toyota Sienna vans produced between model years 1998 and 2003, involved in full-width frontal crashes. For such vehicles, NEI produced data obtained from reports of two tests: DOT 2766, a barrier crash at approximately 35 mph, and Transport Canada 99-144, a barrier crash at approximately 30 mph. The data appeared in a table showing the pertinent test data, and the A and B parameters were calculated for each test by itself. A crash plot was also produced having two data points, presumably reflecting the two tests, plus what appears to be a regression fit. The equation of the regression line, its slope, and its intercept were not given. Thus, it was not possible to trace the derivation of A and B from the crash plot. However, the table indicates values of A and B from the crash plot of 380 lb/in. and 150 lb/in.^{2}, respectively.

So how to check the crash plot numbers? Guidance was sought from the SAE papers referenced on the NEI web site. In the 1992 paper,[6] formulas are presented for analyzing individual tests for A and B, but no discussion on crash plots appears. For an individual crash test, equations and example calculations are included in the 1994 and 1995 papers,[11,12] but again crash plots are not mentioned. The 1996 paper[13] presents vehicle properties that reside in the HVE[14] database, and includes a crash plot for Chevrolet Citations (that includes the equation of the regression line, by the way). It also contains a table with sample stiffness parameters, but the vehicle(s) are not identified; nor are there test data or calculations from which the parameters came. The 1998 paper[15] again presents equations and sample calculations, along with a crash plot (with a regression line equation) for a crushable barrier face, but there is no discussion of how to construct a crash plot for a series of crash tests, and interpret the results. The 1999 paper suggests various crush models and presents a couple of crash plots, again without explanation. In short, how to process a single test is covered. How to construct and interpret a crash plot for multiple tests is not. The attempt to obtain NEI's numbers from a crash plot by studying the cited references was not successful.

Another approach is to construct a crash plot using the methods of this chapter, and attempt to replicate NEI's results. Data for the two frontal tests of Toyota Siennas were obtained from NHTSA as discussed in Chapter 11. The data and analysis are shown in Figure 21.10 wherein it is seen that analyzing Test 2766 by itself, the computed A and B values are 325.5 lb/in. and 107.8 lb/in.^{2}, respectively. By way of contrast, the NEI values are 350 and 130. For Test 3087 analyzed by itself, the computed A

Applicable Crash Tests for Vehicle No. 2: 1998–2003 Toyota Sienna Vans

DOT#	Year	Make	Model	Angle deg	Speed mph	Crush Energy ft-lb	Test Weight lb	Barrier Weight* lb	Restit. Coeff.	C1 in.	C2 in.	C3 in.	C4 in.	C5 in.	C6 in.	C7 in.	C8 in.	C9 in.	Cavg in.	Form Factor*	Multi-plier	Crash Plot Abscissa	Crash Plot Test Data	A lb/in.	B lb/in.²	SMAC K1 lb/in.²	ε
2766	1998	Toyota	Sienna	0	4.50	3,078	45.0	1.000E+09	---										0.00	1.00600	7.0705	0.00	31.82				
3087	1999	Toyota	Sienna	0	34.92	179,983	45.6	1.000E+09	0.147	16.7	19.7	22.0	22.0	20.3	18.1				20.30	1.00561	6.9663	20.41	243.27	325.5	107.8	146.0	0.147
					29.33	125,672	44.4	1.000E+09	0.142	11.7	14.1	15.4	14.8	13.8	10.4				13.84	1.00867	6.9416	13.96	203.59	385.7	152.5	219.0	0.142

(Single-test Calculation columns: A, B, SMAC K1; TC Test 99-144)

* For 3 equally-spaced crush measurements, $C_{avg} = (C_1 + 2C_2 + C_3)/4$; $FF = (C_1^2 + C_1C_2 + 2C_2^2 + C_2C_3 + C_3^2)/(6C_{avg}^2)$

For 5 equally-spaced crush measurements, $C_{avg} = (C_1 + 2C_2 + 2C_3 + 2C_4 + C_5)/8$; $FF = (C_1^2 + C_1C_2 + 2C_2^2 + 2C_2C_3 + 2C_3^2 + C_3C_4 + 2C_4^2 + C_4C_5 + 2C_5^2)/(12C_{avg}^2)$

For 6 equally-spaced crush measurements, $C_{avg} = (C_1 + 2C_2 + 2C_3 + 2C_4 + 2C_5 + C_6)/10$; $FF = (C_1^2 + C_1C_2 + 2C_2^2 + 2C_2C_3 + 2C_3^2 + 2C_3C_4 + 2C_4^2 + 2C_4C_5 + 2C_5^2 + C_5C_6 + 2C_6^2)/(15C_{avg}^2)$

* Use 1.0e9 for fixed barriers

Sample Calculations

C1	C2	C3	C4	C5	C6	C7	C8	C9	Cavg	FF
17.0	19.5	16.1							18.03	1.002440
17.0	17.5	18.0	17.0	16.0					17.25	1.000910
17.9	19.0	19.3	19.5	18.7	17.8				18.87	1.000677

Input quantities are in **BOLD** type
to convert millimeters to inches: divide by 25.4
to convert kilograms to pounds: multiply by 2.2046230
to convert kph to mph: divide by 1.6093440

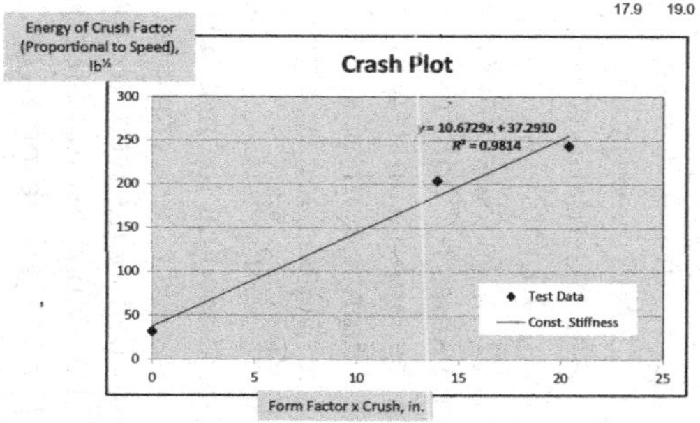

Slope = 10.6729 (lb)^½/in.
Intercept = 37.2910 (lb)^½
A = (slope) × (intercept) = **398.0** lb/in.
B = (slope)² = **113.9** lb/in.²

Crash Plot

$y = 10.672x + 37.2910$
$R^2 = 0.9814$

Energy of Crush Factor (Proportional to Speed), lb^½

Form Factor x Crush, in.

♦ Test Data
— Const. Stiffness

FIGURE 21.10
Crash test data and calculations for Toyota Siennas.

and B values are 385.7 lb/in. and 152.5 lb/in.², respectively, whereas NEI has the values at 430 and 190. Using a crash plot with a constant stiffness model to analyze both tests together results in A and B values of 398 lb/in. and 113.9 lb/in.², respectively. NEI's values are 380 and 150, respectively. Why the differences?

To get to the bottom of this question, changes were made to certain of the parameters and the calculations were repeated. It was found that A and B values could be replicated when analyzing individual tests. However, this author was unable to replicate the NEI crash plot results for a full set of tests. The results are shown in Figure 21.11, with altered parameters outlined with a heavier border.

This exercise leads to the following conclusions:

1. No errors were found in the crush values or test weights used by NEI.

2. NEI's calculation of average crush was correct.

3. NEI utilized the crush widths, called direct damage width, or DDW, reported in the test reports. In Figure 21.11, NEI's crush widths were used, and outlined in a heavier border to indicate the differences from the procedures developed in this chapter. The reported crush widths were different in the two tests, so in effect, the NEI calculations were dependent on the measurement procedures used at the test facility. It is hard to believe that in these two tests at 30 and 35 mph, the entire vehicle width would not have been involved.

4. Despite identifying 4.50 mph as a default value for the damage onset speed, NEI uses 4.2 mph for these tests. This is accomplished in the calculations by setting the speed of the fictitious data point at 4.2 mph, which is outlined with a heavier border in Figure 21.11. In Neptune's 1996 paper, he says "The average and mean values for both front end and rear end structures were found to be approximately 4.5 mph (7.2 km/h). Based on this analysis, a damage offset speed of 4.5 mph (7.2 km/h) was used for both front and rear structures where no low speed data was [sic] available"[13] (p. 256).

5. NEI does not account for restitution effects, which is to say that the energy recovered after a crash is ignored. To simulate this effect, the restitution coefficients were set to zero, and outlined with a heavier border. The actual restitution coefficients on Tests 2766 and 3087 were 0.147 and 0.142, respectively.

6. In the Transport Canada test (3087), integrations of the accelerations did not appear in the test report. Therefore, the restitution coefficient was determined from integrating the acceleration traces as discussed in Chapters 12 and 13. In that test, tri-axial accelerometers were used

Applicable Crash Tests for Vehicle No. 2: 1998–2003 Toyota Sienna Vans

DOT#	Year	Make	Model	Angle	Speed	Crush Energy	Crush Width	Test Weight	Barrier Weight*	Restit. Coeff.	C_1	C_2	C_3	C_4	C_5	C_6	Cavg	Form Factor*	Multi-plier	Crash Plot Abscissa	Crash Plot Test Data	A	B	SMAC K_s	E
				deg	mph	ft-lb	in.	lb	lb	—	in.	in.	in.	in.	in.	in.	in.					lb/in.	lb/in.²	lb/in.²	
2766	1998	Toyota	Sienna	0	4.20	2,306	73.0	3913	1.000E+09	0.000	16.7	19.7	22.0	22.0	18.1		20.30	1.00000	6.5552	0.00	27.53	350.3	126.3	163.2	0.147
3087	1999	Toyota	Sienna	0	34.92	183,059	65.7	4516	1.000E+09	0.000	11.7	14.1	15.4	14.8	13.8	10.4	13.84	1.00000	7.4233	20.30	259.23	433.2	187.3	255.2	0.142
					29.33	128,258	63.0	4464	1.000E+09									1.00000	7.5368	13.84	221.04				
					4.20	2,306	73.0	3913	1.000E+09								0.00	1.00000	6.5552	0.00	27.53	32.776			

TC Test 99-144

* For 3 equally-spaced crush measurements, $C_{avg} = (C_1 + 2C_2 + C_3)/4$, $FF = (C_1^2 + C_1C_2 + 2C_2^2 + C_2C_3 + C_3^2)/(6C_{avg}^2)$

For 5 equally-spaced crush measurements, $C_{avg} = (C_1 + 2C_2 + 2C_3 + 2C_4 + C_5)/8$; $FF = (C_1^2 + C_1C_2 + 2C_2^2 + C_2C_3 + 2C_3^2 + C_3C_4 + 2C_4^2 + C_4C_5 + 2C_5^2)/(12C_{avg}^2)$

For 6 equally-spaced crush measurements, $C_{avg} = (C_1 + 2C_2 + 2C_3 + 2C_4 + 2C_5 + C_6)/10$; $FF = (C_1^2 + C_1C_2 + 2C_2^2 + C_2C_3 + 2C_3^2 + C_3C_4 + 2C_4^2 + C_4C_5 + 2C_5^2 + C_5C_6 + C_6^2)/(15C_{avg}^2)$

* Use 1.0e9 for fixed barriers

Sample Calculations

C_1	C_2	C_3	C_4	C_5	C_6	C_{avg}	FF
17.0	19.5	16.1				18.03	1.002440
17.0	17.5	18.0	17.0	16.0		17.25	1.000910
17.9	19.0	19.3	19.5	18.7	17.8	18.87	1.000677

Note: 2 extra points at zero crush are used in the crash plot

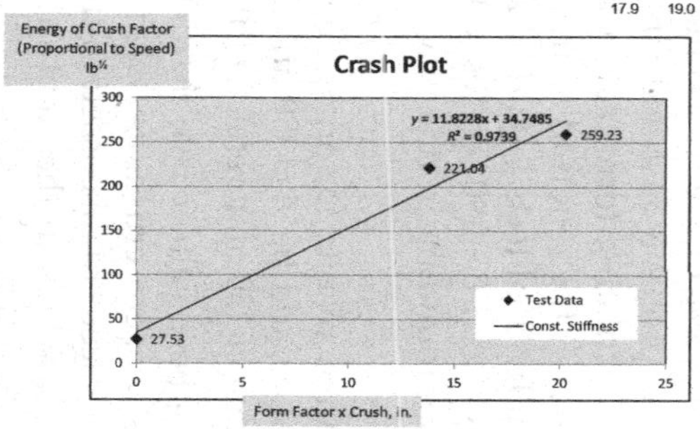

Crash Plot

Energy of Crush Factor (Proportional to Speed) lb^½

y = 11.8228x + 34.7485
R² = 0.9739

259.23
221.04
27.53

Form Factor x Crush, in.

◆ Test Data
— Const. Stiffness

Input quantities are in **BOLD** type
to convert millimeters to inches, divide by 25.4
to convert kilograms to pounds, multiply by 2.2046230
to convert kph to mph, divide by 1.6093440

Slope = 12.0201 (lb)^½/in.

Intercept = 31.2607 (lb)^½

A = (slope) × (intercept) = **375.8** lb/in.

B = (slope)² = **144.5** lb/in.²

FIGURE 21.11
Calculations with inputs altered to obtain approximate Neptune stiffness parameters.

TABLE 21.1
Restitution Coefficient Comparison among the Toyota Sienna Tests

DOT Test No.	Speed (mph)	Accelerometer Location	ΔV (mph)	Restitution Coefficient
2766	34.9	Rear Seats	40.06	0.147
3087	29.3	"CG"	32.65	0.113
3087	29.3	B-Pillars	33.49	0.142

at the bases of the B-pillar, plus a tri-axial package at the "CG," as opposed to the "rear seat cross member" locations used in the NHTSA test 2766. The results of integration are shown in Table 21.1.

The restitution coefficient and ΔV as calculated from the "CG" accelerometers are significantly different from the others. As discussed in Chapter 13, accelerations measured at this location may lead to questionable results in some cases.

7. Taking form factors into account does not appear in the papers cited on the NEI web site. Therefore, the form factors were set to 1.00000 and outlined with a heavier border in Figure 21.11 to simulate this effect. This adjustment has only a slight effect, since full-width flat barrier impacts produce form factors very near unity, anyway.

8. NEI rounds off its calculations to the nearest 10, as noted on its web site. In view of the uncertainties in testing and analysis, such rounding is probably justified (and would completely obscure the effects of calculating form factors for these tests). Using the rounding procedure, the reported A and B values obtained by analyzing the two individual tests would be $A = 350$ and $B = 130$ for Test 2766, and $A = 430$ and $B = 190$ for Test 3087. These exactly match the NEI results, but rounding in this way adds to the difficulty of checking the results.

9. The reported results for analyzing the two tests together, using a crash plot, would be $A = 380$ and $B = 140$. The difference from the NEI B value of 150 may reflect a difference in rounding.

10. To obtain these values, two fictitious data points were used in the crash plot at zero crush, as noted in Figure 21.11. While different from the procedure discussed in this chapter, it is equally justified. However, nowhere in the Neptune papers is there a discussion on this topic, or just what procedure is used.

11. It is also the case that these results were obtained by using the overall vehicle width, rounded to the nearest inch, as the crush width in the fictitious data points. They are outlined with a heavier border in Figure 21.11.

12. The "standard weight" of 3913 lb, as reported on the NEI data sheet, was used for the test weight for the fictitious data points. No explanation was found as to what "standard weight" means, nor where the value came from. Nor was there an indication that NEI used it in their crash plot. The fictitious test weights are outlined with a heavier border in Figure 21.11.

13. In the NEI crash plot, as well as those in Figures 21.10 and 21.11, the deviation of the data points from a straight line suggests that the structure may be constant force instead of constant stiffness, or that there may be some force saturation occurring at 35 mph. However, only constant-stiffness models with no force saturation are discussed in the Neptune references, so those were the only models considered.

In summary, the purchase of stiffness data may be of great assistance to those who may lack the time, skill, or inclination to derive their own parameters from primary sources. However, it behooves the purchaser to develop a full understanding of how the parameters are calculated, which may involve consultations with the vendor.

References

1. Campbell, K., Energy basis for collision severity, SAE Paper 740565, SAE International, 1974.
2. Struble, D.E., Welsh, K.J., and Struble, J.D., Crush energy assessment in frontal underride/override crashes, SAE Paper 2009-01-0105, SAE International, 2009.
3. Marine, M.C., Wirth, J.L., and Thomas, T.M., Crush energy considerations in override/underride impacts, SAE Paper 2002-01-0556, SAE International, 2002.
4. Anderson, G.C., *Vehicle Year & Model Interchange List (Sisters & Clones List)*, Scalia Safety Engineering, Madison, WI, 2009.
5. Strother, C.E., Woolley, R.L., James, M.B., and Warner, C.Y., Crush energy in accident reconstruction, SAE Paper 860371, SAE International, 1986.
6. Neptune, J.A., Blair, G.Y., and Flynn, J.E., A method for quantifying vehicle crush coefficients, SAE Paper 920607, SAE International, 1992.
7. Warner, C.Y., Allsop, D.L., and Germane, G.J., A repeated-crash test technique for assessment of structural impact behavior, SAE Paper 860208, SAE International, 1986.
8. Prasad, A.K., Energy dissipated in vehicle crash—A study using the repeated test technique, SAE Paper 900412, SAE International, 1990.
9. Welsh, K.J. and Struble, D.E., Crush energy and structural characterization, SAE Paper 1999-01-0099, SAE International, 1999.
10. Neptune Engineering, Inc., http://www.neptuneeng.com.

11. Neptune, J.A. and Flynn, J.E., A method of determining accident specific crush stiffness coefficients, SAE Paper 940913, SAE International, 1994.

12. Neptune, J.A., Flynn, J.E., Underwood, H.W., and Chavez, P.A., Impact analysis based upon the CRASH3 damage algorithm, SAE Paper 950358, SAE International, 1995

13. Neptune, J.A., Overview of an HVE vehicle database, SAE Paper 960896, SAE International, 1996.

14. Day, T.D., An overview of the HVE vehicle model, SAE Paper 950308, SAE International, 1995.

15. Neptune, J.A. and Flynn, J.E., A method for determining crush stiffness coefficients from offset frontal and side tests, SAE Paper 980024, SAE International, 1998.

22

Measuring Vehicle Crush

Introduction

The first step in dealing with crush is to distinguish "contact" damage from "induced" damage. As the name implies, contact damage results from contact with another vehicle or object, and generally exhibits surface disturbances in the form of scratches, scrapes, sheet metal folds with the ridges pressed flat, paint removal, or transfers of paint or other material from the crash partner. Induced damage does not exhibit such evidence of contact, but is deformation associated with, or induced by, a contact that does not touch that area of the vehicle. Both contact and induced damage figure into the assessment of crush energy absorbed by the vehicle.

If the reconstruction is to utilize the calculation of crush energy, the vehicle crush must be measured. Even if that sort of calculation is not intended, it is still a good idea to document the crush with some sort of measurement protocol, and note where the boundaries of the contact damage are. This constitutes evidence of the match-up that existed between the collision partners, which is nicely illustrated by a drawing of the vehicles interacting during the crush phase. It is always possible that later in the reconstruction analysis, a crush energy calculation may be desired, even if that was not the original plan.

The concept of crush measurement is simple. The position of a particular point on the exterior surface of the vehicle is measured, both before and after the accident. The crush at that point is simply the vector difference between the two positions. In a crash test, the usual procedure for preparing for crush measurements is to mark the measurement points beforehand, and then position the vehicle so that the area that is to remain undamaged is located at a known distance from a reference surface that is roughly parallel to the area to be damaged. For example, in a frontal crash test, the measurement points would be marked on the front of the vehicle, and the rear of the vehicle would be positioned at a known distance away from a reference surface in front of the vehicle. The vehicle is positioned the same way before and after the test, and measurements are made between the reference surface and the crush measurement locations. The crush is simply the difference between the sets of measurements.

One difficulty that has occasionally reared its head is the crash test personnel needing to identify the crush measurement locations before the crush has occurred. How do the test personnel know that the vehicle component on which the measurement is being made will survive the crash intact and still attached to the vehicle? How do they know that the end points for crush measurement will coincide with the ends of the crush area? As an example, when a high-speed barrier crash is run, how can the reported crush width be less than the full width of the vehicle? When some other kind of test is conducted, producing more contour in the crush, will the maximum crush coincide with a measurement location? In a side impact, where are the limits of crush going to be? At what water line is the deepest penetration going to be? The crush data could be interpreted as deepest crush, or the largest crush measurement in a profile could be interpreted as the maximum crush, or the end points of the measurements could be interpreted as the width of the crush, when in fact one or more interpretations may be incorrect.

The accident investigator has a different set of challenges. The crush has already occurred, so it is easy to decide where to make the measurements. (Crush is often measured with uniform spacing, but we have seen in Chapter 21 that uniformity is an unnecessary restriction.) However, no reference measurements were made before the accident, so an exemplar vehicle must be recruited as a stand-in for the undamaged accident vehicle, and an equivalent set of measurements made. How, then, to ensure that the measurement locations are the same? The answer is inch tape. When sheet metal crushes, it buckles and wrinkles, but for all intents and purposes, it does not stretch. Therefore, inch tape can be applied to the crushed area, as long as one is careful to run the tape through the wrinkles and folds. If the sheet metal is torn, the tape is torn at the same location, just as it would have been if the tape were applied before the crash. The tape is affixed so that its position is known relative to one or two landmarks on the vehicle. Then the same locating procedure is followed on an exemplar vehicle. The desired crush measurement locations are simply marked at the same positions on the inch tape in both cases.

Figure 22.1 shows inch tape affixed to an accident-involved vehicle. One tape line marks a crush profile on the rocker panel; the other marks a profile at the level of the front door beam. In fact, the upper profile measurements are actually on the door beam. In this situation, the outer door panel had been peeled down, so that its outer surface was not accessible. At the same time, it was recognized that the door beam would not be accessible on an exemplar vehicle. This is typical of the sort of challenge presented to the accident reconstructionist. However, foam pads, about ¼ in. thick, had been used when the door was constructed to act as spacers between the outer door panel and the door beam. These are visible as white dashes on the exposed inside surface of the door skin. In this case, the actual crush profile was constructed ¼ in. away from the measurement points on the door beam, to account for the offset. The measurement points were identified by making

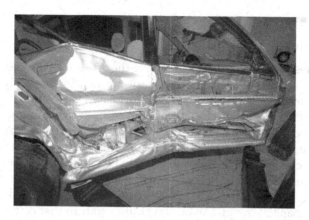

FIGURE 22.1
Crush profiles marked with inch tape.

red marks on the inch tape, and may not be visible in the picture. However, close-up photos were taken wherein the points are visible, in accordance with the earlier discussion.

Figure 22.2 shows those equivalent profiles on an exemplar vehicle. The lower profile was located on the outer edge of the rocker panel, so it was only necessary to position the tape so that at the start of the crush profile, at the aft end of the rocker panel, the tape displayed the same number. The upper profile was a little more of a challenge, though. The tape was started at the aft ends of the front and rear door outer panels. The height of the tape at the two ends of each door panel was set at the same height they were on the accident vehicle. On the rear door, these heights were memorialized with two pieces of inch tape placed vertically, and might be visible in Figures 22.1 and 22.2.

FIGURE 22.2
Crush profile points on an exemplar vehicle.

NASS Protocol

The next question is how to actually make the crush measurements, particularly when vertical reference surfaces are not nearby, patiently waiting to be used. A protocol that was developed for use in the National Automotive Sampling System, formerly the National Accident Sampling System, or NASS, is described by Tumbas and Smith.[1] A prominent feature of the protocol is the establishment of a baseline parallel to the vehicle surface as it existed pre-crash. The baseline is run between stanchions positioned near the corners of the vehicle, and calibrated rods are used for measuring distances. A jig is used to hold the distance-measuring rods, and is visible in the Tumbas paper. One purveyor of a similar fixture that can be disassembled for transport, called the DBD Crush Deformation Jig, is available from WeCARE.* A great advantage of such a jig is that the shape of the deformed vehicle can be captured almost independently of where the vehicle is situated, as long as one has at least six feet or so of space to work in. Thus, the crush data can often be obtained with the vehicle in a wrecking yard, perhaps without moving the vehicle. However, this jig is designed for crush measurements along a straight line, so it is not known whether such a device could be used for a profile with a jog like that seen in Figure 22.2.

The NASS protocol calls for the use of 2, 4, or 6 measurements, evenly spaced, but as mentioned previously, outside the NASS system it is not necessary to use a specified number of measurements, nor that they be equally spaced. Six measurements are enough for a reasonably accurate crush energy calculation in most cases, as long as the measurements are made at such locations that a piecewise linear profile connecting the measurements captures the salient features of the crush and reasonably represents its shape.

If a single crush profile is used, the *de facto* assumption being made is that the crush is fairly uniform in the vertical direction. The Tumbas paper addresses this, and many other issues one encounters when measuring the crush in vehicles involved in actual, or "field," accidents. For frontal and rear impacts, the authors state that the crush measurements should be at the frame level, since that is where the main load paths (structures carrying the highest loads) are. This is a reasonable—and widely followed—procedure. However the authors state, quite without justification, that "if at any station measurement of depth of crush at other than the bumper-frame level exceeds that at the bumper-frame level by 5 or more inches (an underride pattern), the crush measurement at the frame level should be averaged with the greater crush measurement, even if the crush at the bumper-frame level is zero..."[1] (p. 163). Thus, the assigned crush could be halved or doubled if the measured value changed however slightly. However, Tumbas and Smith

---

* WEst Coast Accident Reconstruction Equipment & Education, DBD Crush Deformation Jig, Rudy Degger & Associates, Inc., California.

acknowledge that the total force at the measurement station must be equal to that associated with some crush amount between the two measurements. Some kind of weighted average, perhaps? That subject is taken up in greater detail in Chapter 26.

The Tumbas paper also addresses side impacts. The authors argue that significant structure exists in the door as well as the sill, as long as the hinges, latches, and pillars maintain their integrity. That would be particularly true for vehicles subject to the dynamic FMVSS 214 requirements, which came into effect after the publication of the paper. However, if some of those elements rupture, the authors state—again, quite without justification—that "if the variation [between door and sill measurements] is equal to or greater than 5 in., average the crush between the two levels for those stations where this is the case, even if the crush at the sill level is zero"[1] (p. 164). Again, this "sill averaging" procedure can result in certain assigned crush values being either halved or doubled, a discontinuity that could be triggered by only slight uncertainties in the crush measurement. Either of these averaging procedures could result in a strange looking crush profile, indeed—one that stretches credibility.

A condition addressed in the Tumbas paper and seen in some side impacts is that of "bowing," in which a vehicle struck in the compartment side is seen from above to be bowed, or banana shaped. In this situation, one or both of the ends of the vehicle are displaced laterally, even though there was no contact in the area. Clearly, causing the vehicle to have such end shifts requires additional work, which should be accounted for in the calculation of crush energy. Tumbas and Smith state that the vehicle is considered to be bowed if either of the ends of the vehicle has shifted laterally by four inches or more. Determination of the end shifts depends on some knowledge of the unbowed shape, which may not be all that apparent in the field. "In less severe cases, field personnel have difficulty discriminating 3 vs. 5 in. of end shift and often treat the vehicle as bowed when it is not and vice versa"[1] (p. 175). An investigator would probably be greatly assisted if there were scale drawings depicting the vehicle in its damaged and undamaged condition. Such a mapping procedure is discussed separately in this chapter.

The measurement of crush requires locating the "deflection points," which encompass both the direct and the induced damage, but not remote buckling, and which serve as the end points of the crush profile. The profile is divided into five segments of equal width (again, an unnecessary restriction). The six measurements are made to the surface, from the line between the deflection points. C_1 and C_6 are usually zero (by definition).

If the vehicle is not bowed, that is the end of the procedure; if the vehicle is found to be bowed, some further measurements and calculations must be made. A line is run between the front and rear corners, and the lateral displacements x_1 and x_2 of the deflection points from this second line are measured. To each of the six measurements is added a "bowing constant," which is the average of x_1 and x_2. As is the case with sill averaging, as the crash severity increases and the bowing becomes more pronounced, the use of a bowing constant introduces a discontinuity into the crush measurement procedure.

If some of the NASS inspection procedures seem arbitrary and subject to error, perhaps they are. A sample of the criticism may be found in the 1998 paper by Strother et al.[2]

> The two NHTSA crush measurement procedures which introduce discontinuous changes in the crush of the subject vehicle must be dealt with: the "bowing constant" and "sill averaging" related to "major structural separation." These discontinuities should be eliminated, both in the measurement of test vehicle and subject vehicle crush.
>
> These discontinuities may be avoided through the simple process of mapping the accident vehicle. The vehicle map should include sill crush, deepest crush, and the location of door hinges, latch components, and pillar and bulkhead attachments so that the issue of "major structural separations" can be addressed …. Any such separations will be analytically "repaired" by preparation of a subsequent adjusted crush map of the vehicle. These "repairs" are accomplished during the making of the adjusted crush map by, for example, rotating the front door and perhaps straightening it as well so that its latch now overlays the striker post from which it was separated in the impact …. This procedure is thus continuous, logical and conservative.
>
> Elimination of the measurement discontinuities in the test vehicles is more difficult because of the inadequate documentation of the damage. Unfortunately, there is rarely any other way to deal with "sill averaging" other than to eliminate those data where sill averaging was used to measure test vehicle crush, unless NHTSA photographs and/or other documentation of the test vehicle damage is of sufficient quality to support a decision that the separation was of negligible consequence …. (p. 60)
>
> If the NHTSA routinely published test vehicle crush maps in its reports, it would be possible for the user to refer to the raw crush data to support a more rational crush measurement procedure …. In the case of sill averaging, it is virtually always the case that no information is given in the test report about the nature of the structural separation that led to the decision to apply "sill averaging." (p. 49)
>
> With respect to the bowing constant, it is our contention that the effect of vehicle bowing should always be included in the crush measurements, since the phenomenon is always associated with the dissipation of energy. (p. 60)

Of course, it is the case that the purpose of NASS investigations is to gather data for a large database, the Crashworthiness Data System, or CDS. In that case, there is a need for standardized procedures and the avoidance of systematic errors across a broad range of crashes, whereas the reconstruction of individual accidents, and the adjudication of the disputes resulting therefrom in courtroom settings, may require more attention to detail.[3]

Full-Scale Mapping

The procedure that was actually used in the case illustrated in Figures 22.1 and 22.2 might be called "full-scale mapping," and is the sort of technique suggested above. In this procedure, a full-scale map is made of the vehicle, by placing it over a sheet of plastic affixed to the floor. Then points on the underside of the vehicle, plus the crush profile(s), are located on the map by means of an electronic equivalent of a plumb bob. Alternatively, one could just use a plumb bob. The points are then connected by lines drawn with marking pen. A few of these lines are visible in the bottom portion of Figure 22.1. The same procedure is followed with the exemplar vehicle. The two maps are then overlaid such that the undamaged portions of the vehicle structure line up. In this way, the match-up can consider whole areas of the structure, instead of a small number of exterior points as might be the case with other methods. Moreover, the match-up can take advantage of and take into consideration the stiff structure on the underside of the vehicles, which does not tend to vary with changes in trim packages, bumper covers, and so on.

One characteristic common to probably all passenger vehicles is the architecture and construction of the rocker panels. These are primary structures, of course, and they are made up of an outer panel, an inner panel, and various reinforcements that are spot welded together at various places, including longitudinal joints at the top and bottom. The bottom joint is in the form of a "pinch weld," where upstanding flanges from the various parts are spot welded together. The bottom pinch weld is virtually always visible from under the vehicle, as it does not have to be covered up for cosmetic purposes. Moreover, in the as-manufactured condition, the bottom pinch welds are virtually always straight, and symmetrically arranged in the vehicle. In fact, the pinch welds often form the only truly straight lines of any length on the structure. In all, they make excellent structural references, especially for side impacts.

In the present case, the pinch weld on the struck side was damaged, as is generally the case in side impacts. The opposite pinch was undamaged, however, so the left side pinch welds were matched up between the accident and exemplar vehicles. The accident vehicle was not bowed in this case, as can readily be seen in Figure 22.3, which is the "damage map" that resulted from the overlay procedure. If the vehicle had been bowed, then only the front and rear ends of the left pinch weld would have been matched up.

Here, the accident vehicle is depicted by red lines, and the exemplar vehicle by blue ones. Green is used for auxiliary lines, or to avoid confusion (in this case, between the exemplar right door profile—green—and exemplar right rocker panel profile—blue). Another feature common to unibody vehicles like this one is the front sub-frames extending aft from the main front cross-member. The left front sub-frame could have been used for match-up purposes if for some reason the left pinch weld was not usable.

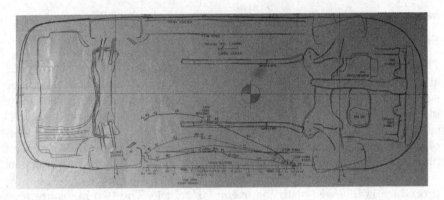

FIGURE 22.3
(See color insert.) Damage map for accident vehicle.

One will notice that on the right of Figure 22.3, the front of the vehicle is depicted with a blue line only, since the front bumper fascia was missing on the accident vehicle. The rear bumper fascias do not quite line up, which is not unusual. The luggage bay area of the underbody does not appear to line up either, but the underbody contours are rounded in that area, so uncertainties in mapping the rounded contours might mask whatever slight bowing there could have been. The match-up of the rear exterior surfaces suggests strongly that there was indeed no bowing.

At this point, all that remains is to draw arrows between comparable crush profile points on the two maps, and measure their lengths and locations, as shown in Figure 22.4.

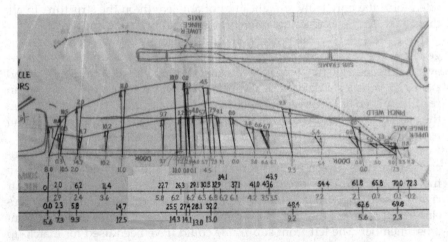

FIGURE 22.4
(See color insert.) Crush profiles.

The crush profile information is drawn and written on a plastic overlay, so as to preserve the clarity of the damage map. The small numbers in red in the upper part of the picture are the inch tape numbers (which repeat every 12 in.), as are the numbers in green and blue near the blue line labeled "DOOR." The numbers in green and purple in the lower part of the picture are the crush measurements for the rocker panel and door profiles, respectively. The black numbers are measurement locations, starting on the left.

On these crush profiles, many more points were utilized than strictly necessary for the computation of crush energy. On the other hand, the measurement locations are set up by merely making marks on the inch tape, so there is hardly any time required to do so. The points were chosen to capture the full geometry of the damaged areas, and the crush vectors are easily measured with a tape measure (in full scale, by the way). All the data are preserved by simply writing them down on a sheet of plastic laid over the damage map, and photographed. If desired, a subset of points can be used for the calculations, as will be demonstrated.

On the door level crush profile, 11 points were measured, the coordinates of which are the purple and black number pairs at the bottom of Figure 22.4. The purple numbers show that none of the crush measurements were zero, despite the fact that the crush profile obviously has to pass through zero at its ends. Therefore, extrapolations back to zero were made at the ends. Creating a piece-wise linear approximation of the crush profile by connecting the crush measurements with straight lines produces the solid-line plot in Figure 22.5, identified as "11 measurements." The measured points are

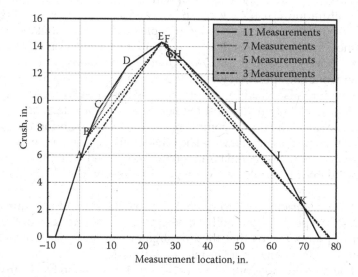

FIGURE 22.5
Crush profile using various numbers of points.

labeled with the letters A through K. Additional points at zero crush were obtained by extrapolating from points A and B at the rear end of the profile, and J and K at the front end.

Points E through H capture some detail in the B-pillar/door cut area, but not all of them are necessary for the crush energy calculation. Points C and I contribute little in describing the shape of the profile. Therefore, a 7 measurement profile was constructed by eliminating points C, F, G, and I from the profile, while retaining the extrapolations. Point E was also retained in order to capture the maximum crush. The result is shown by the line of dots in Figure 22.5, identified as "7 measurements." It is hardly distinguishable from the solid line drawn through 11 measurements.

To simplify the crush profile still further, two more points were eliminated from the profile (J and D). Because of the elimination of point J, the extrapolation had to be made between points E and K. This lengthened the profile somewhat, as seen in Figure 22.5, while decreasing the area under the profile in other regions.

Finally, two more points were eliminated. The loss of point B means that extrapolation would have to occur between points E and A, with an unacceptably large extension of the profile width. Therefore, extrapolation on the front end was eliminated, and the profile was drawn between points A, E, and the extrapolation from point K, as shown by the longer dashes in Figure 22.5. This profile does not start at zero crush, but rather at point A.

So how does using, or not using, detail in the crush profile affect the calculation of crush energy? The precise answer probably depends on the structural model being used, and its parameters. In this case, a constant stiffness model with force saturation was used, with $A = 248.5$ lb/in., $B = 137.3$ lb/in.^{2}, and the saturation crush C_S was 4.6 in. The calculated crush energy was 44,447 ft-lb, using the 11 measured points plus the two extrapolations. Using only 7 measured points resulted in only a 0.63% reduction. Using only 5 measured points produced a 7.22% reduction, and using only 3 measured points with one extrapolation yielded a 13.5% reduction. So while the spreadsheet procedure discussed in Chapter 21 was designed to accommodate up to 19 points and 18 segments, nothing near that level of detail should ever be required. Seven well-chosen points appear to suffice, even for a severe side impact. Other crashes need fewer. The crush in full-width end impacts into a flat barrier, for example, could be adequately described with three points; the point in the middle would be a very good representation of the maximum crush.

This takes us briefly back to the NASS protocol. Instead of requiring the use of 2, 4, or 6 points, it would have been better to require 2, 3, or 5. An odd number of points means that the investigator can more easily capture the maximum deformation, or at least is more likely to.

It will be noticed in Figure 22.4 that the crush vectors, in black, are not all parallel. The outer crush vectors tend to be angled inward toward the middle of the profile, which is typical of side impacts. For all kinds of impacts, it has been this author's practice to measure the crush vector components, shown in green and purple, that are parallel to the longest vector in the profile. This produces a conservative crush measurement, although there was very little effect in this case. The reason for the procedure is that the work done on the structure during the crash is the dot product (or inner product) of the impulsive force and the displacement, which means that only the component of the displacement that is parallel to the force is used in the computation. Of course, the direction of force was not yet known at this stage of the reconstruction, but the longest crush vector can provide a good initial approximation (mostly lateral in this case) for crush energy assessment purposes.

For highly angled crush vectors, as in a corner hit, for example, the baseline, along which the locations are measured, should be rotated so that it is perpendicular to the longest crush vector. Of course, this has the effect of reducing the crush length, which will result in a lower calculation of crush energy. Such a procedure is consistent with that advocated by Asay et al.,[4] who provide an excellent description with other measurement issues associated with pole impacts.

In oblique impacts, what structural parameters, discussed in Chapter 21, should be used—those associated with the end of the vehicle, or the side? The main load-carrying elements of the structure, and the damage pattern, should be examined closely for guidance in this decision. Perhaps some weighted average would be appropriate. For such cases, it may be appropriate to propose a crash test designed to produce a similar crush pattern, so that the structural parameters specific to the crash mode may be obtained directly. It would not be necessary to exactly duplicate the speed or the deformation mode (or to replicate the subject crash); the object would be to exercise the structure similarly, and thereby obtain the parameters that should be used in the reconstruction.

It must be noted that the Crash3 "correction factor" $1 + \tan^2 \alpha$ is not used, because the investigator is not constrained to measure crush normal to the pre-crash surface of the vehicle. Nor is it necessary to measure crush locations along a line parallel to the pre-crash surface (though that is the usual practice). By the time the angle α between the surface and the assumed impulse direction grows to 45° (the Crash3 limit), the "correction factor" grows to 2.0! It is undesirable to have to use a correction factor in the first place, and completely unacceptable to use one that doubles the result. This has been pointed out by other reconstructionists, and discussed in further detail.[4]

The sharp-eyed reader may have noticed that in this severe side impact, the rear door structure lost its integrity, as is reflected in Figure 22.3 and the red dotted line of Figure 22.4. (Actually, the rear striker pulled out of the C-pillar, and the B-pillar ruptured.) This is the very sort of major structural separation that was discussed above—for which the NASS protocol requires the use of sill averaging. Instead, it was decided to make the conservative estimate that the side structure ceased to carry loads, and absorb further energy, once the rupture ruptured. How to implement this assumption?

It was decided to analytically "repair" the separation by taking the structure back to the point at which the separation occurred, and calculating the crush energy at that point. The separation was "repaired" by "shutting the door," in other words, by rotating the door crush profiles (in this case, the entire profile) about the upper front door hinge so that the aft end of the accident vehicle profile would bear the same relationship to the C-pillar striker as it does on an exemplar vehicle. This was done by making a tracing of the profile, and physically rotating the tracing. Anticipating the need to "shut the door," the upper front door hinge axis and the C-pillar striker were mapped during the accident vehicle inspection, though it may not be visible through the overlays in the photograph shown in Figure 22.4. The resulting crush profile is the solid red line seen in Figure 22.4, along with the crush vectors and their measurements. The procedure is akin to that suggested by Strother et al. and leads to crush energy being calculated only up to the point in time at which the structural separation occurred; all additional deformation is assumed to entail no additional crush energy.

It must be said that the mapping procedure is labor intensive, and requires a reasonably smooth surface to work on. However, it provides a full-size illustration of how the structure deformed, and gives all parties to the case, plus the judge and jury, not only a physical representation of what was done, but preserves the profiles so that the crush could be re-measured using some alternate procedure (such as not "shutting the door," for example) if desired. NHTSA reports for severe crashes, particularly vehicle-to-vehicle, offset, underride, and pole impacts, would have their usability enhanced by the inclusion of map photographs, such as Figure 22.3, or perhaps digital equivalents.

Total Station Method

A compromise between full-scale mapping and the NASS protocol would be to establish the locations of the crush profile points and then use a total station to survey them, on both the accident and exemplar vehicles. Of course, undamaged areas of the accident vehicle, and corresponding points on an exemplar vehicle, have to be surveyed as well, so that points in undamaged areas can be matched up. It is probable that points on the underside (such as the pinch weld) could not be used for this purpose, but it may be possible to establish reference points on the ground using a plumb bob, and then survey those points. The profiles can be plotted on a scale drawing using CAD software, which allows the distances between points to be measured. At the cost of some jury understanding of the process, only the survey points need be drawn, as opposed to an entire vehicle. Special operations, such as "shutting the door," can be performed on a CAD drawing as well.

Loose Parts

A challenge can be posed by loose parts, particularly plastic bumper fascia, or bumper covers. These may be off the vehicle completely, or only partly detached. If they are missing, then obviously they cannot be used for crush measurements. Structure that has been revealed by the missing bumper cover is often not accessible on an exemplar vehicle, in which case some other structure, as near to the crush surface of the vehicle as possible, will have to be used for crush measurements. Of course, there is hardly ever an exemplar vehicle available to look at during the course of the accident vehicle inspection. Nevertheless, a choice of points to use for a crush profile will have to be made. For front structures, taking measurements using an alternative structure, such as the upper core support, is discussed further in an SAE paper[5] by this author.

If the component is partly loose, it may be possible to move it into its regular place by matching up points of attachment, and secure it there temporarily with gaffer's tape, or perhaps duct tape (with permission, of course). Again, this is particularly the case with a bumper fascia. Then the crush profile can be established, measured, and photographed, after which the tape can be removed, and the vehicle returned to the condition in which it was found.

This leads us to the issue of bumper cover rebound, or the "air-gap" problem described by Neptune in 1999.[6] This rebound can create a gap between the bumper fascia and the underlying structure post-crash. According to Neptune, personnel should push on the bumper fascia until resistance is felt, and then make the crush measurement. Neptune was referring to NHTSA crash tests, and while this is a reasonable recommendation, it apparently was not being done, at least as of 1999. The effect of Neptune's procedure would be to increase the crush measurements, and reduce the calculated structural stiffness. Neptune has published crush and stiffness resulting from removing the "air-gap," by "employing both the reported damage measurements and the post-test vehicle photographs." While removing the "air-gap" from the crush data is a laudable objective, the need for caution in using the data should be obvious. In a perfect world, the "air-gap" would be removed in the crush measurements of both the crash-tested vehicle and the accident vehicle.

Other Crush Measurement Issues in Coplanar Crashes

The forces that cause a structure to be deformed are, of course, the same forces that contribute to the impulse to the vehicle. Therefore, the damage pattern offers important clues to where the line of action of the impulse intersected the surface of the vehicle. By definition, this line of action must be within the area of contact damage, not induced damage. Since the depth of crush varies with the magnitude of applied forces, the most widely used—and perhaps the best—estimate of the line of action of the impulse is the centroid of the portion of the crush profile due to contact damage. If for no other reason than this, the limits of contact damage should be documented, so that the centroid of that portion can be subsequently calculated. If a mapping process is being used, the ends of the contact damage area can be recorded directly on the map, using a plumb bob.

Reconstruction procedures such as Crash3 require as input the "damage offset," often denoted as D. This is the moment arm of the impulse force, the direction of which must be estimated when ascertaining the offset. While D is often thought of as the offset from the vehicle CG to the centroid of the entire crush profile, it is more appropriate to simply calculate the centroid of the contact damage portion only, and enter its offset directly. Of course, the ΔV vector is a result of the calculation and therefore is not known at the time. Consequently, its direction, which coincides with the direction of the impulse force, is also unknown. D must therefore be estimated, based on the anticipated direction. This is a weakness of the Crash3 program.[3] Having to specify the center of impulse is a feature of any calculation that is based on Newton's Second Law, applied to a single vehicle. Refer to Chapter 18.

Implicit in the entire discussion in Chapter 21 is the assumption that a structure's stiffness parameters are uniform throughout the region of crush. However, vehicle structures are anything but homogeneous with uniform stiffness properties. For end contacts involving the entire vehicle width, this assumption is usually of little or no consequence, since the tests from which the properties are derived are themselves full width. Pole impacts, and almost all side impacts, are different. Only a portion of the structure is engaged, and it is well known that some areas are stiffer—perhaps much stiffer—than others.[7] The computations must take into account, therefore, which portion(s) are included in the crush profile.

As a result, the NHTSA conducted a series of 11 side impacts, using a rigid, contoured barrier as the striking "vehicle." This type of test, as opposed to a vehicle-to-vehicle crash, results in all the crush energy being dissipated in the struck vehicle instead of being distributed between the vehicles according to their stiffness properties:

> Four vehicle models were tested. Occupant compartment impacts at two different impact speeds were done on each to derive stiffness parameters for these vehicles. One of these models was re-tested at a third speed to determine if the derived stiffness parameters were velocity sensitive. This same model was also used in non-compartment impacts, again at two speeds, to explore any impact location sensitivity.[8] (p. ii)

The noncompartment impacts resulted in the A-pillar, wheel, undercarriage, and possibly the transaxle being engaged. For these impacts, NHTSA computed an A value almost twice that of the compartment value, and a B value over twice as high. For side stiffness characteristics, it is once again location, location, location.

Other similar tests of a different set of vehicles were reported by Prasad[9] in 1991. Prasad's tests involved repeated impacts on the side of a single vehicle, a technique introduced by Warner.[10]

Since the crush profile is divided into segments anyway, it is a simple matter to locate crush measurements at the boundaries between stiffness characteristics (compartment and noncompartment, say), as long as one is not constrained by an equal-spacing requirement. Then each structural region can be identified distinctly within the crush profile. One can (and indeed, should) then apply different stiffness parameters, and possibly even different structural models, to the appropriate damage regions.

In the late 1990s, the NHTSA began testing vehicles according to an upgraded version of FMVSS 214 that required crash tests. Unfortunately for the reconstruction community, these are unlike the tests of Willke and Prasad. Instead, they involve a moving barrier with a deformable face, and have the moving barrier angled so as to simulate an intersection collision with both vehicles moving. These factors greatly complicate the extraction of stiffness data from the test result. However, a procedure was presented in 2001 for characterizing the struck vehicle from FMVSS 214D test data.[11]

It is not unusual for impacts with narrow objects (poles for example) to produce such bending in the struck surface that some portions away from the contact actually experience negative crush. At first glance, it may be surprising that negative crush will result in positive crush energy being calculated, but the calculation depends not only on the crush, but also the crush squared. To avoid this counter-intuitive effect, it is customary to ignore all negative crush. As a practical matter, the crush profile is constructed, and the zero intercepts (the locations where the crush changes sign) are calculated. The crush energy calculations include only positive crush regions.

It is also not unusual, particularly in end impacts, for the parts of the nonstruck surfaces to buckle outward (fenders, for example). Since the crush profile does not extend beyond the width of the undamaged vehicle, these deformations are automatically ignored.

This line of thought raises the question: What to do if the crush measurements do not cover the entire damaged area? This can happen on an accident vehicle when parts are missing, or when an element used for the crush profile (such as the upper core support, for example) does not cover the entire crush zone. It can (and in fact, usually does) happen on a test vehicle when the crush profile is ascertained from the bumper, which does not extend the full width of the vehicle. (In the latter case, the full width of the vehicle is crushed, but not all of the crush width gets measured.) The most rational thing to do is extrapolate the measurements to the full width, as long as the extrapolation is not so large and the two measurements being used are not so different as to produce an unreasonable result. If there is not much taper to the crush profile, the best procedure is to simply estimate the crush at the ends of the damage area by setting it equal to the adjacent values.

In full-width frontal and rear flat barrier tests, there is not enough variation in the crush measurements to warrant the effort of extrapolation. Since the testing agencies routinely use uniform crush measurement spacing, and the tests are severe enough to cause the full width to be crushed, it suffices to simply use the overall width as the crush width. Typically, this is a larger number than the reported crush width, so lower stiffness parameters will be calculated. This is of no consequence as long as the full width is also utilized in the field.

A related issue arises when an end impact results in the side sway, or end shift, of the vehicle, as discussed in Chapter 4 regarding the RICSAC tests. As in bowed vehicles in side impacts, work is required to produce such deformations, and special calculations and measurement considerations come into play. The interested reader is referred to the 2006 paper by Welsh et al.[12]

Rollover Roof Deformation Measurements

Crush measurements for rollover accidents impose different requirements on the reconstructionist, because the vertical components are significant (and perhaps of most interest). Some sort of electronic measurement equipment, such as a total station, will be required. Again, it is necessary to locate corresponding points on the accident and exemplar vehicles. As we have previously seen, inch tape can be used for that purpose.

References

1. Tumbas, N.S. and Smith, R.A., Measurement protocol for quantifying vehicle damage from an energy basis point of view, SAE Paper 880072, SAE International, 1988.
2. Strother, C.E., Kent, R.W., and Warner, C.Y., Estimating vehicle deformation energy for vehicles struck in the side, SAE Paper 980215, SAE International, 1998.
3. Struble, D.E., Generalizing CRASH3 for reconstructing specific accidents, SAE Paper 870041, SAE International, 1987.
4. Asay, A.F., Jewkes, D.B., and Woolley, R.L., Narrow object impact analysis and comparison with flat barrier impacts, SAE Paper 2002-01-0552, 2002.
5. Struble, D.E., Crush energy assessment in frontal underride/override crashes, SAE Paper 2009-01-0105, SAE International, 2009.

6. Neptune, J.A., A comparison of crush stiffness characteristics from partial-overlap and full-overlap frontal crash tests, SAE Paper 1999-01-0105, SAE International, 1999.

7. Digges, K. and Eigen, A., Measurements of stiffness and geometric compatibility in front-to-side crashes, *Proceedings of the 17th International Technical Conference on the Enhanced Safety of Vehicles*, Amsterdam, the Netherlands, U.S. Department of Transportation, 2001.

8. Willke, D.T. and Monk, M.W., *Crash III Model Improvements: Derivation of New Side Stiffness Parameters from Crash Tests*, Vol. 2, Report DOT HS 807 353, National Highway Traffic Safety Administration, 1987.

9. Prasad, A.K., Energy absorbed by vehicle structures in side-Impacts, SAE Paper 910599, SAE International, 1991.

10. Warner, C.Y., Allsop, D., and Germane, G., A repeated-crash test technique for assessment of structural impact behavior, SAE Paper 860208, SAE International, 1986.

11. Struble, D.E., Welsh, K.J., and Struble, J.D., Side impact structural characterization from FMVSS 214D test data, SAE Paper 2001-01-0122, SAE International, 2001.

12. Welsh, K.J., Struble, D.E., and Struble, J.D., Lateral structural deformation in frontal impacts, SAE Paper 2006-01-1395, SAE International, 2006.

23

Reconstructing Coplanar Collisions, Including Energy Dissipation

General Approach

This approach recognizes from the outset that the inter-vehicle contact forces are self-equilibrating, due to Newton's Third Law. If, therefore, we define the system as including both vehicles, then the contact forces do not show up when Newton's Second Law is expressed mathematically. The only external forces on the system are tire forces due to friction. As discussed in Chapter 2, the tire forces are in the neighborhood of the vehicle's weight or less, so they produce vehicle accelerations on the order of 1 G or less. In a high-speed crash, vehicle accelerations often exceed 40 G, so the tire forces are negligible by comparison. For collisions at low speeds or with large trucks, tire forces might not be negligible by comparison, so the reconstructionist must remain aware of the assumptions being made. If indeed the tire forces are negligible, then no external impulses are delivered to the system during the crash, and Newton's Second Law reduces to the Conservation of Momentum, which may be expressed as

$$d(m\vec{V}) = 0 \tag{23.1}$$

$$d(I\omega) = 0 \tag{23.2}$$

Equation 23.1 indicates that the change in momentum $m\vec{V}$ is zero; in other words, momentum is conserved (i.e., stays constant) throughout the crash. The magnitude and direction of the system momentum vector is unchanged from impact to separation. Similarly, Equation 23.2 shows that the change in angular momentum is zero; angular momentum is conserved throughout the crash. (Note that the system moment of inertia will change during a crash if the vehicle positions change relative to each other.)

Equations 23.1 and 23.2 are vector equations, involving two components in Equation 23.1 and one component in Equation 23.2. So momentum conservation for a planar system involves three equations. To these scalar equations, we add a fourth, which expresses the conservation of energy:

$$\Delta(KE_1) + \Delta(KE_2) = CE_1 + CE_2 \tag{23.3}$$

where ΔKE is the loss of kinetic energy, CE is the crush energy, and the subscripts refer to the two vehicles. Energy is a scalar quantity, so Equation 23.3 is always a single equation, regardless of the degrees of freedom of the system.

If the run-out has been analyzed for each vehicle, the exit velocities (both linear and rotational) have already been calculated and can be considered known quantities as far as crash analysis is concerned. See Chapter 4. The velocities at impact are not known, however. Each vehicle has three degrees of freedom—two translational and one rotational—so at the instant of impact each vehicle has three velocity quantities: two translational velocity components and one rotational, or yaw, velocity. So the system as a whole has six unknowns.

At this point, we have only four equations with which to find the six unknowns. As we have discussed in Chapter 18, a mathematically rigorous approach would be to introduce some more equations, which turns out to involve some additional assumptions (such as a common velocity) and/or additional unknowns. Instead, in the interest of simplicity we adopt the approach of reducing the number of unknowns, rather than adding equations. Consider that two of the unknowns are the angular velocities (yaw rates) ω_1 and ω_2 of the two vehicles at impact. While yaw rates can be significant when exiting a crash, they are almost never so going into a crash. Rotational kinetic energies due to the impact yaw rates are small compared to the kinetic energies due to the translational velocity components. This is not to say that the impact yaw rates should be ignored; it simply says that they can be estimated and that any errors will not affect the results much. If a vehicle is known to be tracking along a straight path at impact, its yaw rate is known to be zero. If its path is curved and the curve is known from other considerations (e.g., tire marks or a turning maneuver into a particular lane or driveway), the yaw rate can be approximated from an estimate of the vehicle's impact speed and its turn radius, possibly imputed from tire marks. When the analysis is complete and the impact speed is known, the impact yaw rate can be adjusted and the analysis re-run. The process converges rapidly (typically, in one iteration) because of the insensitivity of the results to the yaw rate.

We are then left with four equations and four unknowns: the X and Y components (or, alternatively, the magnitude and direction) of the two impact velocity vectors. In practice, however, the enforcement of angular momentum conservation renders the solution exquisitely sensitive to small variations in the geometry of the crash. (It should be noted that the system yaw moment of inertia changes as the two vehicles move together, which is why it is properly inside the parentheses in Equation 23.2. Even though the ice skater may not know that she is reducing her moment of inertia when she does it, she pulls in her arms to increase her spin rate.)

Consider that the impact geometry is obtained sometimes from physical evidence on the roadway, and almost always from matching up the two damaged vehicles. The angular momentum can be exquisitely sensitive to the impact geometry. If angular momentum conservation is enforced, the computed impact velocities can be significantly affected by what is, after all, user judgment of facts that may be of only secondary interest to those requesting the reconstruction. This is undesirable, to say the least. Experience has shown that it is better to relinquish the use of angular momentum conservation (although angular momenta before and after the crash can be calculated and reported, so that they may be compared). It is worth noting that the source code for Crash3[1] contained angular momentum computations that constituted "orphan code." There was no program entry to this code, because angular momentum conservation was never implemented.

This leaves us with four unknowns but only three applicable equations for the system being analyzed: linear momentum change in two directions, and energy balance. To deal with the situation, one of the vehicles (vehicle 1) is assumed to have a known velocity direction at impact. Again, physical evidence at the scene (or the lack of the same) often tells us the direction that at least one of the vehicles was traveling at impact. In the analysis, vehicle 1 is taken to be the vehicle having the better-known travel direction.

The remaining three unknowns are then the two translational impact velocity components for vehicle 2, and the magnitude of the translational impact velocity for vehicle 1.

There is a trade-off between assuming as known quantities variables that are not actually known (e.g., impulse center, restitution coefficient, or impulse ratio coefficient), or reducing the number of unknowns by using physical evidence or crash dynamics observations to make up the difference (specifying the yaw rates and one of the velocity directions). The choice is up to the reconstructionist. However, simplicity, understanding the analysis parameters, and utilizing physical evidence all have their merits, as does taking advantage of the insensitivity of the results to reasonable impact yaw rate variations.

As Al Fonda points out,[2] it is not necessary to assume that the collision is instantaneous. It suffices to enforce momentum conservation and energy balance between the two points—impact and separation—regardless of how much vehicle motion occurs between the two points (again, as long as tire forces are negligible). Taking this approach, we avoid introducing errors into the run-out analysis that otherwise occur by assuming that run-out begins at the point of impact. At the same time, some additional outputs are provided that serve to refine, and improve the accuracy of, the final results. These considerations are discussed in Chapter 24.

It is also the case that there is no mention, in either the momentum conservation or the energy conservation principle, of the requirement for a common velocity, or the location at which a common velocity must occur. Looking for such a mention is rather like turning to the U.S. Constitution to find a discussion of the Senate filibuster rule. Instead, the two principles of mechanics apply whether or not there is a common velocity achieved, and even when there is no collision at all. In the oft-cited example of a sideswipe collision, it suffices to ensure that the energies are accounted for, particularly the energy dissipated in friction (or snagging) between the two vehicles. If there is no energy dissipated, no momentum is transferred, and the vehicles continue on their way unimpeded by each other, like two ships passing in the night.

The results of applying the energy and momentum principles are invariant with respect to coordinate systems, as long as Newtonian (i.e., unaccelerated) coordinates are used. This means that the impact velocity vectors are physically the same, though obviously the description of the vectors is dependent on the coordinate system chosen. In particular, we can choose a coordinate system fixed to the ground but oriented so that its abscissa is parallel to the linear momentum vector. This has the advantage of simplifying the equations and allowing derived results that apply to the degenerate condition of a uniaxial collision. Recall that near-uniaxial collisions are a problem area in Crash3.

Development of the Governing Equations

In the current analysis, the issue is attacked head-on by rotating the coordinate system to coincide with the linear momentum vector. No difficulties have been encountered with uniaxial crashes. The axes in this momentum-oriented coordinate system are denoted as ξ (xi) and η (eta), whereas the (arbitrarily chosen) axes used to survey the scene are X and Y. The ξ-axis is chosen to coincide with the linear momentum vector. The angle φ of this axis from the X-axis is found as follows:

$$\phi = Arc\tan\left[\frac{m_1 V'_{Y1} + m_2 V'_{Y2}}{m_1 V'_{X1} + m_2 V'_{X2}}\right] \tag{23.4}$$

Then the separation velocity components may be transformed into this rotated coordinate system, as follows:

$$V'_{\xi K} = V'_{XK}\cos(\phi) + V'_{YK}\sin(\phi) \tag{23.5}$$

$$V'_{\eta K} = -V'_{XK}\sin(\phi) + V'_{YK}\cos(\phi) \tag{23.6}$$

where $K = 1, 2$, for Vehicle 1 or Vehicle 2. Once the impact velocity components are computed in this coordinate system, it is understood that they may be transformed back into the original X–Y coordinate system. In the following, V is the linear velocity, m is the mass, W is the weight, I is the yaw moment of inertia, ω is the angular velocity, L is the linear momentum, CE is the crush energy, and $KE = \frac{1}{2}mV^2 + \frac{1}{2}I\omega^2$ is kinetic energy, all expressed in consistent units. Primed quantities denote values at separation, those without primes are quantities at impact.

We start by enforcing conservation of linear momentum in the η direction. Since this direction is perpendicular to the resultant momentum vector, by hypothesis this component of momentum is identically zero; that is,

$$L'_\eta = m_1 V'_{\eta 1} + m_2 V'_{\eta 2} = L_\eta = m_1 V_{\eta 1} + m_2 V_{\eta 2} \equiv 0 \tag{23.7}$$

Thus,

$$V'_{\eta 2} = -\frac{m_1}{m_2} V'_{\eta 1} \tag{23.8}$$

$$V_{\eta 2} = -\frac{m_1}{m_2} V_{\eta 1} \tag{23.9}$$

The second of our equations is momentum conservation in the ξ-direction, as follows:

$$L'_\xi = m_1 V'_{\xi 1} + m_2 V'_{\xi 2} = L_\xi = m_1 V_{\xi 1} + m_2 V_{\xi 2} \tag{23.10}$$

from which we have

$$V_{\xi 2} = V'_{\xi 2} - \frac{m_1}{m_2}(V_{\xi 1} - V'_{\xi 1}) \tag{23.11}$$

Our third equation is energy conservation, equating the kinetic energy loss for the two-vehicle system, to the energy dissipated in deforming the two vehicles, as follows:

$$\frac{1}{2}m_1\left[(V_{\xi 1})^2 + (V_{\eta 1})^2\right] + \frac{1}{2}I_1(\omega_1)^2 + \frac{1}{2}m_2\left[(V_{\xi 2})^2 + (V_{\eta 2})^2\right] + \frac{1}{2}I_2(\omega_2)^2$$

$$-\frac{1}{2}m_1(V'_1)^2 - \frac{1}{2}I_1(\omega'_1)^2 - \frac{1}{2}m_2(V'_2)^2 - \frac{1}{2}I_2(\omega'_2)^2 = CE_1 + CE_2 \tag{23.12}$$

where CE_K is the crush energy in the Kth vehicle, and the exit velocity in the Kth vehicle is given by

$$(V'_K)^2 = (V'_{\xi K})^2 + (V'_{\eta K})^2 \qquad K = 1, 2 \tag{23.13}$$

Substituting Equations 23.11 and 23.9 into Equation 23.12 results in

$$\frac{1}{2} m_1 \left[(V_{\xi 1})^2 + (V_{\eta 1})^2 \right] + \frac{1}{2} m_2 \left\{ \left[V'_{\xi 2} - \frac{m_1}{m_2} (V_{\xi 1} - V'_{\xi 1}) \right]^2 + \left[-\frac{m_1}{m_2} V_{\eta 1} \right]^2 \right\}$$

$$= (keq)_1 + (keq)_2 + CE_1 + CE_2$$

$$\tag{23.14}$$

where the kinetic energy quantity $(keq)_K$ is defined by

$$(keq)_K = \frac{1}{2} m_K (V'_K)^2 + \frac{1}{2} I_K \left[(\omega'_K)^2 - (\omega_K)^2 \right] \tag{23.15}$$

for $K = 1, 2$. Collecting like terms, multiplying through by $2/m_2$, and defining the mass (or weight) ratio r as

$$r = \frac{m_1}{m_2} = \frac{W_1}{W_2} \tag{23.16}$$

the energy equation becomes

$$r(1 + r)\left[(V_{\xi 1})^2 + (V_{\eta 1})^2 \right] - 2r\left[V'_{\xi 2} + rV'_{\xi 1} \right] V_{\xi 1} + \left[V'_{\xi 2} + rV'_{\xi 1} \right]^2$$

$$= \frac{2}{m_2} \left[(keq)_1 + (keq)_2 + CE_1 + CE_2 \right] \tag{23.17}$$

This is a single equation that appears to have two unknowns, $V_{\xi 1}$ and $V_{\eta 1}$. However, since we wish to prescribe the direction of vehicle 1's impact velocity vector, in the interest of reducing four unknowns to three, $V_{\xi 1}$ and $V_{\eta 1}$ are not independent. The angle of that velocity vector θ_1 (treated as a known quantity) is related to the impact velocity components for vehicle 1 as follows:

$$\tan(\theta_1) = \frac{V_{\eta 1}}{V_{\xi 1}} \tag{23.18}$$

and therefore

$$(V_{\xi 1})^2 + (V_{\eta 1})^2 = \frac{(V_{\xi 1})^2}{\cos^2(\theta_1)} \tag{23.19}$$

Substitution into the energy equation 23.17 and dividing by r results in

$$\frac{1+r}{\cos^2(\theta_1)}(V_{\xi 1})^2 - 2\left[V'_{\xi 2} + rV'_{\xi 1}\right]V_{\xi 1} + \frac{1}{r}\left[V'_{\xi 2} + rV'_{\xi 1}\right]^2$$

$$- \frac{2g}{rW_2}\left[(keq)_1 + (keq)_2 + CE_1 + CE_2\right] = 0 \tag{23.20}$$

where g is the acceleration due to gravity. This equation is quadratic in $V_{\xi 1}$, of the form

$$a(V_{\xi 1})^2 - 2bV_{\xi 1} - c = 0 \tag{23.21}$$

where

$$a = \frac{1+r}{\cos^2(\theta_1)} \tag{23.22}$$

$$b = V'_{\xi 2} + rV'_{\xi 1} \tag{23.23}$$

$$c = \frac{2g}{rW_2}\left[(k_{eg})_1 + (k_{eg})_2 + CE_1 + CE_2\right] - \frac{1}{r}\left[V'_{\xi 2} + rV'_{\xi 1}\right]^2 \tag{23.24}$$

There are two solutions to this quadratic equation, as follows:

$$V_{\xi 1} = \frac{b \pm \sqrt{b^2 + ac}}{a} \tag{23.25}$$

Once $V_{\xi 1}$ is found, then the remaining impact velocity components are found as follows:

$$V_{\eta 1} = V_{\xi 1}\tan(\theta_1) \tag{23.26}$$

$$V_{\xi 2} = V'_{\xi 2} + r\left(V'_{\xi 1} - V_{\xi 1}\right) \tag{23.27}$$

$$V_{\eta 2} = -rV_{\eta 1} \tag{23.28}$$

These quantities can be transformed back into the coordinate system in which the scene was drawn. This is done as follows:

$$V_{XK} = V_{\xi K}\cos(\phi) - V_{\eta K}\sin(\phi) \qquad (23.29)$$

$$V_{YK} = V_{\xi K}\sin(\phi) + V_{\eta K}\cos(\phi) \qquad (23.30)$$

where $K = 1, 2$.

The Physical Meaning of Two Roots

As opposed to the linear equations that result from impulse–momentum considerations, the inclusion of energy conservation gives rise to a quadratic equation because kinetic energy is proportional to the square of the velocity. The ± sign in the quadratic equation and the resulting two solutions (instead of the one solution that would result from linear equations) reflect the fact that the mathematics are the same regardless of which vehicle is the bullet and which is the target. For example, the front of Vehicle 1 can strike the rear of Vehicle 2 and have the same momentum and energy conservation equations as if the front of Vehicle 2 struck the rear of Vehicle 1. The solutions are different, though. In the example situation, the signs of the closing velocity would be opposite; the vehicle having the larger velocity would swap places with the vehicle with the smaller velocity. The calculated speeds have to be examined to determine whether the positive or the negative root is appropriate for the case being studied.

Extra Information

A key advantage of the current analysis is the use of distinct positions for impact and separation. The analysis begins with the vehicles positioned at separation. As mentioned previously, the ability to analyze vehicle run-outs is improved because impact and run-out conditions are not co-mingled. More importantly, the vehicles are positioned relative to each other on the basis of their conditions and physical evidence available to the investigators post-accident, whereas any evidence of their initial engagement is often wiped out during the ensuing crash. Again, the separation conditions (and the velocities) are the known quantities, as opposed to the impact conditions, which to some degree or other are the unknowns.

Starting from their positions at separation, their headings and center of mass positions at impact can be predicted on the basis of their velocities and the crush duration.

As discussed in Chapter 18, the duration tends not to be a function of speed (the quantity being sought). Therefore, a reasonable estimate of duration can be made independent of the analysis. When the center of mass impact positions are predicted from the reconstruction results, the vehicles should be just in contact—not overlapping and not distant from each other.

Similarly, the vehicle headings at impact can be predicted from the yaw rates at impact and separation, the vehicle heading at separation (known from run-out analysis), and the crash duration. Since the direction of vehicle 1 impact velocity was assumed (usually from scene evidence or circumstances, as discussed earlier), the crab angle of vehicle 1 is predicted when the analysis is done. The actual crab angle is often known within a few degrees (not uncommonly, to be near to or at zero). The crash duration can be adjusted within reason to have the crab angle prediction come out correctly, but then the duration also affects the predicted positions of both vehicles at impact, including the impact heading of the other vehicle. Therefore, the ability to adjust the duration is usually strictly limited, but such considerations help form the reconstructionist's final opinion as to what the duration was, without the necessity of running a time-forward simulation to obtain such an estimate.

One could point out that this solution approach, having assumed an impact velocity direction for Vehicle 1 and only retaining three unknowns, is only marginally better than the analysis of central collisions that was discussed in Chapter 20. However, the inclusion of kinetic energy in this analysis means that the contribution from vehicle rotations is retained. See, for example, Equation 23.12. This element is entirely absent in the analysis of momentum conservation for central collisions. Moreover, the crush energy is also included in the current analysis. As we have seen, the total crush energy is directly connected to the severity of the crash. A prediction of the crash severity while leaving out crash damage or crush energy could constitute a neglect of key evidence.

As discussed in Chapter 18, eccentric collisions can be analyzed through the use of effective masses for the two vehicles. The effective mass values would be substituted for m_1 and m_2 in Equations 23.16 through 23.28, and then the results from these equations would be interpreted in terms of the centers of mass.

Sample Reconstruction

The governing equations for coplanar collision analysis are sufficiently simple that their implementation on a spread sheet is fairly straightforward. A sample reconstruction is self-explanatory, and is shown in Figure 23.1.

Case Name = Sample Reconstruction

Location of point at which separation (exit) begins

		X-Component of Exit Momentum, slug*ft/sec = 9608	−3192 = Y-Component of Exit Momentum, slug*ft/sec
X_sep =	**−30.00** ft	Angle B of Momentum Vector, deg = −18.38	
Y_sep =	**5.11** ft	Mass Ratio = $r = m_1/m_2$ = 1.4966	
Angle β of V_i vector, CCW from global X axis =	**−10.0** deg	$(1+r)/Cos^2(\beta-\theta)$ — ACoeff = 2.5508	
Crush Duration dt, from impact to separation =	**0.100** sec	BCoeff = 95.804 $V_{Q}+rV_{Q}$	
Root choice =	**pos** POS/NEG 1)	CCoeff = −2081.2 $2[(keq)_i + (keq)_j + CE_i + CE_j]/(rW_i) - (V_{Q}+rV_{Q})^2/r$	

Input quantities are in **BOLD** type
Quantities from linked spread sheets

	Vehicle 1	Vehicle 2	
	2001 Chevrolet Suburban	2000 Mitsubishi Galant ES	ρ = Sqrt 0.92[(OAL^2 + OAW^2)/12]
Radius of Gyration ρ, inch =	64.52	55.35	$m = W/g$
Vehicle Mass m, slug =	158.17	105.68	$I = m\rho^2$
Yaw Moment of Inertia I, slug-ft² =	4572.7	2248.5	
Yaw Rate at Exit ω', deg/sec =	−2.09	−221.63	
Exit speed V', ft/sec =	**0.00**	**0.00**	
Yaw Rate at Impact ω, deg/sec =	34.01	45.22	
Departure Angle θ', deg =	−13.99	−23.32	
X-Component of Exit Velocity V'_x, ft/sec =	33.00	41.52	
Y-Component of Exit Velocity V'_y, ft/sec =	−8.22	−17.90	
Exit Veloc. Parallel to Mom. Vector V'_i, ft/sec =	33.91	45.05	
Exit Veloc. Perpendicular to Mom. Vector V'_j, ft/sec =	2.60	−3.90	

			keq = ½MV² + ½I($\omega'^2 - \omega^2$)
Exit Trans. Energy + Rot. Energy Gained, ft-lb =	91,489	124,867	
Crush Energy CE, ft-lb =	19,944	84,111	Sums to 104,056 ft-lb
Exit Angular Momentum, slug*ft²/sec =	15,282	25,477	Sums to 40,760 slug*ft²/sec

Impact Speed Parallel to Mom. Vector V_i, ft/sec =	61.95	3.09	V_{Q} = (BCoeff + Root*Sqrt(BCoeff2 + ACoeff*CCoeff))/ACoeff
Impact Speed Perpendicular to Mom. Vector V_j, ft/sec =	9.12	−13.65	$V_{Qi} = V_{Qj}*tan(\beta-\theta)$
X-Component of Impact Velocity V_x, mph =	42.04	−0.93	
Y-Component of Impact Velocity V_y, mph =	−7.41	−9.50	
Resultant Impact Velocity V, mph =	**42.69**	**9.55**	
Approach Angle θ, deg =	−10.0	−95.6	$V_{Q} = V_{Q} + r*(V_{Q} - V_{Q})$
X-Component of Velocity Change ΔV_x, mph =	−19.54	29.25	$V_{Qi} = -r*V_{Qi}$
Y-Component of Velocity Change ΔV_y, mph =	1.81	−2.70	
Resultant Velocity Change ΔV, mph =	**19.62**	**29.37**	Calculated Distance Between CGs at Impact, ft = 12.46
Direction of Velocity Change $\Delta\theta$, deg =	174.7	−5.3	Centers of Mass Approach at 61.27 ft/sec = 41.77 mph
			Centers of Mass Separate at 11.06 ft/sec = 7.54 mph
Kinetic Energy Lost, ft-lb =	218,565	−114,509	"Restitution Coefficient" at Centers of Mass: 0.180
Impact Angular Momentum, slug*ft²/sec =	4,692	37,194	Closing Speed = 43.0 mph
Calculated X-Cordinate of Position at Impact, ft =	−37.12	−24.90	Sums to 104,056 ft-lb
Calculated Y-Cordinate of Position at Impact, ft =	6.06	3.67	Sums to 41,886 slug*ft²/sec
Calculated Vehicle Heading ψ at Impact, deg =	−10.60	278.58	
Calculated Crab Angle at Impact, deg =	−0.60	14.19	
Prin. Dir. Of Force, relative to impact headings, CW deg. =	−5	104	

X-Component of Impact Momentum, slug*ft/sec = 9608   −3192 = Y-Component of Impact Momentum, slug*ft/sec

FIGURE 23.1
Sample reconstruction.

References

1. Source Code Listing for Crash3, National Highway Traffic Safety Administration, Washington, DC.
2. Fonda, A.G., Computer implementation of momentum and energy solutions: Computer solutions, refinements and graphical extensions to the CRASH treatment, In: *Forensic Accident Investigation: Motor Vehicles*, T.L. Bohan and A.C. Damask, Editors, Michie Butterworth, Charlottesville, VA, 1995, pp. 351–394.

24

Checking the Results in Coplanar Collision Analysis

Introduction

The solution of three simultaneous equations that derive from the physics model is not only difficult, but it is also far from sufficient to reconstruct an accident. There are always measurement or observation uncertainties of some sort, and there are almost always missing data. Of course, these affect the results. As a consequence, in any given crash, there are myriad possible solutions. So the problem is not simply to get results, but to get results that comport with all that is known, which may include physical evidence that is not quantifiable. Certainly the results should reflect a position and velocity at impact that is physically achievable by the vehicle and driver. Fact witness testimony should be taken into account (though not necessarily relied on!). The analysis is set up to allow the reconstructionist to check the results, which include parameters that can be examined for reasonableness. One can then make adjustments in the parameters that are not known with that much certainty, and thereby tune the results so that all reasonableness checks are passed. Some of these will be discussed below.

Sample Spreadsheet Calculations

There is simply too much information and calculations in a reconstruction for it all to be contained on a single spreadsheet. Even if such a thing were possible, the spreadsheet would be far too large to print on a paper size that could be accommodated by most printers. So the calculations are set up with linked spreadsheets, incorporated in a single workbook. The spreadsheets are as follows:

Weight and dimensions (one for each vehicle)

Lock fractions and run-out trajectory (one for each vehicle)—see Chapters 2 and 4

Crash tests, with crash plot calculations (one for each vehicle)—see Chapter 21

Momentum conservation—see Chapters 19 and 20

Crush energy calculations (one for each vehicle)—see Chapter 21

Momentum and energy conservation—see Chapters 23 and 24

Momentum and energy conservation for coplanar impacts was discussed in Chapter 23, and a sample spreadsheet of the calculations was presented in Figure 23.1. It is reproduced here as Figure 24.1, and will serve to facilitate the discussion of how to check the results. Quantities entered directly into the spreadsheet are in bold type and highlighted with dark gray; those based at least in part on linked spreadsheets are highlighted with light gray.

Choice of Roots

The first task is to check whether the analysis represents the crash at hand, or one in which the roles of the vehicles are swapped, as discussed in Chapter 23. This is most easily done by checking the signs of the impact velocity components. Generally, if one or more of the signs are wrong, the indication is that the root choice (entered here as "pos") may also be wrong. Changing the root choice will usually change all the velocity components. In this case, switching the root choice to "neg" changes the signs, and the magnitudes, of all four impact velocity components, and effectively has vehicle 2 sliding mostly down the road in the opposite direction sideways at a high speed and striking a slower vehicle 1 in the rear with its side! This results in the same crush energy and momentum vectors, but it is clearly a different crash. Therefore, the negative root choice must be resoundingly rejected for this example.

Crash Duration

As discussed in Chapter 19, the crash duration is largely independent of speed; hence, it is largely independent of the reconstruction calculations. It is mostly a function of the types of structures involved, and the way they engage each other. Thus, the crash duration can be estimated before the reconstruction analysis is done.

Case Name = Sample Reconstruction

		X-Component of Exit Momentum, slug*ft/sec =	**9608**	−3192	= Y-Component of Exit Momentum, slug*ft/sec
Location of point at which separation (exit) begins		Angle ß of Momentum Vector, deg =	−18.38		
X_sep =	−30.00	ft	Mass Ratio = r = m_2/m_1 =	1.4966		
Y_sep =	5.11	ft				
Angle ß of V_1 vector, CCW from global X axis =	−10.0	deg	ACoeff =	2.5508	(1+r)/Cos²(ß-θ)	
Crush Duration dt, from impact to separation =	0.100	sec	BCoeff =	95.804	V_a + V_b	
Root choice =	pos	POS/NEG (1)	CCoeff =	−2081.2	2[(keq)_1 + (keq)_2 + CE_1 + CE_2](V_aW_2) − (V_a+rV_a1)²/r	

Input quantities are in BOLD type
Quantities from linked spread sheets

	Vehicle 1	Vehicle 2				
	2001 Chevrolet Suburban	2000 Mitsubishi Galant ES				
Radius of Gyration ρ, in. =	64.52	55.35	ρ = Sqrt[0.92(OAL² + OAW²)/12]			
Vehicle Mass m, slug =	158.17	105.68	m = W/g			
Yaw Moment of Inertia I, slug ft² =	4572.7	2248.5	I = mρ²			
Yaw Rate at Exit ω', deg/sec =	−2.09	0.00				
Exit speed V', ft/sec =	34.01	45.22				
Departure Angle θ', deg =	−13.99	−23.32				
X-Component of Exit Velocity V_x', ft/sec =	33.00	41.52				
Y-Component of Exit Velocity V_y', ft/sec =	−8.22	−17.90				
Exit Veloc. Parallel to Mom. Vector V_l', ft/sec =	33.91	45.05				
Exit Veloc. Perpendicular to Mom. Vector W', ft/sec =	2.60	−3.90				
Exit Trans. Energy + Rot. Energy Gained, ft-lb =	91,489	124,867	keq = ½mV² + ½I(ω'²−ω_a²)			
Crush Energy CE, ft-lb =	19,944	84,111	Sums to **104,056**	ft-lb		
Exit Angular Momentum, slug*ft²/sec =	15,282	25,477	Sums to	40,760	slug*ft²/sec	
Impact Speed Parallel to Mom. Vector V_l, ft/sec =	61.95	3.09	V_l1 = (BCoeff + Root*Sqrt[BCoeff² + ACoeff*CCoeff])/ACoeff			
Impact Speed Perpendicular to Mom. Vector V_w, ft/sec =	9.12	−13.65	V_a1 = V_l1 *tan(ß−θ)		V_l2 = V_a +r(V_a1− V_l1)	
X-Component of Impact Velocity V_x, mph =	42.04	−0.93			V_a2 = −rV_a1	
Y-Component of Impact Velocity V_y, mph =	−7.41	−9.50				
Resultant Impact Velocity V, mph =	**42.69**	**9.55**	Calculated Distance Between CGs at Impact, ft =	12.46	ft	
Approach Angle θ, deg =	−10.0	−95.6	Centers of Mass Approach at	61.27	ft/sec	41.77 mph
X-Component of Velocity Change ΔV_x, mph =	−19.54	29.25	Centers of Mass Separate at	11.06	ft/sec	7.54 mph
Y-Component of Velocity Change ΔV_y, mph =	1.81	−2.70	"Restitution Coefficient" at Centers of Mass:	0.180		
Resultant Velocity Change ΔV, mph =	**19.62**	**29.37**	**Closing Speed =**	**43.0**	mph	
Direction of Velocity Change Δθ, deg =	**174.7**	**−5.3**				
Kinetic Energy Lost, ft-lb =	218,565	−114,509	Sums to	**104,056**	ft-lb	
Impact Angular Momentum, slug*ft²/sec =	4,692	37,194	Sums to	41,886	slug*ft²/sec	
Calculated X-Cordinate of Position at Impact, ft =	−37.12	−24.90				
Calculated Y-Cordinate of Position at Impact, ft =	6.06	3.67				
Calculated Vehicle Heading ψ at Impact, deg =	−10.60	278.58				
Calculated Crab Angle at Impact, deg =	−0.60	14.19				
Prin. Dir. Of Force, relative to impact heading, CW deg =	−5	104				

		X-Component of Impact Momentum, slug*ft/sec =	**9608**	−3192	= Y-Component of Impact Momentum, slug*ft/sec

FIGURE 24.1
Momentum and energy conservation calculations.

However, the crash duration affects the calculation of the center of mass positions, vehicle headings, crab angles, and the relative alignments of the vehicles at impact. So it is not good practice to adjust the estimated crash duration, except by small amounts, because it affects so many parameters at once. If one or a few of the impact conditions do not seem quite right, it is better to look elsewhere for the problem first.

The assumed crash duration of 0.100 sec is shown near the top of Figure 24.1, in bold type and highlighted in dark gray, as a user input. If one were to build a time line from the point of impact to rest, the duration would have to be added to the time values shown on the trajectory spreadsheets (where the time clock starts at separation).

Selecting Which Vehicle Is Number 1

If one of the vehicles has a known trajectory at impact, that vehicle should be vehicle 1 in the analysis, since its impact velocity vector angle is directly entered into the spreadsheet by the user. The angle of the velocity vector for vehicle 2 is a product of calculations by the analysis.

In the sample case, a Mitsubishi Galant turned left in front of a Chevrolet Suburban, which was heading straight, according to the police report. Accordingly, the Suburban was selected as vehicle 1 in the reconstruction analysis, shown in Figure 24.1. No tire marks on the road were reported, but the police investigation was minimal at best, and outright wrong at worst. (The wrong location was provided for the accident scene.) In keeping with the level of investigation, there were no photographs taken at the scene, so the Suburban travel direction could not be independently verified.

Yaw Rate Degradation

As discussed in Chapter 4, some sort of reasonable decrease in the yaw rates from separation to rest should be expected. If a splining operation is used to analyze the run-out, it is often the case that the yaw rates will not look just right the first time a solution is attempted. Do not despair. Undesirable yaw rate fluctuations can be changed by slight adjustments in the vehicle headings during run-out, consistent with physical evidence (such as tire marks). It may be that additional points on the run-out trajectory will need to be specified, particularly if actual yaw rates are high. Generally, no more than 90° of vehicle rotation should occur between key points, so that the effect of crab angle on deceleration can be properly modeled.

Yaw Rates at Impact

In Figure 24.1, the impact yaw rate was assumed to be zero for both vehicles. In fact, a more realistic yaw rate for vehicle 2 would be about 10 or 15°/sec (positive for a left-turning vehicle because a mathematical coordinate system was in use). Since the vehicle was turning left, accident scene geometry, fact witness testimony (as to whether tire squealing was heard, for example), or just plain physical evidence could be used to make an estimate of the pre-crash turn radius. However, no such information was available in the sample case. If such information were available, the reconstructed impact speed could then be used to make a refined estimate of pre-impact yaw rate.

In the sample case, changing the impact yaw rate for vehicle 2 to 15°/sec changes its impact heading, and crab angle, by only about 0.7°. (This is yet another instance of the initial yaw rate not influencing the crash much.) A what-if question like "What is the effect of initial yaw rate?" is easily answered through the use of a spreadsheet.

Trajectory Data

The trajectory analysis provides the distance in each segment, the path angle in the segment, the vehicle rotation during the segment, the computed crab angle, the average deceleration, and the time in the segment. The vehicle rotation should be checked to see if it is more than 90° between key points.

Looking at the crab angle during run-out is useful in explaining variations in the vehicle drag factor. Questions about the vehicle exit velocities can often be answered by checking the crab angles.

Vehicle Center of Mass Positions

In Figure 24.1, the variables X_sep and Y_sep serve as a common location, in scene coordinates, for the reconstruction calculations. (This is not necessarily the "point of impact," but could be if the user desires. It could also be some landmark, such as a manhole cover.) The position of each vehicle center of mass at separation is located (by a formula contrived by the user) relative to X_sep and Y_sep, and is shown on the trajectory spreadsheet. The workbook is set up this way as a convenience to permit easily moving the whole collision complex without having to recompute the position of each vehicle separately. This could happen when one needs to place the collision on roadway evidence, for example. Also, the sensitivity of the results to assumed impact location can be explored. The inputs are at the top of the spreadsheet, and are in bold type and highlighted in dark gray to indicate they are input directly by the user.

For both vehicles, the positions of their centers of mass at impact are calculated. The X and Y coordinates are shown six lines up from the bottom.

Impact Configuration Estimate

The vehicle positions at separation are provided by the reconstructionist as inputs, usually based on matching up the damaged vehicles to each other, and to any scene evidence, on a scale drawing. The reconstructionist also inputs the yaw rates at impact, usually based on crude estimates such as what we have already seen. The estimates can be refined once the impact velocities have been calculated. The vehicle velocities at separation (including yaw rate) are obtained by analyzing the run-out trajectories. The vehicle velocities at impact are calculated by the physics model. These are the necessary ingredients for calculating the average translational and rotational velocities during the crush phase.

The crash duration is input by the reconstructionist. When multiplied by the average velocities, and the results are added to the vehicle positions at separation, the vehicle positions (translational and rotational) at impact are calculated.

Scale drawings of undamaged vehicles can be placed in this configuration on a scene drawing to see if they are in contact with each other. Any gaps or overlaps may be correctable by adjusting the crash duration, though again that adjustment should be done carefully. It should be noted that the impact configuration estimate is the only area in the analysis where the crash duration is used. Changes in the crash duration have no effect on the physics model, or on the resulting speed calculations.

The drawings of the undamaged vehicles at impact can be used to directly measure the distance between their centers of mass, taking into account the rounded shapes of the vehicles. This number can be compared to the calculated distance that is produced by the analysis, and printed out on the spreadsheet just to the right of the approach angle values (see Figure 24.1). The primary means of adjusting the estimated crash duration is to get an acceptable match with the measured distance. The Excel "Goal Seek" function can be used for this purpose, but the vehicle headings and crab angles should be checked immediately afterwards to ensure that they are still in accordance with physical evidence and vehicle performance limits.

Vehicle Headings at Impact

The calculated vehicle headings ψ at impact are shown near the bottom of the spreadsheet (see Figure 24.1).

In Chapter 23, it was pointed out that the impact headings are calculated from the yaw rates at both separation and impact. It is useful to remember that when there is a significant yaw rate at separation, variations in the crush duration will have a significant effect on the calculated heading at impact. However, one is limited in adjusting this parameter by its effect on the distance between the centers of mass at impact, as discussed above. Variations in the exit yaw rate can also have significant effects, as will be seen in later discussion.

Adjustments of vehicle heading at separation would appear to directly affect the heading at impact, though again limitations are imposed by the necessity of having the vehicles match up. However, such adjustments will also affect the exit yaw rate calculated in the trajectory analysis, so the heading at separation has a nonlinear effect on the calculated heading at impact. In fact, it should be said that yaw rate degradation, and hence vehicle heading at separation, should be taken care of before worrying about the impact heading, for just this reason.

Alternatively, one can rotate the whole collision complex, particularly if the changes are favorable for both vehicles. This can be tricky, however, because of the nonlinear effects cited above.

Getting the impact headings and crab angles right can take a lot of patience. The calculations are simple enough, and variations can be easily explored with a spreadsheet. But a lot of details are made available to the analyst— details that cry out to be pinned down. Once an apparent mathematical solution has been obtained, the positions of the vehicles should be compared with each other and with any evidence at the scene (such as a kink in a tire mark indicating impact location).

Crab Angles at Impact

In many cases, a vehicle is known to be tracking at impact. For example, the vehicle may have been engaged in everyday driving at the time of impact, without accident avoidance maneuvers. A vehicle that is known to be tracking at impact should have a calculated impact crab angle of 2–3° or less. Alternatively, a vehicle might be leaving multiple tire marks on the pavement during impact, from which information of both the vehicle heading and its velocity vector direction can be found. In such cases, the crab angle at impact is approximately known.

If the crab angle for either vehicle at impact is known from physical evidence, the analysis should be checked to ensure that the calculated crab angle is a reasonable match to the known crab angle. The crab angle at impact is therefore printed out on the spreadsheet. See Figure 24.1, near the bottom, just below the calculated vehicle headings at impact. If a crab angle appears to be out of whack, probably the vehicle heading at impact will have to be re-calculated using a change of inputs.

The crab angle of the Suburban at impact could not be ascertained or estimated independently from tire marks, because none would have remained until the author's scene inspection, and because there were no photographs and no police measurements. The reconstruction indicates it was near zero (less than a degree). However, the crab angle of the Galant, vehicle 2, at impact (about 14°) is suspiciously high, even though the vehicle was turning. What is the significance of this result? If 14° is considered too high, how did it get that way?

In Chapter 23, the crush duration was pointed to as a potential culprit because it figures into the crab angle calculation. However, it is also used in the prediction of both vehicle heading angles ψ, as well as the X and Y coordinates of both vehicles at impact. Therefore it is not a knob to be adjusted carelessly. There is nothing in the present results to indicate anything amiss with the crash duration, so no adjustments were made to it.

We note that the exit yaw rate of –222°/sec in Figure 24.1 comes from calculations that employed only one intermediate point. The actual accident vehicle had a highly curved trajectory, so three intermediate points should have been used. Making this change results in an exit yaw rate of –156°/sec—which seems more realistic. This value produces a crab angle for vehicle 2 at impact of about 5.6°—a much more reasonable result in this instance.

Approach Angles

The approach angles θ are the angles of the velocity vectors at impact. For vehicle 1, the angle is specified by the user and is therefore in bold type and highlighted in dark gray in Figure 24.1. For vehicle 2, the approach angle is not specified. That angle is determined from the analysis. Be aware, however, that if either the X or Y component of impact velocity is small, small changes in the component can have disproportionately large influences on the magnitude of the angle, and can even change its sign, without affecting the resultant velocity much. In this situation, concerns about this variable can (and should) be deferred until the other reasonableness checks are satisfied.

Restitution Coefficient

The analysis provides a calculation of the rate of approach of the vehicle centers of mass at impact, and their rate of separation at the end of the crash. The rate of separation is the rate at which the distance D between the centers of mass of vehicles A and B increases. The distance D is given by

$$D = \sqrt{(X_A - X_B)^2 + (Y_A - Y_B)^2}$$

(24.1)

Differentiating with respect to time yields

$$\frac{d(D)}{dt} = \frac{1}{2}\left[(X_A - X_B)^2 + (Y_A - Y_B)^2\right]^{-\frac{1}{2}}$$

$$\times \left[2(X_A - X_B)\left(\frac{dX_A}{dt} - \frac{dX_B}{dt}\right) + 2(Y_A - Y_B)\left(\frac{dY_A}{dt} - \frac{dY_B}{dt}\right)\right]$$

$$= \frac{(X_A - X_B)(V_{XA} - V_{XB}) + (Y_A - Y_B)(V_{YA} - V_{YB})}{D}$$

(24.2)

The position and velocity values in Equation 24.2 reflect conditions at separation. To obtain the rate of approach at impact, one simply swaps the roles of vehicles A and B (so as to produce the desired sign reversal) and substitutes values pertinent to impact. The analysis then computes the ratio of the approach and separation rates, and labels it the "restitution coefficient."

It should be pointed out that the term "restitution coefficient" is not strictly correct, since we are not calculating the relative velocities at the point where the vehicles are in contact, normal to the contact surface. The distinction has to do with the degree to which the normal to the contact surface varies from a line drawn between the centers of mass, and with the vehicle yaw rates, which could make the velocities at the point of contact different from those at the centers of mass. Besides, vehicle contact occurs along a whole surface, not just at a point, so fussiness about the actual coefficient of restitution is not likely to produce an increase of understanding.

All that said, the purpose of this check is to warn the reconstructionist if the vehicles continue to move into each other (i.e., the centers of mass continue to approach each other) after impact, the solution is likely to be problematic.

It is interesting to note that despite the fact that the restitution coefficient must be specified (as a known quantity) in the Brach's Planar Impact Mechanics,[1] in Woolley's IMPAC program,[2] and in Smith's PLASMO program,[3] a negative "restitution coefficient" in the present analysis has nothing to do with the impact conditions or the impact analysis. Rather, it simply means that the exit velocities fail to have the vehicles actually separating, but continuing to move into each other. Generally, this is not because one or both of the exit velocity vectors are in the wrong direction, but simply that they are of inconsistent magnitudes. So the system momentum vector, determined on the basis of the run-out trajectories, is probably only slightly in error. The culprit in a faulty "restitution coefficient" is almost certainly in the run-out analysis. The negative "restitution coefficient," or even a "restitution coefficient" outside the bounds of experience—0.0 – 0.3, say—needs to be cured. The cure indicates necessary adjustment(s) in one or both of the run-out trajectories, not in the collision itself. Crab angles, and the attendant drag factors, are usually the primary suspects. Friction coefficients should also be looked at.

It is also interesting to note that the approach speed is not the same thing as closing speed, which is given by

$$V_{cl} = \sqrt{(V_{XA} - V_{XB})^2 + (V_{YA} - V_{YB})^2} \qquad (24.3)$$

This quantity deals only with the velocity vectors—not the positions of the centers of mass. The difference between closing speed and approach speed is easily seen by considering two vehicles traveling in opposite directions at constant speeds, on opposite sides of the road, not colliding. The vehicle velocity components, and the closing speed, remain constant before, during, and after the time they are abreast of each other.

By contrast, the approach speed will very gradually start to decline as the angle of the line between their centers of mass starts to deviate from the velocity vectors. This decline will become noticeable as the vehicles come abreast of each other. When they are abreast, the approach speed will suddenly change signs (as it should), and asymptotically approach its original magnitude. This effect can be seen by driving at a constant speed past a radar-operated speed warning sign.

Principal Directions of Force

If we apply Newton's Second Law, Equation 18.7, to each collision partner separately, and assume that its mass remains constant, we have

$$m d\vec{v} = \vec{F} dt \qquad (24.4)$$

If we integrate across the crash duration Δt, we have

$$m(\vec{V}_S - \vec{V}_I) = \int \vec{F} dt = \vec{F}_{AVG} \Delta t \qquad (24.5)$$

or

$$m\Delta\vec{V} = \vec{F}_{AVG}\Delta t \tag{24.6}$$

where the subscripts S and I indicate separation and impact, respectively, $\Delta\bar{V}$ is the velocity change (a vector), and $\bar{F}_{AVG}$ is the average force (also a vector) applied during the crash phase. From the above, we realize the important result that the ΔV vector is collinear with the average force vector. The principal direction of force, or PDOF, is therefore the same as the direction of the ΔV vector.

All this may seem to be dwelling on the obvious. However, it is worth examining in detail because some reconstruction methods (Crash3 and its derivatives) treat the PDOF as a known quantity, to be specified by the user. This is like specifying an important part of the answer, one typically of high interest, before the reconstruction has begun. It may be necessary in Crash3 because of the way it handles eccentric impacts (not unlike the effective mass concept discussed in Chapter 18), but in the present analysis such an assumption is unnecessary. The requirement to assume a PDOF and other topics regarding Crash3 are discussed in papers by Woolley et al.[4] and Struble.[5] In fact and in the current analysis, the PDOF is determined by the direction of the ΔV vector. Since the analysis has the velocity components of both the impact and separation velocities, it can, and does, calculate the PDOF. It is printed out as a clockwise angle from the local (vehicle) x-axis, consistent with widespread usage. The analysis assumes that for the purpose of calculating PDOF, the vehicle heading is a counterclockwise angle, and adjusts accordingly. Therefore, the spreadsheet formula for PDOF will need to be changed if the user is working in an SAE coordinate system, with clockwise angles. The reconstructionist can convert the angles to clock position, if desired.

The PDOFs are shown at the bottom of Figure 24.1. If such angles were reported in scene coordinates (relative to the X–Y-axis), they would be 180° apart, in accordance with Newton's Third Law. However, they are reported with respect to the vehicles, each of which has its own (x–y) coordinate system.

Energy Conservation

The kinetic energy lost in the collision is calculated, and reported in the first line of numbers near the bottom of the spreadsheet (see Figure 24.1). In the sample case, the Suburban was moving about 43 mph at impact and about 23 mph (34 ft/sec) at separation, losing about 219,000 ft-lb of kinetic energy in the process. The Galant was moving about 10 mph at impact, but at separation it had a speed of about 31 mph (45 ft/sec) and was rotating at about

150°/sec. Thus, the Galant gained about 115,000 ft-lb of kinetic energy when it got hit on the side. These changes produced a total loss in kinetic energy of about 104,000 ft-lb, as reported on the spreadsheet.

At the same time, the Suburban absorbed about 20,000 ft-lb of crush energy. The value was calculated in a linked spreadsheet and reported about half-way down in Figure 24.1. The Galant, on the other hand, absorbed about 84,000 ft-lb of crush energy, for a total crush energy of about 104,000 ft-lb.

The decrease in kinetic energy and the increase in crush energy (from zero) are both reported in the spreadsheet, in bold type face. The numbers are identical. Proof positive is thus provided that there is an energy balance during the crash phase.

Momentum Conservation

The X- and Y-components of the momentum vector at separation are obtained from the vehicle velocity vectors at separation, as calculated in the two trajectory spreadsheets and copied to Figure 24.1. They are shown in bold type face at the top of the spreadsheet. The angle of the resultant momentum vector is about 18° clockwise from the X-axis (or –18° in the mathematical coordinate system used in this reconstruction).

The X- and Y-components of the momentum vector at impact are calculated from the impact velocity vectors of the two vehicles, and are printed at the bottom of Figure 24.1. At impact, the X-component of momentum is about 9600 slug-ft/sec, and comes mostly from the Suburban. Not only is the Suburban much more massive than the Galant, it has an X-component of velocity at about 42 mph. At impact, the Y-component of momentum is about –3200 slug-ft/sec, due to both vehicles having a negative Y-component of velocity. (Even though the Suburban was traveling "straight ahead," the roadway was not quite parallel to the X-axis at the point of impact, so with its mass and much higher speed, it also contributed.) The momentum components are reported at the bottom of the spreadsheet.

Again, both the X and Y components of momentum at impact and separation are identical. This is proof positive that momentum is conserved during the crash phase.

These two conservation checks can be used to demonstrate that the principles upon which the physics model is based are satisfied. It matters not what the resulting equations were, or how they were solved. With three equations and three unknowns making up a quadratic system, there are two, and only two, solutions. By choosing the root to be positive or negative, we have chosen one of them, but both solutions satisfy the principles of conservation of energy and momentum. In that sense, it is not necessary to get bogged down in the mathematics. If the principles of physics are satisfied, the solution is the same, independent of whether advanced mathematics or a Ouija board was used to find it.

Direction of Momentum Vector

This quantity is printed out by the analysis, in the scene coordinate system that the user chose when determining the trajectory inputs. In a very real sense, it is determined entirely by the run-out analysis, before momentum conservation is enforced during the crash phase. If this angle appears suspect, the run-out conditions need to be examined.

In a coarser sense, it is useful to look at the general directions the vehicles take from separation to rest to get a feel for how the momentum vector is oriented. If we have a general idea of how the vehicles approached the crash, we can get an idea of how their relative speeds contributed to the momentum vector. This is particularly true in intersection crashes. If the vehicles are traveling in orthogonal directions, one of the vehicles will supply all of the momentum in the direction it was traveling, and the other will do likewise. We can thus readily see what sort of adjustments are required in vehicle exit velocity magnitudes or directions, to swing the momentum vector one way or another.

Speaking of exit velocity directions, the analysis draws a line between the separation position and the first point on the run-out trajectory, and points the exit velocity vector in that direction. If the vehicle's trajectory is curved, it would be erroneous to simply draw a straight line between separation and rest, as that would lead to an incorrect direction for the exit velocity vector.

Momentum, Crush Energy, Closing Velocity, and Impact Velocities

The closing velocity is calculated by the analysis, as described previously, and printed out. One can think of the closing velocity as being parceled out to the vehicles as ΔVs, with a small increase due to the restitution coefficient, according to their masses. Equations 19.11 and 19.12 provide this insight, even though they were developed for uniaxial, rather than coplanar collisions. Equation 19.20 also shows us the direct connection between closing velocity and crush energy. All these quantities exist without any reference to global (scene) coordinates (though obviously the descriptions of velocity vectors, in terms of components and direction, depend on the coordinates they are referenced to).

The tie-in to scene coordinates occurs when the ∆Vs are connected to the separation velocities, allowing the determination of impact velocities. Since the separation velocities determine the system momentum, we can think of this connection as the place where momentum and crush energy meet. Any imbalance between momentum and crush energy, so to speak, will show up in the impact velocities.

Suppose, for example, that a stopped vehicle is hit from behind. If the analysis has the struck vehicle moving forward at impact, we have two possible indications:

1. The system momentum is too high. We can lower it, without affecting the crush energy and having minimal effects on the closing velocity, by finding a way to lower one or both of the separation velocities (keeping an eye on the restitution coefficient, of course). This entails an examination of the run-out trajectories. The effect of lowering the system momentum would be a lowering of both impact velocities, which is the indicated "fix" for the struck vehicle.

2. The closing velocities, and hence the crush energy, are too low. We may think of a higher closing velocity as requiring a higher impact velocity by the striking vehicle, but it can also be achieved by a lower impact velocity by the struck vehicle. The latter would be the objective of adjustments in the analysis. This could be achieved, with minimal effects on the momentum, by calculating a lower crush energy.

This example is somewhat over-simplified, in that we are generally dealing with coplanar collisions rather than uniaxial ones. It is also the case that most crashes do not involve stopped vehicles. However, one might be checking the reasonableness of a possible solution against some parameter other than impact speed, such as the ∆V for airbag deployment threshold on the striking vehicle. Nevertheless, looking at a crash in this way (balancing crush energy against momentum) is useful in guiding one toward a reasonable solution.

The analysis does not speak English (or any other spoken language), but still manages to offer consistent hints at the direction one must go to find a reasonable answer. It does this by presenting parameters than can be compared (such as forces, or impact angle, or crab angle, for example). In the process of making adjustments to satisfy the various reasonableness checks within the analysis, one becomes aware of these hints, and where they lead.

Angular Momentum

As discussed previously, the analysis does not conserve angular momentum. However, knowledge of the vehicle positions and velocities at impact and separation allows the angular momentum to be calculated at those two points in time. They can be compared with each other, though as a practical matter it is difficult to know what to make of the differences between them.

Angular momentum can be calculated about any (fixed) point. It could be done about the location of the system center of mass at separation, for example, if that were deemed helpful in understanding the crash. For simplicity, however, the analysis does its calculations about the origin of scene coordinates. Therefore, it is helpful to set up the scene coordinate system so that its origin is as close to the impact point as practical, consistent with finding landmarks at the scene that are permanent, and that allow the coordinate axes to be parallel to some feature of the roadway.

In the sample case, the exit angular momentum summed to about 41,000 slug-ft^2/sec, and the impact angular momentum summed to about 42,000 slug-ft^2/sec, as seen in Figure 24.1.

Force Balance

A force balance (Newton's Third Law) is meaningful only for the contact damage portions of the crush profiles. Therefore, the analysis allows the specification of the beginning and end of that portion of each crush profile, and restricts force calculation to only the crush profile segments between those points.

The analysis allows for multiple crush profiles, and for the use of weighting factors where two or more profiles are used to describe the damage. Damage to separate structures—possibly with different stiffness parameters and/or structural models—may all carry weighting factors of unity, meaning that the forces from the various structures are simply added together, as are the crush energies. On the other hand, underride/override crashes may produce separate crush profiles at different water lines. The forces and crush energies from the different profiles are then combined using weighting factors based on force levels seen in load cell barrier crashes, as explained in Chapter 26. The weighted force values, and the sum of the weighted values, are calculated and printed out in a linked spreadsheet—one for each vehicle.

In the sample case, one crush profile was developed for the front of the Suburban (with a weighting factor of 1.000, obviously). A constant stiffness model (see Chapter 21) produced a total weighted force of about 66,300 pounds. A single crush profile was developed for the side of the Galant. Again, a constant stiffness model was employed having a saturation crush of 4.6 in. The total weighted force was about 66,800 lb.

As with angular momentum, the analysis does not enforce a force balance. (In any case, the forces are not dynamic real-time contact forces generated during the crash, but forces associated with residual crush measured after the fact, as discussed in Chapter 21.) However, the total weighted forces over the colliding contact areas (where match-up occurs) should be in reasonable proximity. Significant imbalance would indicate a need to reconsider the elements going into the force calculation.

Vehicle Inputs

A linked spreadsheet contains all of the weight and dimensional properties that pertain specifically to each vehicle. Crash test data, including weights, speeds, crush, and stiffness calculations, are contained in another spreadsheet, which also contains the calculated raw and weighted forces and crush energies for each crush profile (where weighting factors are applied to multiple crush profiles).

Another spreadsheet for each vehicle contains the lock fraction and trajectory calculations. Finally, a spreadsheet containing the physics model includes the angle of the vehicle 1 velocity vector, the crush duration, and the root choice. Of course, all the inputs are echoed somewhere, at least once, in the workbook.

Final Remarks

This chapter should illustrate that reconstruction is neither exact, nor a science. It is engineering—in fact, reverse engineering—which not only takes full cognizance of physics and mathematics, but also relies on observation, measurement, knowledge of vehicle design and construction, and the ability to deal with errors and unknowns. It is also an art, significantly shaped by experience and intuition. The analysis was designed with the objective of placing information and tools in the hands of the reconstructionist so that he or she can deal effectively with uncertainties, and most easily apply the engineering and the art to the problem of quantifying what happened in a crash.

References

1. Brach, R.M. and Brach, R.M., *Vehicle Accident Analysis and Reconstruction Methods*, SAE International, 2005.
2. Woolley, R.L., The 'IMPAC' program for collision analysis, SAE Paper 870046, SAE International, 1987.

3. Smith, G.C., Conservation of momentum analysis of two-dimensional colliding bodies, with or without trailers, SAE Paper 940566, SAE International, 1994.
4. Woolley, R.L., Warner, C.Y., and Tagg, M.D., Inaccuracies in the CRASH3 program, SAE Paper 850255, SAE International, 1985.
5. Struble, D.E., Generalizing CRASH3 for reconstructing specific accidents, SAE Paper 870041, SAE International, 1987.

25

Narrow Fixed-Object Collisions

Introduction

Impacts with narrow fixed objects such as poles, trees, retaining wall ends, and other obstacles require special consideration from the reconstructionist. For example, real-world poles may be fixed (or not, if they are mounted on a slip base), but they are not infinitely rigid. Nor are they infinitely massive. They can be separated from their base, pushed over in the ground, or totally or partially fractured (splintered). While they are not infinitely massive, their mass can be significant relative to that of the vehicle. Trees usually are not broken unless they are fairly small, but they can bend and sometimes recover. Trees can dissipate energy aerodynamically if their foliage whips about in the air, and by vibrating. Trees are living things and can grow and heal; some can even straighten out somewhat after the accident.

Then again, the very narrowness of objects has effects on the vehicle. By definition, narrow objects are narrower than the vehicle. Unlike the barriers used to characterize the structure as was discussed in Chapter 21, narrow objects do not engage the entire vehicle structure. The part that is engaged is exercised differently than in a barrier crash. As in all crashes, the vehicle structure is inhomogeneous. In a barrier crash, the behavior of its more compliant parts gets averaged with the action of the stiffer parts, but such "averaging" does not occur when something narrow is hit. Overall structural behavior can depend on which part carries the brunt of the load. In any case, frontal pole crash tests are comparatively few in number, effects are not well understood, and ignorance abounds. Lord Kelvin would surely note the plethora of theories and the paucity of numbers.

Nevertheless, the reconstructionist is occasionally asked to wade into this swamp of unknowns and render an opinion. This chapter is intended to help out by providing an overview of the state of knowledge.

Wooden Utility Poles

Many thousands, or perhaps hundreds of thousands, of utility poles have been erected for the purpose of carrying overhead wires for electric power, telephone, cable television, and so on. It should not be surprising to find that, as engineering structures, their characteristics are governed by standards. Of foremost interest is American National Standards Institute (ANSI) Standard 05.1-1979,[1] *Specifications and Dimensions for Wood Poles.*

The large end of the pole is called the "butt." Though not regulated by the standard, typical installation practice appears to consist of burying the butt 6 ft below ground line. Wooden utility poles are identified by class, which is based on their circumference measured 6 ft from the butt. The specified circumference varies with the wood species, because a pole of a given length and class is intended to have a particular load-carrying capacity when a horizontal load is applied 2 ft below the top, regardless of species.

Appendix B of the standard does not specify any requirements, but is included for design purposes. It shows the horizontal loads used in the calculations for each pole class. The calculations assume the pole is a simple (tapered) cantilever beam, with the maximum bending stress occurring at ground line. An average taper is used in the analysis. A Class 4 pole is often encountered; its horizontal design load is 2400 lb.

The standard also specifies how the pole shall be marked. Poles are often branded, but a metal tag may be used as well. The bottom line on the pole label includes a hyphenated number of the form C-LL, where C is the pole class and LL is the length. For example, "4-40" indicates a Class 4 pole 40 ft long (of which 6 ft can be assumed to be below ground). Other identification lines (that may be omitted on the marking) indicate the manufacturer, plant location and year of treatment, and species and preservative. Other tags may be placed by the utility company. It is good practice to record this information (e.g., by photography, notes, and/or on a sketch of the scene), so that there is no confusion about which pole is being looked at, and what its size is. It is not uncommon for police to use a pole as a permanent point in their investigation, and to photograph or otherwise record the label information.

Early attempts to assay the dynamic properties of wooden posts appear in papers by Michie et al.[2] and Wolfe et al.[3] In the Wolfe research at Southwest Research Institute, Class 4-40 creosote-treated southern pine poles were modified with drilled holes and groove patterns, installed, and impacted by a 4000-lb pendulum at 20 mph, at bumper height. These basic experiments have not led to the adoption of weakened wooden utility poles for roadside use, but the tests did yield fracture energy data, which have been used for reconstruction purposes. Unfortunately, the results from the weakened poles are questionable (e.g., stress concentrations), and the only pole that was not weakened was not fractured. In contrast to crashing a vehicle into a pole, the full-scale pendulum testing procedure seems elegant, and allows the fracture energy to be more readily isolated from other energy dissipation mechanisms. However, it is probably difficult to summon enough kinetic energy in a pendulum to achieve fracture of a nonweakened pole.

In 1981, Mak et al. published a reconstruction procedure[4] using the Wolfe data, based on three assumptions: (a) a vehicle-to-pole impact occurs in three phases that are independent of each other; (b) pole fracture is associated with a certain critical level of breakaway fracture energy, or BFE; and (c) once fractured, the segmented pole structure is accelerated to the velocity of the vehicle at separation. The procedure was "validated" with five full-scale crash tests into slip-base poles, in which the CRASH program, with its "poor correlation for velocity change due to vehicle crushing," was used to compute the vehicle's crush energy in the crash. User beware.

For reasons not explained in the Mak paper, the BFE for complete fracture of poles <26 in. in circumference was reported to be a constant 20,000 ft-lb, independent of pole size. This seems counterintuitive. In at least some copies of the paper, the three remaining equations for BFE for wooden utility poles appear to have some incorrect powers of 10. The set of equations in Table 1 of the paper should read:

$$\text{Complete Fracture BFE} = 20,000 \tag{25.1}$$

$$C \leq 26 \text{ in.}$$

$$\text{Partial Fracture BFE} = \tfrac{1}{2}[20,000 - (1.4 \times 10^{-2})C^{4.38}] \tag{25.2}$$

$$\text{Complete Fracture BFE} = (1.4 \times 10^{-2})C^{4.38} \tag{25.3}$$

$$C > 26 \text{ in.}$$

$$\text{Partial Fracture BFE} = \tfrac{1}{2}[(1.4 \times 10^{-2})C^{4.38} - 20,000] \tag{25.4}$$

where BFE is the breakaway fracture energy in ft-lb and C is the pole circumference (measured at the ground plane?). The coefficients and exponents in these equations appear to be empirical fits to the Wolfe data, but no explanation or derivation is provided.

If the BFE for complete fracture is to be a continuous function of C, then the right-hand side of Equation 25.2 must be equal to 20,000 ft-lb at $C = 26$ in. Setting Equation 25.3 equal to Equation 25.1 yields a transition circumference C of 25.4 in. Close enough. However, the partial fracture BFEs in Equation 25.2 decrease from 10,000 ft-lb at zero circumference to zero at 25.4 in., and then increase from zero in Equation 25.4 as the circumference increases. This is a bizarre result.

Perhaps a rigid contoured moving barrier, like the one used in NHTSA side impacts[5] to characterize side structures, could be impacted into wooden utility poles with a mass and speed sufficient to produce pole fractures. This would be an interesting exercise, and might produce some usable results.

In 1987, Morgan and Ivey[6] analyzed wooden pole impacts, primarily with an eye to whether the pole would fracture or not. Statements are made with little, if any, explanation or foundation. The results of two full-scale vehicle crash tests into wooden utility poles (Class 4-40) were presented in a table. One vehicle was a 2600 lb Chevrolet Vega moving at 29.4 mph, which did not fracture the pole. The other vehicle was a 4325 lb Dodge St. Regis moving at 61.5 mph, which did fracture the pole. The authors assert that "any automobile impacting a pole at <30 mph will be stopped at impact." A "failure boundary" consistent with this statement is presented for Class 4-40 poles.

In 1998, Kent and Strother[7] explored the energy required to fracture a wooden utility pole when hit by a vehicle. Scale model poles, 1/8 size, were constructed of various woods, some of which were configured to correspond with poles tested by Wolfe et al. These were impacted with an instrumented pendulum, and the fracture energies measured. For 20 hemlock specimens, "A second-order polynomial fit most accurately represents the relationship between the energy to completely fracture the pole specimens and the specimen moment of inertia, calculated about a horizontal axis perpendicular to the velocity vector ($R^2 = 0.9758$)." Birch and poplar specimens had fracture energies in this range, but green cottonwood and pine energies were considerably lower.

The fitted curve for the hemlock poles was scaled up so as to match the fracture energy for Pole NP-1 in the Wolfe study. The relationship between the fracture energy (*FE*) and cross-section moment of inertia then became

$$FE = 39.782 \, I_c - 0.0051 \, I_c^2 \tag{25.5}$$

where *FE* is the fracture energy in ft-lb, and I_c is the cross-section moment of inertia in inches.[4] It is unfortunate that a modified pole like NP-1 had to be used for the scaling-up process (since no full-size unmodified poles were fractured in the Wolfe study). As it is, Kent and Strother state that "fracture energies found [by Wolfe et al.] are probably understated relative to those that would have been found if unmodified poles of similar section moduli had been used"[7] (p. 37). Because of the scaling procedure, similar considerations were thus passed on to their work.

As a southern yellow pine pole was used for the scaling, Equation 25.5 represents that wood species, as tested by Wolfe et al. If the pole is known to be made of some other kind of wood, the Kent paper presents a procedure for taking that fact into account.

According to Kent and Strother, the fracture energies presented "by earlier authors may have overestimated fracture energy because they relied upon extrapolations past known data." They acknowledge that poles were assumed to act as classic cantilever beams; that is, with completely rigid support at the base and complete freedom at the top. Thus yielding of the soil, the mass of equipment such as transformers, and constraining effects of wires connected to other poles or the ground were ignored.

As for trees, Kent and Strother point out that "live trees can be expected to require less fracture energy than dry poles of the same size and type, if they completely fracture," because of moisture contents that are significantly higher than in utility poles. Incomplete fracture, or—to use a medical term—"green stick fracture," may be more the rule than the exception for living trees.

Poles that Move

Yielding of the soil is another mechanism of energy absorption. For wooden utility poles, a reasonable energy accounting proposed by Daily et al.[8] holds that the loss of kinetic energy loss between impact and separation is the sum of the crush energy absorbed by the vehicle, the fracture energy (if any) of the pole, and the work done in moving the pole in the ground. In this last item, the force required to move the pole is equal to the average crush force in the vehicle, which is calculated as half the peak crush force, which in turn is assumed equal to the pseudo-force calculated using a constant-stiffness crush model for the vehicle structure. The question of how well such a structure model works for narrow-object impacts is not discussed.

The pole movement is treated as an angle change in the ground, with a measured displacement of the contacted area on the pole. The angles of the pole before and after the impact figure into the calculations. If the pole has fractured, the authors recommend taking circumference measurements equidistant above and below the fracture surface, and averaging the results. All this means that it is essential to document the geometry of the impacted pole before it is removed and/or replaced.

Daily et al. report that three Class 6 poles, without wires attached, were impacted by late 1990 vintage Ford Taurus and Mercury Sable passenger cars at speeds in the range of 43–49 mph. The poles fractured in all three cases.

An energy accounting was performed, and impact speeds were reconstructed, for all three impacts. Pole fracture energies were calculated using the procedures of both Mak et al. and Kent and Strother. While the Mak fracture energies were in the range of 31,000–44,000 ft-lb, the Kent and Strother numbers were in the 12,000–16,000 range. However, the total energies (including post-crash vehicle run-out) were in the 177,000–246,000 ft-lb range, so the difference in fracture energies did not influence the speed calculations very much. On the other hand, the energy required to move the pole was calculated at 29,000–133,000 ft-lb.

The speed estimates using the Mak fracture energies were within the measurement uncertainty of the radar guns used to record impact speed; calculations using the Kent and Strother energies were about 2 mph lower.

However, the fact that the sum of the parts is correct does not mean that the parts themselves are correct. As pointed out by Nystrom and Kost,[9] using A and B values derived from barrier impacts to calculate the vehicle speed into a rigid pole tends to understate the speed by about 33%. It follows that the Daily calculations, for both the vehicle crush energy and the force exerted on the pole, are likely to be low as well. If true, the results would tend to counterbalance overestimations resulting from the Mak equations.

Considerable attention is given in the Daily paper to what the authors call "uncertainty analysis." None of this analysis deals with the suitability of using barrier crashes to characterize the structure and calculate the vehicle crush energy, nor with the use of the pseudo-force obtained thereby to calculate an average force required to move the pole. These assumptions would appear to be of primary importance. However, the authors claim a precision of 8–15% on impact speed, with the most significant contributions to the uncertainty coming from the Mak equation exponent and the vehicle stiffness coefficients.

Crush Profiles and Vehicle Crush Energy

We have seen in Chapter 21 how to calculate crush energy for a vehicle displaying an arbitrarily shaped crush profile. In 1987, Smith et al.[10] applied such a procedure to narrow object impacts, using maximum crush (instead of average crush) as the measure of crush magnitude, and deriving shape factors α and β to take nonuniformities into account. The approach was similar to that in Chapter 21, except that use of maximum instead of average crush caused the definition of the β quantity to be different. (α does not appear if average crush is used in the analysis.) Six NHTSA pole impacts were reconstructed, along with a repeated pole test series that was also reported in the paper. Using stiffness coefficients derived from flat barrier tests, and using the original rather than the deformed crush widths, impact velocities were predicted with errors that ranged from 79% to 104% (on the low side for all impacts but one). However, for any structural model to work correctly, it must be based on tests that have exercised the structure in a fashion similar to the crash being reconstructed. This condition is violated if one attempts to reconstruct narrow object impacts using models based on barrier crash tests.

To illustrate the situation, 14 vehicles with frontal pole impact tests were found in the NHTSA crash test database. Many of these vehicles were subjected to only a single pole impact. However, six of the 14 vehicles were subjected to pole impacts using the repeated test procedure discussed in Chapter 21: a 1985 Ford Escort,[11] a 1985 Ford Tempo,[12] a 1985 Nissan Sentra,[13] a 1987 Hyundai Excel,[14] a 1987 Ford Taurus,[15] and a 1992 Chevrolet Caprice.[16] In pole impacts such as these, the procedure for establishing the crush width is important, but was not described in the reports. The vehicles were not mapped, as was suggested in Chapter 22, but detailed measurements of the undeformed and deformed shapes were recorded around the perimeter of approximately the front halves of the vehicles. It was not clear how these measurements related to the six crush measurements C1 through C6 or the reported crush widths. However, the *Accident Reconstruction Journal* did its own analysis, and issued revised crush profiles for some of the tests at the higher speeds.[17] These crush values were used in the current analysis. It was assumed that the same points on the bumper were used for each successive crush profile. Since each vehicle was impacted repeatedly, the original spacing of the outside measurements was used for the crush width throughout each vehicle's test series.

An excellent discussion on measurement issues associated with pole impacts, particularly the assessment of crush width, is presented in the 2002 paper by Asay et al.[18]

A typical crash plot from the repeated impact series, in this case for the Caprice, is shown in Figure 25.1. These repeated impacts had very low initial severities, so it was not necessary to add a fictitious data point to represent the no-damage condition. In this case, the A value represents a damage onset speed of 3.21 mph.

We see that the excellent fit of the test data to a straight line confirms the appropriateness of a constant-stiffness model for this structure. This was also true for the Escort, the Sentra, and the Excel. Even though the R^2 values for the Tempo and the Taurus were 0.9644 and 0.9440, respectively, there was enough of a deviation from a straight line to ask whether a constant-force model might fit the data as well. Indeed, repeated tests of a Taurus into a flat barrier[19] plus individual barrier tests of Tauruses and Mercury Sables in the same production run were analyzed by Struble et al.[20] A constant-force model for those tests was found to produce an even better fit to the data. This question was not explored for the pole tests.

Given the existence of frontal flat barrier and pole tests on structurally identical vehicles, one is led to ask how the stiffness coefficients compare. Are they the same, or if not, is there some kind of linkage between them? Application of a constant-stiffness model to all the vehicles, whether they were crashed into a flat barrier, a centered pole, or an offset pole, produced the results shown in Table 25.1.

For the repeated pole impacts, a fictitious no-damage data point was not needed (or used) in the crash plot, but was calculated anyway to see what the no-damage threshold speed would need to be to produce an A value that matched that obtained from the crash plot. Generally, these damage onset speeds were 4 mph or less. This is reasonable in view of the flat barrier tests

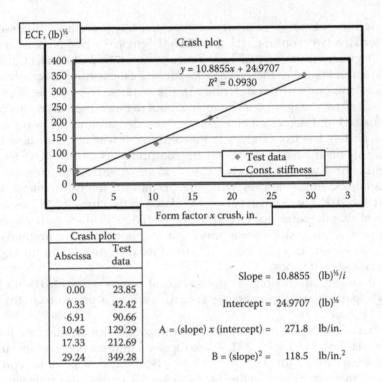

FIGURE 25.1
1992 Chevrolet Caprice frontal pole impacts.

analyzed by Neptune.[21] In the individual pole tests, the fictitious data point was based on an onset speed of 3 mph.

That said, the crash plots for the Tempo and Sentra had negative intercepts and thus negative A values. Negative no-damage onset speeds are not physically reasonable, so the values were set to zero in Table 25.1.

In setting up the offset pole tests for the Omni, Accord, and Fuego, an effort was evidently made to evaluate stiffness variations along the front of the vehicle. However, Table 25.1 indicates that for the Omni and the Accord, the lateral location of a pole impact had no effect at all on the crush stiffness, and only a moderate effect for the Fuego. These tests were also analyzed by Smith et al.

A search for overall trends in Table 25.1 suggests that any resemblance between the barrier and pole stiffnesses is purely coincidental. One might expect the pole impact values to be lower, but the calculated values are often higher. Perhaps the reason is that even though the recommended procedure was followed in using pre-crash spacing to establish the crush width, it was always smaller than the vehicle overall width (OAL) used for barrier crashes.

TABLE 25.1

Summary Barrier-to-Pole Stiffness Comparison

ID	Pole Test DOT No.	Vehicle	OAL in.	OAW in.	Flat Barrier		Individual Tests				Repeated Tests	
							Centered Pole		Offset Pole		Centered Pole	
					A lb/in.	B lb/in.²	A lb/in.	B lb/in.²	A lb/in.	B lb/in.²	A lb/in.	B lb/in.²
1	0698	1973 Chevrolet Vega	170	65	171	53			167	60		
2	0760	1979–91 Ford LTD	220	78	180	48	209	82				
3	0700	1988–83 Honda Civic	148	62	201	66			275	192		
4	0846, 0876	1978–90 Dodge Omni	165	67	179	59	169	67	171	70		
5	0819, 0873	1982–85 Honda Accord	175	65	166	45	160	63	155	62		
6	0847, 0872	1983–85 Renault Fuego	177	67	190	59	248	125	217	101		
7	0862	1977–90 Chevrolet Impala	212	75	220	55	239	90				
8	0614	1980–84 Volkswagen Rabbit	155	63	138	61			284	239		
9	1836–1840	1981–85 Ford Escort	164	66	227	59					251	89
10	1846–1850	1984–94 Ford Tempo	176	66	344	74					0	142
11	1841–1845	1982–86 Nissan Sentra	165	64	135	58					0	97
12	1648–1651	1986–89 Hyundai Excel	161	63	257	64					47	33
13	2170–2174	1986–91 Ford Taurus	188	71	375	84					344	68
14	2185–2189	1991–96 Chevrolet Caprice	214	77	379	62					272	119

On the basis of the linearity of the crash plots, one can safely use pole tests for a given vehicle to reconstruct narrow-object impacts of a similar vehicle at severities within the range tested. But the real moral of the story is that one cannot use barrier stiffness values to reliably reconstruct pole impacts, and vice versa. Given the small number of pole tests available for study, any variation of stiffness values across car lines, vehicle sizes, and so on remains unexplored.

It is worth noting that as of this writing, the newest car for which frontal pole test data are available is 20 years old. Studies have shown that vehicles have grown stiffer, particularly with the advent of offset frontal crash tests into barriers, both rigid and deformable. It can be reasonably expected that frontal pole stiffness of current vehicles will also be higher. Perhaps research or governmental organizations will be persuaded to conduct a new round of frontal pole tests.

Maximum Crush and Impact Speed

A different approach is to ignore the underlying structural model and the crush profile, and proceed directly to an empirical relationship between maximum crush and impact speed. Because of the paucity of pole impact tests, this approach requires that all existing tests be lumped together in the analysis, thus blurring any distinctions between different structures, engine compartment geometries, and so on. In addition to being simpler, however, the use of just the maximum crush neatly sidesteps the nettlesome issues regarding crush width. If they are instructed to measure the maximum crush, crash test personnel are likely to do so with greater consistency as compared to an entire complex crush profile.

The approach of using maximum crush only was taken by Nystrom and Kost[9] in 1992, who analyzed 19 tests in the NHTSA database. For the most part, these tests deal with the first eight vehicles in Table 25.1. They used a regression analysis to develop the equation

$$V = BP0 + BP1 \times CRM \tag{25.6}$$

where V is the pre-impact velocity, BP0 is the no-damage threshold speed, and CRM is the maximum crush on the vehicle. Correlations between the quantity BP1 and various vehicle parameters were sought by using regression analysis. With BP0 set to 5.0 mph, the preferred regression was on vehicle weight, with

$$BP1 = 0.964 - 3.51(10)^{-5} \times W \tag{25.7}$$

where V is in mph, BP0 = 5.0 mph, CRM is the maximum vehicle crush in in., and W is the vehicle weight in lb. In SI units,

$$BP1 = 1.53\text{--}2.47(10)^{-4} \times W \tag{25.8}$$

where V is in kph, BP0 is 8.0 kph, CRM is the maximum vehicle crush in cm, and W is the vehicle weight in N. The fit to the 19 pole tests produced an rms error of 1.8 mph.

Despite the goodness of fit, it must be cautioned that the test data analyzed by Nystrom and Kost[9] are even longer in the tooth than the 14 vehicles analyzed using crush profiles. At this writing, the newest vehicle in the Nystrom and Kost data set is nearly three decades old and virtually extinct from the vehicle fleet. Again, pole impact tests on modern vehicles are needed.

Side Impacts

Whatever difficulties are entailed in quantifying damage and reconstructing frontal pole impacts are compounded when the side of a vehicle strikes a pole. As we saw in Chapter 22, side impacts in general can involve issues of bowing and intrusion that influence how crush is measured, and the crush energy assessed. These issues are amplified when a narrow object is involved, since the concentration of impact forces virtually guarantees that bowing will be involved in a severe crash.

On the positive side, the side impact occupant protection standard—FMVSS 214—was modified in 2007 to add pole impacts to the matrix. The intent was to so configure the test that an occupant's head would be directly contacted by the pole unless some form of restraint system intervened. As a result of this rulemaking action, the NHTSA began conducting side pole impacts in the model year 2010. Consequently, a radically different situation exists, compared to frontal pole impacts: a growing body of crash tests is emerging on late model vehicles. In these tests, the vehicle is impacted at 20 mph at an angle of 75° into a rigid pole in the driver's door.

Measurements from a longitudinal vertical reference plane are made at 75 mm (3 in.) spacing before and after the test, at five water lines. The difference is reported as crush. Since the reference plane is located a specified distance from vehicle center line at the front and rear, bowing deformations are automatically included, but are mostly slight in the tests. Yaw motions also appear to be slight, meaning that rotational energies can probably be ignored.

The body of test reports has become sufficient to allow a reconstruction methodology to be constructed and put into practice. Perhaps an enterprising reader will do so.

References

1. Specifications and Dimensions for Wood Poles, ANSI Standard 05.1-1979, American National Standards Institute, New York, 1979.
2. Michie, J.D., Gatchell, C.J., and Duke, T.J., *Dynamic Evaluation of Timber Posts for Highway Guardrails*, Highway Research Record 343, Highway Research Board, 1971.
3. Wolfe, G.K., Bronstad, M.E., and Michie, J.D., *A Breakaway Concept for Timber Utility Poles*, Transportation Research Record 488, Transportation Research Board, 1974.
4. Mak, K.G., Labra, J.J., and Noga, J.T., A procedure for the reconstruction of pole accidents, *Proceedings, American Association for Automotive Medicine*, 1981.
5. Prasad, A.K., Energy absorbed by vehicle structures in side-impacts, SAE Paper 910599, SAE International, 1991.
6. Morgan, J.R., and Ivey, D.L., Analysis of utility pole impacts, SAE Paper 870607, SAE International, 1987.
7. Kent, R.W., and Strother, C.E., Wooden pole fracture energy in vehicle impacts, SAE Paper 980214, SAE International, 1998.
8. Daily, J.G., Daily, J.S, and Rich, A.S., A method for vehicle-wooden utility pole impact speed reconstruction, *Accident Reconstruction Journal*, September/October, 2009, 41–51. (Victor Craig).
9. Nystrom, G.A., and Kost, G., Application of the NHTSA crash database to pole impact predictions, SAE Paper 920605, SAE International, 1992.
10. Smith, G.C, James, M.B., Perl, T.R., and Struble, D.E., Frontal crush energy and impulse analysis of narrow object impacts, ASME Paper 87-WA/SAF-5, American Society of Mechanical Engineers, 1987.
11. Markusic, C.A., *Final Report of a 1985 Ford Escort 3-Door Hatchback Into a Pole Barrier in Support of Crash III Damage Algorithm Reformulation*, DOT HS 807 913, DOT 1836-40, U.S. Department of Transportation, National Highway Traffic Safety Administration, 1992.
12. Johnston, S.A., *Final Report of a 1985 Ford Tempo 4-Door Sedan Into a Pole Barrier in Support of Crash III Damage Algorithm Reformulation*, DOT HS 807 915, DOT 1846-50, U.S. Department of Transportation, National Highway Traffic Safety Administration, 1992.
13. Markusic, C.A., *Final Report of a 1985 Nissan Sentra 4-Door Sedan Into a Pole Barrier in Support of Crash III Damage Algorithm Reformulation*, DOT HS 807 926, DOT 1841-45, U.S. Department of Transportation, National Highway Traffic Safety Administration, 1992.
14. Markusic, C.A., *Final Report of a 1987 Hyundai Excel Into a 0° Frontal Pole Barrier in Support of Crash III Damage Algorithm Reformulation*, DOT HS 807 775, DOT

1648-51, U.S. Department of Transportation, National Highway Traffic Safety Administration, 1991.

15. Looker, K.W., *Final Report of a 1987 Ford Taurus Into a 30.5 CM Diameter Pole Barrier in Support of Crash3 Damage Algorithm Reformulation*, DOT HS 808 230, DOT 2170-74, U.S. Department of Transportation, National Highway Traffic Safety Administration, 1994.

16. Looker, K.W., *Final Report of a 1992 Chevrolet Caprice Into a 30.5 CM Diameter Pole Barrier in Support of Crash3 Damage Algorithm Reformulation*, DOT HS 808 213, DOT 2185-89, U.S. Department of Transportation, National Highway Traffic Safety Administration, 1994.

17. Craig, V., Repeat pole impact tests of three more vehicles, *Accident Reconstruction Journal*, 7(5), Victor Craig, 1995.

18. Asay, A.F., Jewkes, D.B., and Woolley, R.L., Narrow object impact analysis and comparison with flat barrier impacts, SAE Paper 2002-01-0552, 2002.

19. El-Habash, N.A., *Final Report of Frontal Barrier Impacts of a 1886 Ford Taurus 4-Door Sedan in Support of Crash III Damage Algorithm Reformation*, DOT HS 907 350, DOT 1201-05, U.S. Department of Transportation, National Highway Traffic Safety Administration, 1988.

20. Struble, D.E., Welsh, K.J., and Struble, J.D., Crush energy assessment in frontal underride/override crashes, SAE Paper 2009-01-0105, SAE International, 2009.

21. Neptune, J.A., Overview of an HVE vehicle database, SAE Paper 960896, SAE International, 1996.

26

Underride/Override Collisions

Introduction

In Chapter 25, we dealt with a class of collisions that are sufficiently unlike barrier crashes that the traditional methods, while providing answers, turn out to be unreliable. That is the worst kind of situation, because those who want so badly to know what happened in a crash can be led down the garden path to an answer that may seem right, but in fact is not.

In this chapter, we encounter another such situation—underride or override collisions. These may be between two vehicles, such as when an automobile runs part way under the back or side of a truck trailer. Or a high-clearance vehicle may run up behind, and attempt to climb into the trunk of, a passenger car, or may strike another vehicle in the side and have its bumper penetrate above the rocker panel. But some fixed object crashes fit into this category, too. As we shall see, a vehicle may be deliberately crash-tested into an overhanging barrier. Or a tractor/trailer may be so massive and rigid, and so far above the ground, that it acts like an overhanging barrier.

All of these types of collisions have in common with pole impacts the fact that only a portion of the structure is exercised, unlike the situation with a full-width barrier. In this case, the stiffer parts of the structure are not engaged because they do not match the height of the stiff parts of the collision partner. We call this phenomenon "geometric incompatibility" between the colliding structures. A related subject is "architectural incompatibility," which gives rise to a mismatch of the stiffer parts of the colliding structures because the vehicle load paths are arranged so differently. The mismatches may be exacerbated by "mass incompatibility," when one collision partner is much more massive than the other, and "stiffness incompatibility," when they differ so much in stiffness.

The distinguishing feature of all underride/override crashes is that the primary load paths of the vehicles are engaged only slightly, if at all, because of height mismatches. In such crashes, the crush at one water level of a vehicle differs significantly from the crush at another. Usually, the bumper of one vehicle has been overridden, forcing the crush into areas above the bumper level, away from the primary load paths. In this way, underride/override crashes are different from crashes into barriers and poles, and different from most vehicle-to-vehicle collisions.

Of course, it is still desirable, for reconstruction purposes, to calculate the crush energy. As we have seen, flat barrier test data are widely used for this purpose in other kinds of crashes. Since there is no reason to expect underride/override crash tests to be performed on a routine basis any time soon, our objective here is to tap into that same wealth of data for frontal flat barrier crashes, and find a way to apply the information to frontal underride/override crashes.

Rear crashes by flat moving barriers exist but are not nearly so common. Perhaps an inquisitive reader will research the situation with rear override crashes, and develop a corresponding reconstruction procedure for those. Side impacts with overrides are another area of potential research, although the improvement of side impact countermeasures may render side impact incompatibilities less problematic in the future.

NHTSA Underride Guard Crash Testing

In 1993, as part of its efforts to upgrade the Crash3 program, the NHTSA conducted a series of crash tests of vehicles into a rigid truck underride guard style indenter mounted on a fixed barrier, at a height calculated to just miss the bumper. The indenter was square in cross section, about 4 in. on a side. To generate data over a range of crash severities, repeated impacts utilizing the procedure of Warner et al.,[1] discussed in Chapter 21, were employed.

Similar to the NHTSA repeated pole impacts discussed in Chapter 25, measurement points for the vehicle shapes were established pre-test and were distributed around the front half of the perimeter of the vehicles. For these tests, however, two sets of vehicle shapes were generated—one at vehicle bumper height, and one at underride guard height. An apparently independent set of crush measurements at each of these heights was also reported for each test. As in the pole impacts, the crush measurement locations do not appear to correspond to the shape measurements.

One of the vehicles tested was a 1990 Ford Taurus,[2] about which there is a plethora of crash test data. It was subjected to impacts at 10, 15, 15, and 35 mph, in sequence. As noted in Tavakoli's analysis of the tests,[3] the upper crush profile measurements did not coincide with the maximum penetration of the indenter; therefore, the upper crush was under-reported.

Another vehicle studied was a 1993 Honda Civic.[4] This time, the nominal test speeds were 5, 15, 15, 20, and 35 mph. Again, it appears that the upper crush measurements were located not at maximum crush, but at some water line above that level.

Synectics Bumper Underride Crash Tests

In 1996, Synectics Road Safety Research Corporation ran a series of single and repeated impact underride tests at various water lines and crash severities.[5] Three similar vehicles were crashed into a horizontal metal beam that was attached to a rigid frame, and mounted to a rigid barrier. In three of the tests, the underride barrier was positioned so as to directly attack the vehicle's greenhouse structure—an interesting scenario that will not be treated here. Six other tests had the barrier positioned so as to contact the fronts of the vehicles just above their bumpers.

In the first test, the underride barrier yielded (and absorbed an unknown amount of energy), thus negating the usefulness of the test for calculating vehicle crush energy. The barrier was subsequently rigidized. The next four tests were a repeated impact on a 1987 Plymouth Reliant, but no vehicle exit speeds were reported; combined with uncertainties in the impact speeds, crush energy calculations are considered to be unreliable. That left one test amenable to analysis—a single impact involving a 1987 Plymouth Reliant in which the impact speed was measured with a speed trap.

Analyzing Crush in Full-Width and Offset Override Tests

In 2000, Croteau et al.[6] reported on crash tests in which a stationary passenger car was struck in the rear by a heavy truck. The authors proposed that the total crush energy in the struck vehicle be calculated as

$$DE_T = F_1 DE_1 + F_2 DE_2 \qquad (26.1)$$

where DE_T is the total damage energy, DE_1 is the damage energy based on bumper level crush, and DE_2 is based on upper structure crush. Both DE_1 and DE_2 were based on a constant-stiffness structural model derived from flat barrier testing. When used with suitable values of F_1 and F_2, the resulting sum would be the correct value of DE_T, known from an energy balance in the test (assuming that the truck absorbed no crush energy). Setting F_1 equal to 1.0, the authors found that for this particular struck vehicle, F_2 was 0.46 for a full-width override, and 0.56 for an offset override. This implies weighting factors of $1.0/(1.0 + 0.46) = 0.68$ and $0.46/(1.0 + 0.46) = 0.32$ for the lower and upper structures, respectively, for this vehicle in a full-width override crash. In an offset override, the lower/upper weighting factors would be 0.64 and 0.36, respectively. The trend between these two test conditions may seem counterintuitive, but Equation 26.1 does suggest some kind of subdivision of crush energy between the lower and upper structures, whose deformation was documented in the two crush profiles.

The NHTSA Tests Revisited

In a 2001 paper, Marine et al.[7] generated crash plots for the NHTSA underride tests. The authors reported the need for engineering judgment in assessing the crush width, as was the case with pole impacts. In their analysis, they adopted the method proposed by Strother et al.,[8] for full-width barrier tests, which is to use the front track width of the vehicle plus 6 in. for the crush width.

The authors did not use the published residual crush values C_1–C_6 for these tests, since "these measurements were apparently not made to the same points on the vehicle from test to test"[7] (p. 199). (The reported crush widths varied among the tests.) Instead, Marine et al. used the pre- and post-test shape measurements to generate their own crush profiles.

For the Taurus, there is considerable discussion in the Marine paper that the measurement points established pre-test do not all reflect the deepest penetration due to the underride guard. Instead, the "equivalent uniform crush calculations may underestimate the direct impact residual crush by approximately three inches in the most severe impact"[7] (p. 205). This illustrates some of the pitfalls of establishing the measurement locations before the test, as discussed in Chapters 22 and 25.

A crash plot was generated for each vehicle, using the upper crush profile only. The one created for the Taurus was reasonably linear, with the points being slightly above the regression line in the middle and slightly below it at the ends. Crash plot points from full-width frontal barrier tests were added for comparison. Remarkably, a straight line drawn through the full-width points would be indistinguishable from the underride regression line. The implication is clear: the upper structure is just as stiff as the entire structure, including the bumper and longitudinal rails. This seems entirely counterintuitive, which the authors point out.

One could cite the measurement issues as a cause for this result. However, the authors state that "even if the crush values for all the above-bumper impact data points were increased by three inches, the data would still be quite similar to the full-frontal barrier test data"[7] (p. 205).

In the crash plot for the Honda Civic, the regression line for the underride tests also fits the data points well. However, the full-width barrier test points are all significantly above the regression line, which is to be expected (although the authors did not seem to think so).

The authors close by recommending that the upper radiator core support be used for upper-structure crush measurements, since that is the most significant structure in the region. Of course, the structure is not accessible if the hood is jammed shut. In that case, the hood edge may provide a perfectly acceptable line. For accident reconstructionists, another procedure is to use some inch tape along the line of deepest penetration (again, if the hood remains latched), as discussed in Chapter 22.

For crash test personnel taking pre-test measurements on a vehicle about to be crashed, and who cannot necessarily anticipate all the nuances of the vehicle damage that is about to ensue, it is highly recommended that some alternative pre-test measurement locations and measurements be kept in the "back pocket" in case some of the planned measurements cannot be made. For example, in frontal tests, points on the upper core support can be documented in case the hood latch is ruptured, and points on the hood edge can be documented in case the hood cannot be opened. Of course, the usual even spacing between crush measurements may need to be abandoned, but at least the crush information will not have been lost. The same considerations apply to other components that are not infrequently loose on the vehicle, destroyed altogether, or removed post-crash. Bumper fascias, radiator grilles, and doors are prime examples. It is highly recommended that Neptune's procedure[9] be used on both crash test and field accident vehicles to eliminate any "air gaps."

More Taurus Underride Tests

In the 2001 Marine paper, one can find the origins of a follow-on Marine paper[10] written in 2005. In fact, a schematic drawing of a vertically offset rigid barrier in the 2001 paper is remarkably like a photograph of the real thing in the 2005 paper. As if to investigate whether the puzzling results from the NHTSA tests were due to measurement uncertainties or actual structural behavior in underrides, Marine et al. subjected a 1990 Ford Taurus to a series of repeated impacts into an overhanging barrier. The overhanging portion was flat, and extended 12 in. out from the main barrier face. The bottom of the overhanging barrier was 20.5 in. above the ground. Some bumper contact was thus permitted, but the residual crush in the upper structure would still be greater than that in the bumper, and "crush profiles of this type are frequently encountered in ... real-world accidents"[10] (p. 211).

Impact speeds were 10.5, 10.9, 14.4, and 18.3 mph, resulting in total crush energies corresponding to single-impact speeds ranging from 10.3 to 27.5 mph. Detailed crush measurements were recorded for both the front bumper and the hood edge. Upper radiator core support measurements were also made before the test series, and after the third and fourth impacts (when the upper core support was sufficiently accessible). The result of this test series was the valuable addition of reliable crush measurements to the sparse collection of underride test data.

The test methodology should be extended to other vehicles. As the authors stated, "Continued testing involving different barrier geometries and other vehicle models is warranted. Furthermore, investigation of override/underride crush energy involving irregular crush profiles (something other than profiles from flat barriers) would also provide useful, and practical, information"[10] (p. 216).

Using Load Cell Barrier Information

The override tests performed by the NHTSA, Synectics, and Marine all have the effect of forcing the upper structure to crush more, relative to the bumper, than do flat barrier tests. Even in the latter kind of tests, however, it is well known that the force levels in the upper structure are less than those seen at the bumper.[11] This knowledge comes from the fact that in the frontal flat-barrier tests at 35 mph, the barrier face is instrumented with an array of load cells, as discussed in Chapter 14.

The demarcation line, or the so-called "boundary," between upper and lower structures is arbitrary, but the usual practice is to assign all forces and displacements below the top of the bumper to the lower structure; forces and displacements above the top of the bumper are assigned to the upper structure. Determining the location of this boundary should be done carefully and consistently, according to the layout of the vehicle structure. The procedure set out by Izquierdo[12] contains useful guidelines.

In flat barrier tests, it is possible to calculate the contributions of the lower and upper structures to the total crush energy, as pointed out in a 2009 paper by Struble et al.[13] This is possible because at each instant of time, integration of the longitudinal accelerations reveals the vehicle velocity, which enables the calculation of how much kinetic energy has been lost up to that time. A second integration of the accelerations produces the vehicle displacement at that time, and how much the displacement has increased since the prior integration step. The product of the average force during the time step, and the displacement increase during that time, yields the work done during that time by the force, and the crush energy absorbed by the structure that generates that force. The cumulative energy absorbed can be calculated by adding to the partial sum from the previous time step.

According to Struble et al., summation of the load cells that contact the lower structure permits the calculation of the total force in that structure, and the crush energy absorbed therein. Similar calculations pertain to the upper structure. The sum of the lower and upper crush energies can be compared to the loss in kinetic energy at each time step in the integration process.

Struble et al. applied this computation procedure to the so-called "Volpe Tests,"* a series of 10 NHTSA impacts of various vehicles into load cell barriers (LCBs) at 30 and 35 mph. These tests were selected because the 30 mph impacts utilized a load cell barrier, which is not usually the case. The seven Volpe Tests that have adequate LCB data included Toyota Camrys at 30 and 35 mph, Honda Accords at 30 and 35 mph, and a Jeep Liberty, a Honda Odyssey, and a Chevrolet Avalanche pickup at 30 mph. All were 2004 models. The data from five of these tests produced crush energy calculations from LCB data that matched the kinetic energy loss, within the energy equivalent of 0.7 mph. The other two tests had energy calculation discrepancies equivalent to 1.2 and 2.3 mph, respectively.

The proportions of crush energy absorbed in the upper structure ranged from about 12% to 28% in these seven tests. The possible effects of test speed on these numbers could be examined only for the Honda Accord and the Toyota Camry. Between 30 and 35 mph, the upper structure crush energy proportion varied by three percentage points for the Accord, and one percentage point for the Camry.

These results suggest a procedure whereby LCB tests (typically, only at 35 mph) be used to estimate the proportions of upper and lower crush energies for a particular vehicle. All available barrier tests for that vehicle can be used to characterize the structure (e.g., calculate A and B). Then, for a similar accident vehicle with underride or override damage, separate crush profiles can be measured for the upper and lower structures. The structural parameters for the vehicle can be used with each crush profile to calculate raw crush energies in the upper and lower structure. Then the upper and lower proportions can be used as weighting factors to calculate a weighted crush energy average for the entire vehicle.

This procedure has some conceptual similarities to that proposed by Croteau et al., except that the proportions are determined from publically available barrier crash test data, instead of special-purpose crash tests. For some accident scenarios, of course, a special-purpose crash test may be necessary anyway.

Shear Energy in Underride Crashes

The Volpe tests, from which the upper and lower proportions were calculated, were all full-width flat barrier impacts. Might not the proportions be different for underride crashes? Unfortunately, we know of no underride

---

* DOT Tests 5136– 5140, 5158, 5212–5213, and 5215–5216, U.S. Department of Transportation, National Highway Traffic Safety Administration, 2004.

crashes with instrumented barrier faces. A future research program with an offset barrier having the upper and lower faces mounted on load cells would be interesting, indeed. It would probably suffice to have load cells only at the four corners of the upper and lower barrier faces, respectively, if they are sufficiently rigid. Of course, there may be legitimate concerns about off-axis loads on the load cells.

In the meantime, the proof of the pudding will have to be in the eating. In other words, how well does the proposed procedure predict the energy of an underride test?

We know that the deformation mode in underride crashes is different from that in barrier crashes. Instead of the entire structure being forced to deform into a planar surface (at least dynamically), the fronts of the upper and lower structures are offset from one another. These differences cause the structure in the immediate vicinity of the boundary to undergo shear deformations, just as a flowing fluid does in the transition area between a stationary pipe wall and the moving middle of the stream. In a fluid, this shear deformation is resisted by its viscosity. In a homogeneous solid, shear deformations are resisted by the shear modulus, which is related to the modulus of elasticity by the well-known equation

$$G = \frac{E}{2(1 + v)} \tag{26.2}$$

where v is Poisson's ratio—usually about 0.3 for most metals. Thus, the material shear stiffness is its compressive stiffness, multiplied by a factor of about 0.38.

Surely, a vehicle structure must have an analogous resistance to the shear deformations that occur in underride/override crashes, which has been discussed by other researchers[14,15] with regard to pole impacts, for example. Therefore, one cannot expect the sum of the weighted crush energies to account for all the crush energy in an underride/override crash. Some additional energy must be dissipated in the creation of shear deformations in the boundary region; that is, shear energy. The shear energy is the product of the shear force times the shear deformation, which is the difference in deformation between the upper and lower structure. The shear force is the average of the upper and lower forces, multiplied by a factor that Struble et al. found from a limited number of tests to be about 0.35.

There remains the question posed above: Do the same crush energy proportions apply in an underride test? The definitive answer must await impacts into an instrumented offset barrier. In the mean time, the proposed methodology can be tested against the few underride tests, described previously, into rigid indenters that resulted in underride crush patterns.

Reconstructing Ford Taurus Underride Crashes

One of the vehicles so tested is often-crashed 1986–91 Ford Taurus/Mercury Sable. Struble et al. found that the best characterization for that vehicle in flat barrier tests was a constant force model in which the saturation force F_S was 1367 lb/in., and the saturation crush C_S was 3.182 in. See Chapter 21 for a discussion of such structures, and how to characterize them.

Of the 13 barrier tests[*] at a variety of speeds that were used for the characterization of this vehicle, four had sufficient load cell and vehicle attitude data to allow their upper and lower crush energy proportions to be assessed. The proportion for the upper structure ranged from 18% to 26%, with an average of 23.8%.

In these four 35 mph tests, the accuracy of the method of calculating the crush energy from load cell data was checked by comparing the results to the kinetic energy losses. The calculated crush energies matched the kinetic energy losses within the equivalent to 0.5 mph of velocity. On the other hand, reconstructions using the constant force model in the structural characterization produced crush energy calculations that approximated the kinetic energy losses within the equivalent of 2 mph. (The reconstructions were not exact because the regression line in the crash plot of the 13 tests did not pass exactly through the four points for these tests.)

To apply the results to the NHTSA underride tests of the Taurus, Struble et al. noted the under-reporting of the maximum penetration of the above-bumper structure by the underride fixture. The hood edge, along which the upper crush measurements were reported, protruded into the underride test fixture past the horizontal cross beam, and into the vertical uprights supporting the cross beam. This is apparent in the shape of the upper structure recorded in the NHTSA tests. The configuration of the uprights and crossbeam can be seen in Figures A.1 and A.9 of the Taurus report,[†] and Figure A.11 of the Honda report[‡]—unlike the fixture shown in Figure 13 of the 2002 Marine paper. In the NHTSA Taurus tests, the true deepest penetration above the bumper was thus a straight line, offset from the penetration of the uprights by the 4-in. depth of the horizontal member.

For each test, the crush energies in the upper and lower structures were calculated by applying the structural characterizations to the two profiles, and then multiplying by the weighting factors of 0.238 and 0.762, respectively, as obtained from the load cell barrier tests. The crush force in each structure was calculated from the respective crush profile, using the same

---

[*] DOT Tests 0944, 0949, 1103, 1104, 1177, 1201 - 05, 1385, 1403, and 1600, U.S. Department of Transportation, National Highway Traffic Safety Administration.
[†] Looker, DOT 1939–42.
[‡] Looker, DOT 1943–47.

weighting factors. The shear force was estimated at 0.35 times the average of the upper and lower structure forces. The shear deformation was equal to the absolute value of the difference between the average crush in the upper and lower structures. The shear energy was then the product of the shear force and the shear deformation. The total energy of deformation was the sum of the upper and lower structure crush energies, plus the shear energy.

This energy (CE) was then compared to the cumulative kinetic energy lost up to that point, using the repeated-impact methodology proposed by Warner. The results were expressed in terms of energy equivalent speed (EES). The EES for the crush energy assessment is the speed at which the vehicle would have a kinetic energy equal to the crush energy, and thus may be found by substituting CE for KE in Equation 1.6. The EES for cumulative kinetic energy loss comes from Equation 21.90.

For Test 1 in the NHTSA series, the methodology over-predicted the actual EES by 1.2 mph (12%). In the remaining tests, the EES was under-predicted by amounts ranging up to 4.1 mph (in the fourth test), and 12.6% (in the third test) (see Figure 26.1).

Note that the cumulative kinetic energy loss in the NHTSA Taurus tests was about 41 mph, whereas the tests on which the structural characterization is based ranged up to 35 mph. In particular, the last two underride tests may have involved an engagement of the top of the engine that would have increased the resistance to further upper structure crush. Or measurement uncertainties could be at work.

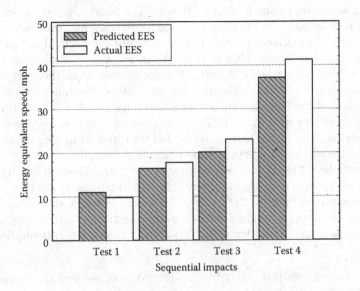

FIGURE 26.1
Predictions compared to NHTSA Taurus tests.

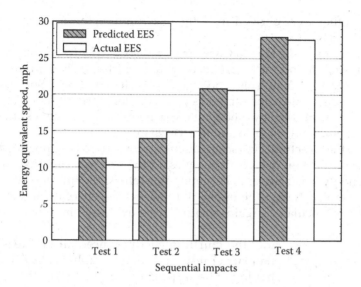

FIGURE 26.2
Predictions compared to Marine Taurus tests.

A parallel analysis was done of the Marine Taurus tests, again predicting the crush energy using the constant force model developed from full-width flat barrier tests of the 1986–1991 Ford Taurus. In this case, the Marine crush data were used as published. The predicted crush energies are compared to the actual (cumulative) values, again expressed in terms of EES and shown in Figure 26.2.

Again, the methodology overpredicted the crush energy in the first test, this time by the equivalent of 0.9 mph (11%). In the second test, the EES was underpredicted by 0.9 mph (6%). In the third and fourth test, the EES was within 1.3%.

Note the narrower range of energies in the Marine test series. Throughout this series, the EESs were within the range of the flat barrier test data, obviating any concerns about extrapolations.

Reconstructing Honda Accord Underride Crashes

The test procedure and measurement protocol for the NHTSA Honda Accord underride crashes paralleled those of the NHTSA Taurus underride crashes. Crush documentation was carried out in the same way. Again, it appears that the upper crush measurements were not at the deepest penetration, but at some water line above that level. Therefore, the crush profiles were derived from the existing data in the same way they were in the Taurus tests.

To characterize the Honda Civic Structure, six barrier tests,[*]—all at 30 and 35 mph, were found for its production run, which was 1992 through 1995. The three tests at 35 mph also contained load cell barrier data. Both constant-force and constant-stiffness models were developed from these data, but a better fit was achieved by modeling the structure with constant stiffness. Its parameters were $A = 0$ lb/in. and $B = 119.25$ lb/in^2.

Again, the structural characterization was applied to the upper and lower structures to obtain raw crush energies. In the first four impacts, the lower crush energy was zero, only becoming nonzero in the fifth test. Weighting factors were applied to the upper and lower crush energies, and the shear energy was calculated. Again, energies were expressed in terms of equivalent speeds.

In the test series, only the fourth impact produced an EES reasonably close to those of the flat barrier tests at 30 and 35 mph used to characterize the structure. It should not be surprising that the EES of that test was most closely predicted by the analytical procedure—0.9 mph, or 3.3%. The first test, at 4.7 mph, was the farthest away from the flat barrier speeds in severity, and resulted in the largest EES prediction error—3.5 mph (see Figure 26.3).

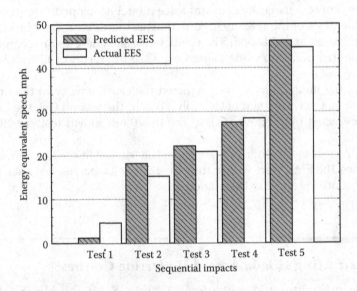

FIGURE 26.3
Predictions compared to NHTSA Civic tests.

---

[*] DOT Tests 1725, 1801, 1822, 1892, 2000, and 2066, U.S. Department of Transportation, National Highway Traffic Safety Administration.

Reconstructing the Plymouth Reliant Underride Crash

For the 1987 Plymouth Reliant that was subjected to a single underride test by Synectics, five flat barrier tests* of vehicles that were representative were found. A constant force model with saturation force $F_S = 1140$ lb/in. and saturation crush $C_S = 12.69$ in. were used to characterize the structure. The structural boundary height was not known, but an estimate was made from the test photographs, and the upper structure was found to absorb about 36% of the crush energy.

Again, the lower structure was found to not contribute to the crush energy absorption in this test. The EES of the test was underestimated by about 3.8 mph.

Conclusions

Underride/override crashes cannot be reliably reconstructed using flat-barrier data applied only to the bumper level crush. Some accounting must be made of the differences between the upper and lower structure crush and structural behavior.

A method has been developed to calculate the crush energy in this type of crash. The method can be applied to any vehicle for which frontal load cell barrier data are sufficient to determine how much of the crush energy is absorbed by the upper and lower structures. However, the method is more complicated than that for full-height frontal crush patterns, and entails the calculation of shear energy. Crush measurement is also more complex for underride/override crashes, both for the test personnel and accident reconstructionists.

Very few underride/override tests are available with which to check the applicability and accuracy of the method. More tests, with accurate measurements of speed and crush, are needed.

References

1. Warner, C.Y., Allsop, D.L., and Germane, G.J., A repeated-crash test technique for assessment of structural impact behavior, SAE Paper 860208, SAE International, 1986.

---

* DOT Tests 0207, 0502, 0593, 0794, and 0941, U.S. Department of Transportation, National Highway Traffic Safety Administration.

2. Looker, K.W., *Final Report of a 1990 Ford Taurus into Heavy Truck Rigid Underride Guard in Support of CRASH3 Damage Algorithm Reformulation*, U.S. Department of Transportation, National Highway Traffic Safety Administration, DOT HS 808 231, DOT 1939–42, 1993.

3. Tavakoli, M.S., Valliappan, P., and Pranesh, A., Estimation of frontal crush stiffness coefficients for car-to-heavy truck underride collisions, SAE Paper 2007-01-0731, SAE International, 2007.

4. Looker, K.W., *Final Report of a 1993 Honda Civic DX into Heavy Truck Rigid Underride Guard in Support of CRASH3 Damage Algorithm Reformulation*, U.S. Department of Transportation, National Highway Traffic Safety Administration, DOT HS 808 228, DOT 1943–47, 1993.

5. Navin, F.P.D., *Bumper Underride Crash Test Series: 1996 Test Results*, Synectics Road Safety Research Corporation, Vancouver, BC, 1997.

6. Croteau, J., Werner, S., Habberstad, J., and Golliher, H., Determining closing speed in rear impact collisions with offset and override, SAE Paper 2001-01-1170, SAE International, 2001.

7. Marine, M.C., Wirth, J.L., and Thomas, T.M., Crush energy considerations in override/underride impacts, SAE Paper 2002-01-0556, SAE International, 2002.

8. Strother, C.E., Woolley, R.W., and James, M.B., A comparison between NHTSA crash test data and CRASH3 frontal stiffness coefficients, SAE Paper 900101, SAE International, 1990.

9. Neptune, J.A., A comparison of crush stiffness characteristics from partial-overlap and full-overlap frontal crash tests, SAE Paper 1999-01-0105, SAE International, 1999.

10. Marine, M.C., Wirth, J.L., Peters, B.W., and Thomas, T.M., Override/underride crush energy: Results from vertically offset barrier impacts, SAE Paper 2005-01-1202, SAE International, 2005.

11. Digges, K., and Eigen, A., Measurements of stiffness and geometric compatibility in front-to-side crashes, *Proceedings of the 17th International Technical Conference on the Enhanced Safety of Vehicles*, Amsterdam, the Netherlands, U.S. Department of Transportation, 2001.

12. Izquierdo, F.A., and Ayuso, F.J.P., *Structural Survey of Cars: Measurement of the Main Resistant Elements in the Car Body*, INSIA—University Institute for Automobile Research, Madrid, Spain, 1999.

13. Struble, D.E., Welsh, K.J., and Struble, J.D., Crush energy assessment in frontal underride/override crashes, SAE Paper 2009-01-0105, SAE International, 2009.

14. Nilsson-Ehle, A., Norin, H., and Gustafsson, C., Evaluation of a method for determining the velocity change in traffics, *Proceedings, Fourth International Technical Conference on Experimental Safety Vehicles*, Kyoto, Japan, U.S. Department of Transportation, 1973.

15. Buzeman-Jewkes, D.G., Lovsund, P., and Viano, D.C., Use of repeated crash-tests to determine local longitudinal and shear stiffness of the vehicle front with crush, SAE Paper 1999-01-0637, SAE International, 1999.

27

Simulations and Other Computer Programs

Introduction

In previous chapters, we have talked about time-backward approaches to solving certain problems in dynamics—specifically, starting with the known conditions at the ends of trajectories and working backwards in time toward the unknowns: the initial conditions at their beginnings. Such approaches are often facilitated by balancing energy (and/or momentum) at the two end points of the process.

A more typical approach to dynamics problems is to work forward in time and traverse the entire time span before getting to the final conditions. This is usually done by drawing a free-body diagram, enumerating the forces, and using Newton's Second Law to derive the differential equations of motion, which have time as an independent variable. The equations may be solved—perhaps in closed form, but most often numerically—by starting the clock at the beginning of the event, applying the initial conditions (admittedly, unknown) and letting the process move forward in time. In general, the end conditions obtained from solving the initial-value problem do not match the facts that the accident investigation has revealed. Therefore, new and improved estimates of the initial values must be made in an iterative fashion until reasonable convergence is achieved with the (known) conditions at the end.

Computer programs that solve initial-value problems are known as simulations, since the results simulate how the system responds to the specified initial conditions and the stimuli introduced during the run. Strictly speaking, they are not reconstructions, which attempt to show what did happen; they simply show what *could* happen if certain conditions pertain. Most simulations require multiple inputs, so the final results reflect that particular combination of inputs. Another set of inputs may produce the same end results, or a set of results that is so similar as to constitute an equally valid explanation of what could have happened in the accident being investigated.

Nevertheless, simulations can flesh in details not available otherwise. They can shed light on whether a particular hypothesis is even physically possible.

They can forecast the outcomes of certain experiments, such as crash tests. And simulations can answer various "what if" questions.

It is not the purpose of this chapter to impart knowledge of how to design and build simulations, or other types of programs, for that matter. Rather, the purpose is to provide an overview of selected simulations and other programs that are available to the reconstructionist. The list of such programs cannot be all inclusive. Rather, discussion is targeted at a selection of programs that are frequently used by reconstructionists, widely known, publicly available, or otherwise likely to be encountered during a reconstructionist's career. Discussion of any program does not constitute a preference for, approval of, or an endorsement of, that program.

CRASH Family of Programs

CRASH

None of the CRASH family of programs is a simulation. Instead, the original CRASH program (Calspan Reconstruction of Accident Speeds on the Highway)[1] was not intended to be a detailed, highly accurate reconstruction program. Instead, it was developed as a precursor to the SMAC program, which is a simulation of two vehicles colliding on a flat plane. SMAC is an initial-value problem, in which the governing differential equations are solved, subject to specified initial conditions. CRASH was intended to help SMAC users make a first estimate of the initial conditions to be used in launching a SMAC run.

Making this first estimate involved performing a preliminary reconstruction, of course. Two basic approaches were used. The first of these was to analyze the vehicle run-out trajectories post-crash, to allow a determination of the separation conditions as the vehicles exited the crash. Of course, information was needed about the accident site to make this determination. As discussed in Chapter 20, the separation conditions could then be used in a momentum conservation analysis, which would result in the vehicle impact velocities (and ΔVs, once the velocities were known). This trajectory-based approach worked well for oblique (intersection and other nonhead-on) collisions. However, if the pre-impact velocity directions were within $\pm10°$ of a collinear (head-on) configuration at impact, the momentum calculation became very sensitive to the impact angle. Again, see Chapter 20.

> Advice to the Crash3 user is "The introduction of sideslip angles creates an increased variety of impact configurations in which the initial velocities are nearly parallel [while the vehicles may be anything but parallel.] Therefore, the user should remember when running cases with side slip or near head-on or rear-end collisions to examine the results carefully."[2] (p. 3-11)

The second approach was to analyze the vehicle crush damage, and determine the ΔVs based on the structural model proposed by Campbell[3] (see Chapter 21). Of course, no impact speeds could be obtained from damage information alone—only the closing speed and ΔVs. When the trajectory solution was within ±10° of being uniaxial, the damage-based ΔVs were automatically used.

At a minimum, the second approach was used. If both solutions were performed, it was hoped that the ΔVs would bear a reasonable resemblance to one another.

The CRASH programs made several assumptions, such as: (a) the accident site is flat and horizontal; (b) there are no vertical load transfers to the tires during braking or acceleration; (c) steering and braking inputs are open loop (no feedback to the driver); (d) the structure has a constant stiffness behavior, with no force saturation (see Chapter 21); (e) the damage profile is uniform vertically; (f) no sideswipes; (g) negligible tire forces during impact; and (h) negligible restitution. Another way of phrasing the last assumption is to say that crush energy is analyzed only up to maximum crush, or during the loading phase, and not beyond.

In the mid-1970s, the NHTSA announced its intention to perform the National Crash Severity Study, or NCSS,[4] in which field accidents would be analyzed by crash investigation teams, in enough depth to establish relationships between crash severity and injury severity. The selected measure of crash severity was ΔV, which meant that the investigation teams would have to perform reconstructions.

Lessons learned from the NCSS program resulted in the massive (and still ongoing) National Accident Sampling System (now called the National Automotive Sampling System, or NASS), which is a nationwide stratified sampling program intended to gather comprehensive, yet representative, accident data. To obtain comprehensive information, the Crashworthiness Data System, or CDS, was set up within NASS. CDS investigations including vehicle inspections, measurements, photographs, and analysis accompany other information such as police reports. Most importantly, the investigation teams would use CRASH to perform reconstructions, and thus use a standardized method to obtain ΔVs. The purpose was not so much to reconstruct specific accidents accurately, but to standardize procedures and avoid systematic biases in the collection of data. The program was updated and revised several times in the 1980s to a widely used and distributed mainframe version: Crash3 (sometimes written as CRASH3 or CRASH III). The program was ported to a DOS-based PC platform with the name CRASHPC. No algorithm changes were made in the transition.[5]

Crash3 and EDCRASH

With the advent of personal computers, reconstructionists everywhere now had the resources to run programs like Crash3. A 1200 baud dial-up connection to a timeshare computer was no longer needed. Additionally, Engineering Dynamics Corporation (EDC) converted the public-domain Crash3 code to compiled BASIC and offered it for sale under the name EDCRASH. Since the computational methods were left unchanged, the names Crash3 and EDCRASH could be used interchangeably when talking about reconstruction methods, and that will be done here. However, EDC developed a more user-friendly interface for both inputs and outputs of EDCRASH. Whether or not there was a perceived government imprimatur attached to the use of Crash3 by the NASS teams, Crash3 and EDCRASH became arguably the best-known computer programs for reconstruction calculations.

Crash3 does not distinguish between impact and separation. Instead, the vehicle run-out trajectories are defined between impact and rest. Three basic path options are offered: straight path, curved path, and a straight or curved path with end of vehicle rotation, beyond which the vehicle travels in a straight line to rest with zero crab angle. The trajectory coordinates at the (optional) intermediate point, the (optional) end of rotation, and rest position are specified by the user.

The trajectory is analyzed by a procedure called Spin2, which calculates the post impact velocities based upon the distance between the points of impact and rest, the surface friction, the average rolling resistance of the tires, the direction of rotation, the number of revolutions (maximum of two), and the curvature of the CG path. The trajectory is not a time forward stepwise calculation as in SMAC, but an analytical procedure based on a theory of vehicle spin-out proposed by Marquard,[6] which can be explained as follows:

Suppose a vehicle is translating and rotating on a uniform flat surface with some or all of wheels having reduced lock fractions that allow some or all of the tires to roll. When the vehicle is parallel to its direction of travel, the translational deceleration is reduced, depending on the lock fractions, as can be seen in the spreadsheets developed in Chapter 4. The tires that can roll are not saturated longitudinally and can still generate lateral forces. Since the only lateral tire movement at this point is due to vehicle yaw rate, all the lateral tire forces are in a direction opposed to the yaw velocity, and thus work together to generate the maximum available resistive yaw moment.

When the vehicle is lateral to its direction of travel, the tires are sliding mostly sideways, whether the wheels are rolling or not. Thus, the force levels are little affected by the lock fraction, though the force directions will be altered to a degree, depending on any wheel spin rates. The translational component of motion causes the forces on the front and rear tires, which are at their maximum level, to be directed the same way: toward the trailing side of the vehicle. These forces due to translation create the maximum possible translational deceleration, and the yaw moment from one pair of tire forces that tends to balance out that from the other pair, creating very little overall resistive moment. Further lateral tire forces to oppose the rotational motion component cannot be generated, because the tires are already saturated. Thus, the maximum available yaw moment cannot be developed, and the yaw deceleration assumes a much-reduced value.

We see, then, that the translational and rotational decelerations fluctuate between limits as the vehicle rotates. When one is a maximum, the other is a minimum, and vice versa. Thus, the Marquard theory holds that the velocity–time curves for translation and rotation show alternating shallow slopes. See Fonda[7] for further details.

The areas under the linear and angular velocity curves are the linear and angular displacements. However, in the CRASH programs, McHenry modified Marquard's procedure to recognize that the vehicle may stop spinning before it stops translating and thereafter "roll out" or "track" to rest. If desired, the user can specify where in the trajectory this "end of rotation" occurs.

In simulations, such a situation is not uncommon, and it may be encountered in real life. To analyze such behavior, McHenry fitted a fifth-order polynomial function to the ratio of linear and angular velocities observed in simulations at the time of separation. The solution involved five coefficients, $\alpha 1$ through $\alpha 5$, which are defined by polynomials expressing their relation to the initial radius from the instantaneous center of rotation. The constants in the polynomials were based on a set of 18 SMAC runs, not known to be published, so neither the data nor the fitting technique are known, according to a 1987 paper by Fonda.[8] The spin solution starts with an estimate of the initial instantaneous center radius and then iterates until convergence is achieved, resulting in the separation conditions needed to calculate post-impact momentum.[9]

Alternatively, the user can choose the SMAC TRAJ simulation routine to analyze the run-out phase. Again, the user specifies the locations of the impact, an (optional) point on curve, an (optional) end of rotation, and point of rest. At end of the simulation, the computed results are compared to user-specified trajectory data. An error term is computed for each specified path position and heading. If the error terms are sufficiently small, the trajectory simulation is said to have "converged." Otherwise, linear velocity components and yaw rate are adjusted, and simulation is tried again. A maximum of five tries is allowed. If there is still no convergence at the end of the process, the trajectory simulation with the smallest errors is chosen.

The momentum calculation uses the assumption that there is a common velocity at one point in the mutual crush zone (or common CG velocity in the case of a collinear collision). The centroid of the crush volume of each car is selected as the common point. The collision force is directed along the line of action which passes through the common point and has the direction of the PDOF, which is specified for each vehicle by the user. This specification is necessary to calculate the eccentricity of the impact for each vehicle. Supposedly, the user is clever enough to discern the PDOF from the crush pattern, which is a dubious proposition even if the reconstructionist is clever. Unfortunately, specifying the PDOF *a priori* is tantamount to guessing at half the answer before the problem is solved, and removes the objectivity that was an original goal of reconstruction procedure.

Momentum conservation was originally envisioned to include both linear and angular momentum. However, the angular momentum solution was abandoned. Consider that the Crash3 calculation procedure considers only the geometry of impact, and not separation. In effect, the vehicles are not allowed to move during the collision, as they are in the real world. Thinking of a high-speed offset frontal crash between two vehicles, the total angular momentum is sensitive to the alignment of their CGs, and it stands to reason that unless vehicle movement is accounted for, the post-crash total angular momentum calculation could be adversely affected. In hindsight, the angular momentum solution was doomed at the outset by the very structure of the CRASH analysis.

For uniaxial collisions, the closing speed can be obtained by knowing the total crush energy, and the closing velocity can be apportioned into ΔVs using the restitution coefficient and vehicle masses, as discussed in Chapter 19. For oblique collisions, the ΔVs can be found if the analyst is also able to estimate the PDOF relative to the site, the point of application of the resultant force on the vehicles, and the offset distances of the force from the CGs.[10] As mentioned, these estimates come with some difficulties and uncertainties. When the damage-only solution was studied by NHTSA, the estimation of "this important parameter" by field investigators was reported to have 95% confidence limits of ±20°, leading to confidence limits in the calculation of ΔV ranging from ±14% at the higher severities to ±18% at the lower severities.[11]

The damage analysis is based on a (linear) constant-stiffness structural model with no force saturation (see Chapter 21). The crush is measured perpendicular to undeformed sides, at 2, 4, or 6 equally spaced locations (see Chapter 22). Moreover, a correction factor is applied because in Crash3, the crush is measured perpendicularly to the undamaged surface, and the force can be at some angle to the perpendicular. Mathematically, the factor becomes indefinitely large as the angle approaches 90°. Consequently, the angle is limited to 45°, so the factor is limited to "only" 2.0. This potential doubling of the crush energy calculation by applying a correction factor has been harshly criticized,[12,13] although NHTSA cited the correction factor as leading to improved ΔV estimates compared to those obtained from CRASH2.

WinSMASH

In the mid-1990s, NHTSA re-wrote the Crash3 program in the Delphi programming language in the Microsoft Windows environment, making several enhancements to it and integrating it with the NASS/CDS data entry system.[5] Enhancements included (a) a reformulated damage algorithm, (b) updated stiffness coefficients, (c) input fields for over-writing default stiffness coefficients, (d) a new algorithm for reconstructions with a missing vehicle, and (e) estimation of barrier equivalent speed. The last item is defined as "the speed with a vehicle would have to collide with a fixed barrier in order to absorb the same amount of energy or produce the same amount of crush to the vehicle as in the crash." WinSMASH retains all the simplifying assumptions of Crash3, including no restitution. The trajectory analysis used in both programs is identical.

According to Sharma,

> Due to the difficulties associated with the scene data collection, the trajectory option is rarely used by the NASS researcher. Less than one per cent of the coded ΔVs in NASS/CDS are computed using the trajectory algorithm. Since the trajectory option is rarely used, no initiative was taken to update this portion of the algorithm in WinSMASH.[5]

In the damage analysis, the Crash3 stiffness coefficients A and B have been replaced by d_0, the intercept of the crash plot, and d_1, the slope of the crash plot. Refer to Chapter 21. Like Crash3, generic vehicle size and stiffness categories are used, but at least the user can now enter vehicle-specific stiffness coefficients. According to NHTSA, the use of vehicle-specific dimensions, inertial properties, and updated stiffness coefficients improves the estimations of ΔVs, which is to be expected.

Where there is no low-speed crash test of a particular vehicle, the stiffness parameters d_0 and d_1 are determined using by adding a fictitious data point that corresponds to no damage at 12 kph (7.5 mph) for frontal impacts, and 16 kph (10 mph) closing speed for rear moving barrier impacts. For side impacts, the value of d_0 is assumed to be 63.3 $N^{1/2}$ (which is equivalent to an approach speed of approximately 16 kph (10 mph) in a crabbed moving barrier impact like that prescribed in FMVSS 214). WinSMASH also contains generic stiffness parameters for various vehicle categories. The use of these numbers for specific reconstructions should be avoided if at all possible.

The damage width, also known as the Field L, includes contact and induced damage. When such damage in the front and rear impacts extends across the entire vehicle width, the damage width is taken as the distance between the right and left bumper corners on an exemplar vehicle. Nevertheless, when high-speed barrier tests are analyzed for vehicle characterization, the overall vehicle width (OAW) is recommended for crush width.

The Field L–D is the distance from the vehicle coordinate axis to a point that is identified as the "center of the Field L" in Sharma's text, but appears to be the centroid of the damage area in the accompanying illustration. The damage offset is

> the distance from the center of the direct damage width to either the vehicle's damaged end plane center or the damaged wheelbase center ... measured along the general slope of the damage plane.

However, Sharma's Figure 3 seems to indicate a measurement parallel to a coordinate axis.

According to Sharma,

> The ΔV computed by WinSMASH is most sensitive to PDOF and yet it is the most difficult measurement to obtain.

The PDOFs on the two ehicles must be within 15° of being collinear, or WINSMASH displays an error message.

> For 2000–2005 NASS/CDS cases, 53 per cent of the highest severity impacts (by vehicle) have ΔV values. The other unknown ΔVs could not be computed for reasons including non-horizontal impacts, side swipe, rollover, severe over-ride, overlapping damage, and no vehicle inspection .... Of those coded ΔVs, about 58 per cent are calculated using the standard or barrier option

according to Sharma.

In WinSMASH, the correction factor for obliquity

> causes Crash3 to over-predict the value of ΔV in the oblique side impacts. In reconstructing an oblique side impact that has [an obliquity of] 45 deg, eliminating the correction reduced the ΔV error to less than 10 per cent from 40 per cent,

according to Sharma. In WinSMASH, the use of the correction factor is optional.

Increasingly, vehicles have electronic data recorders that calculate and document the ΔV if they are involved in frontal crashes. NHTSA has evaluated the accuracy of these data in staged collisions, and compared them to the results of WinSMASH reconstructions in NASS cases. Excluding those cases with questionable reconstructions, NHTSA found that WinSMASH underestimates longitudinal ΔV by about 20%, on average. Failure to consider restitution may be contributing to that difference.

According to Sharma,

> The WinSMASH was not designed to be a simulation program but rather a consistent, uniform method of judging accident severity in terms of the change in velocity. It should be emphasized that the WinSMASH program, as Crash3, should be statistically valid for a large number of cases; it may not provide accurate results in a particular case. The software should only be used with caution for individual cases .... The software is currently being tested and will be made available to the public once all features are implemented. As with Crash3, NHTSA maintains that WinSMASH is intended as a statistical tool to identify and isolate problems in motor vehicle safety, not as a simulation program, and should be used accordingly.

SMAC Family of Programs

SMAC

The computer program Simulation Model of Automobile Collisions (SMAC) was developed by Raymond R. McHenry and others at Calspan Corp. under a series of contracts from the NHTSA.[14,15] The goal was to generate a representation of an automobile that was relatively simple (so as not to overtax the computer resources of the day), but yet capable of not only being a deformable crash partner, but also a realistic representation of vehicle behavior as it traveled on its wheels over the ground. That was a tall order, indeed, that required sophisticated computer programming, yet resulted in a remarkably useful analysis tool.

Each car in a car-to-car crash was represented as a three-degree-of freedom (translation and yaw rotation in a level plane) rectangular solid that rested on four wheels that could be steered (perhaps due to crash damage) and/or braked, through the use of time-based input tables. Thus, an open-loop driver system was employed (no feedback so as to represent a path-following task). Tire/roadway forces were calculated from a friction circle for constant normal forces (no load transfer due to braking or cornering), with cornering coefficients specified for each tire.[16]

Assumptions made in the program included: (a) the accident scene was a flat, horizontal road surface; (b) no rollover was allowed; (c) all external forces were applied at the tires; and (d) the vehicle had a homogeneous exterior, meaning that the same linear force–deflection characteristic was used all the way around the periphery of the vehicle.

To model crash behavior, the periphery of the rectangular solid was divided into equally spaced intervals. Each of these intervals formed a pie-shaped wedge extending radially from the CG. When two of these vehicle representations contacted each other, the position of each was known relative to the other, so that the wedges, or RHO vectors, that could be in contact were identified. These RHO vectors would radially compress variable amounts until the normal pressures between the deformed surfaces achieved a reasonable balance.

Summation of these forces dictated the motion of each vehicle due to the collision, through the use of Newton's Second Law, integrated twice to obtain the velocity and position of each vehicle. The process was continued for each collision time increment until the vehicles were no longer in contact. Separate integration step times were used for the collision, separation, and trajectory phases.

Recommended values for these phases were 0.001, 0.010, and 0.050 sec, respectively. The program switched between integration time-step values automatically, depending on which phase of the crash the simulation was in. Because a trajectory phase was assumed at the start of the run, it was important to ensure the vehicles were not already in contact at that time, so that the program could switch to the collision phase, and the concomitant time step, and have the opportunity to stabilize the integrations, when contact was detected. Often, this meant a slightly greater vehicle separation than would be indicated by placing scale drawings of the actual vehicles together, because of the rounding of the actual vehicle contours, compared to the rectangular boxes used in SMAC.

Of course, the time steps could all be set to that of the collision phase, but that would tend to exceed the capabilities of the output storage arrays. On the other hand, the run could be truncated deliberately (the run time could be set to 500 msec, say) to concentrate on just the collision and separation phases. This is a good practice if a secondary "side slap" impact occurs, so that the program can handle the second impact with a time step fine enough to allow the integrations to stabilize and capture the accelerations with suitable precision.

The components of the RHO vector crush normal to the undamaged surface became the vehicle dynamic crush. Residual crush depths and separation velocities were calculated from a rebound model based on the use of energy rather than velocity as a separation criterion. Tangential forces between the vehicles were calculated based on a circumferentially uniform friction coefficient.

The result of the program was a time-based tabulation of translational and angular position, velocity, and acceleration of the two mass centers. Trajectory information was shown by graphics of the planar outlines of the vehicles, in addition to any tire marks. Vehicle shapes with residual crush were also shown graphically.

EDSMAC

In the mid-1980s, Engineering Dynamics Corporation (EDC) ported SMAC to compiled BASIC, and offered the resulting program for sale as EDSMAC.[17] Generally, all EDSMAC calculations were identical to those in SMAC, so references to the calculations in one program generally apply to those in the other. However, the use of the braking and steering tables, and of the terrain boundary, was slightly altered. According to EDC, three minor errors and the associated error messages were eliminated. EDSMAC utilized enhanced input, output, and graphics capabilities to take advantage of the personal computer environment.[18]

As a simulation, EDSMAC solves an initial-value problem and works in a time-forward fashion, in contrast to EDCRASH, which is a reconstruction program. In fact, EDCRASH creates a complete input file for use in EDSMAC, in keeping with the original intention of CRASH.

EDSMAC generates a scale drawing with a rectangular representation of each vehicle. Gray shading illustrates both crush areas, which are determined from the final length of the RHO vectors. If a crush area has narrow spikes, or appears to have narrow slivers taken out of it, it is due to the way the displacements move toward or away from the CG as the RHO vectors change length; the jaggedness indicates that adjacent vectors have apparently attained significantly different lengths. This is a clue that EDSMAC's representation of the structure may not be suitable for the crash being analyzed (if the RHO vectors in the two vehicles are perpendicular, for example). Such a circumstance could arise in an acute angle collision, in which the front of one vehicle strikes the side of another, near the front or back end.

On the same output display of the crush patterns are the vehicle ΔVs. These are computed from the integral of the accelerations that are calculated during the collision routine, as opposed to performing a vector subtraction of the velocities that exist at initial contact and separation. It is very useful (and in fact, strongly recommended) to examine the data output tables and independently determine the times of initial contact and separation, based on the accelerations being indistinguishable from those due to tire forces. (EDSMAC defines separation in terms of the accelerations being below a specified value, but the user may well have a different opinion.) The user can then: (a) look up the (longitudinal and transverse) velocity components at those times; (b) transform the longitudinal and lateral components at impact and separation into scene coordinates; (c) subtract the scene-oriented components at impact and separation to obtain the ΔV components in scene coordinates; (d) transform the components back into vehicle coordinates; and (e) compute the resultant. Transforming the components back and forth is necessary to account for any vehicle rotations that may occur, although they are generally small during the collision phase. The user will emerge with a much better understanding of how the ΔV vector is oriented with respect to the vehicle, and may also find some disagreement with the EDSMAC output display (particularly if a different separation time is used). If the calculations have been done correctly, the user should not be abashed.

The calculations outlined above can be performed easily with a pocket calculator that can convert back and forth between rectangular and polar coordinates (by converting the orthogonal components into polar coordinates, changing the angle by the requisite amount, and converting the polar coordinates back into (rotated) rectangular components). Alternatively, one can use a spreadsheet program, in which case it can be re-used when the need arises.

One can extend the above procedure to examine the velocities and ΔVs at points in the vehicle other than the CG (such as the occupant's seated position, location of interior contact, etc.) If there is significant vehicle rotation, the effect of location in the vehicle on velocities and ΔVs could be significant. Moreover, with a little more programming effort, one could trace the path of a free particle during the collision, as an indication of what direction an unrestrained occupant could move. This could be of interest to those dealing with occupant kinematics and injury exposure.

As a simulation program, EDSMAC generates acceleration values for both vehicles, and reports the results in tabular form, as functions of time. A crash pulse is thereby obtained. However, the results are a long way from being comparable to measurements from a crash test. While the program performs admirably, the structural representation is necessarily highly idealized. While the calculated accelerations may be representative in a gross sense, they are not representative in detail. The best way to get a crash pulse is to perform an instrumented crash test, keeping in mind the considerations discussed in Chapter 13.

While its ability to predict a crash pulse is very limited, EDSMAC is very useful as a precursor to a crash test, particularly if the test is oblique, or if the test conditions have been changed from the actual circumstances (due to test facility limitations, for example). Such simulations are valuable in reducing test uncertainties, such as the vehicle travel post-impact, the possibility of a side-slap, the ΔV that will be achieved, and so on. With regard to side slap, though, it must be remembered that EDSMAC retains the general rectangular shape of the vehicle. The side sway that is seen in the front of a vehicle when it strikes a moving vehicle in the side, or the bowing of the struck vehicle, cannot be represented. Such deformation, not simulated by EDSMAC, affects the timing of the side slap and the match-up of the two vehicles when they re-engage one another.

EDSMAC has a terrain boundary, in which the flat accident scene can be divided into two regions by a straight line. Where the origin is located, one friction coefficient can be specified; on the other side of the line, a separate coefficient is possible. It is not well known that in the original SMAC program, a second boundary was also created, with polar symmetry to the specified boundary. At the origin, one friction coefficient existed; on the other side of either line existed the second friction coefficient. The clever reconstructionist could then set up the scene coordinates so that the two terrain boundaries represented the edges of the paved road (or perhaps the edges of the unpaved median). The analyst who was unaware of this feature could get some puzzling results. In EDSMAC, this mirror image feature of the terrain boundary was eliminated.

Another factor affecting tire forces is the cornering stiffness, which is the amount of lateral force produced per unit slip angle of the tire, measured in lb/deg or lb/rad. Cornering stiffness data are difficult to measure, and published data are rare. The original SMAC documentation[19] indicated a typical cornering stiffness of 1/6 the static normal force on a fully inflated tire, 80% of that value if the tire is partly inflated, and 20% if it is flat. Nowhere in the EDSMAC program documentation is there found a recommendation to set the cornering stiffness to zero to simulate an airborne vehicle (which is clearly outside the bounds of program applicability), although such practices are known to have occurred.[20]

Finally, the observations of Day and Hargens are worth noting:

A characteristic of all simulation programs, including EDSMAC, is the fact that a single simulation cannot be used to describe the only way an accident could have occurred. Experience has shown that it is easy to create different scenarios (sometimes significantly different!) by making small changes in the input data .... Reconstructions attempt to find the only way an accident could have occurred, while simulations can identify many possible ways an accident could have occurred.[18] (pp. 89 and 91)

EDSMAC4

EDSMAC4 incorporates a number of enhancements by EDC into its EDSMAC program, and is discussed in a 1999 paper.[21] Among these enhancements are the following:

a. The program is not limited to two vehicles; any number of vehicles and vehicle interactions can be accommodated. In addition to improved collision detection, discussed below, this enables the simulation of multi-vehicle chain-reaction collisions.

b. The program has improved collision detection, so that the simulation time step is decreased before the collision occurs, rather than at the first sign of overlap. This avoids excessive vehicle overlap before the time step gets reduced. Previously, users had to avoid the fatal error resulting from this situation by manually reducing the time steps during the trajectory and separation phases.

c. The structure has a constant-stiffness force–deflection characteristic like that in Crash3, employing the same A and B constants, instead of a force–deflection characteristic that goes through the origin as in EDSMAC. The paper claims an improved representation of the structure and a more realistic crash pulse, but no comparison with actual crash data is offered.

d. The vehicle structure has different stiffnesses for front, side, and rear, instead of uniform stiffness around the periphery.

e. Up to 360 RHO vectors are allowed per vehicle—up from 100 previously. Damage can exist around the periphery of the vehicle. This is claimed to help in the simulation of narrow objects, but no validation is offered. No penetration is allowed beyond the CG (since that is where the RHO vectors originate).

f. For each RHO vector, up to 3000 adjustments are allowed in an attempt to balance the surface pressure between vehicles at that point.

g. In keeping with advancements in computing power, the RHO vector adjustment increment, delRHO, is reduced by a factor of 10 and the number of interactions has been increased by a factor of 10. The paper claims these changes to be effective, but no details are offered.

h. The paper claims improved support for pole and barrier impacts. No validation is offered, however.

i. Restitution appears to be modeled the same way as previously, although the restitution algorithm is bypassed if the three restitution constants (C0, C1, and C2) are all zero.

j. Load transfers due to X–Y plane accelerations are handled quasi-statically, which is to say that tire force increments are calculated using the CG height, wheel positions on the vehicle, and accelerations. As previously, suspension motions are not modeled. To distribute lateral load transfer between the front and rear axles, the vehicle's roll couple distribution is used.

k. The program supports articulated vehicles (tow vehicle and trailer), including pinch forces when the articulation becomes too large. Forces and yaw moments due to pinching are included, but no validation is cited. Since the vehicle segments do not have a degree of freedom in roll, roll moment transfer at the connections is not modeled.

l. Explicit support of tandem axles and dual tires is included, instead of the user having to obtain the effects of these features by representing the wheel locations and tire forces as a single wheel and tire. No validation is cited in the paper.

m. Displacement of the wheels in the vehicle's x–y plane can be specified by the user, but no validation of such effects is cited in the paper.

n. The paper claims support for tire blow-out, though the implementation consists of user-specified changes in the cornering stiffness as a function of time—not a full-blown 3-D representation of the vehicle and tire.

o. Using the HVE Event Editor, trajectory target positions can be set up that do not influence the simulation, but assist in determining how far away from the measured path the simulated path is.

p. Accelerometer locations can be specified for each vehicle.

q. The collision surface can be flat or on irregular 3-D faces. How this is possible for a yaw-plane model is not explained in the paper.

Like EDSMAC, the new version models motion in the plane; no vertical, roll, or pitch motions are included. Thus, there is no roll steer, or bump steer. The calculation procedure for ΔV appears to remain unchanged. As previously, even though crush is reported perpendicular to the undamaged vehicle surface, the actual displacements of points on the surface are along the direction of the RHO vectors—toward the CG. This can be expected to differ, at least for some points, from the deflection directions on an actual crashed vehicle.

PC-CRASH

PC-CRASH is a program combining a momentum-based impact analysis with a time-forward simulation of pre- and post-impact trajectories. The user can specify vehicle positions and velocities at impact, make an estimate of vehicle positions at separation, allow the program to perform a momentum-conservation analysis of the impact (at the separation configuration) to determine the velocities, and then watch how the trajectory simulation predicts vehicle motions during run-out. The cycle can be repeated until the user is satisfied with how the simulation predicts tire marks and rest positions.

Unlike the SMAC program family, PC-CRASH does not simulate the impact itself. Unlike the CRASH program family, PC-CRASH does not utilize an empirically based "spin" analysis; nor was it developed to take any account of vehicle stiffness, crush energy, or vehicle deformation, other than to match up the vehicles. Developed in Austria, the momentum-only impact analysis could be viewed as a response to the fact that in Europe, not only are manufacturer test data proprietary, EuroNCAP test data are not made available to the public, either.[22] In the United States, manufacturer test data are also proprietary, but Government NCAP and Standards Enforcement test data are publicly available, so structural characteristics can be obtained or developed for almost any passenger vehicle, or at least a sister or clone vehicle. Reconstructionists have not been so fortunate with vehicles not imported to North America.

PC-CRASH uses a momentum-based two- or three-dimensional model that relies on restitution, rather than vehicle crush or stiffness coefficients. The collision is treated as instantaneous, occurring at a single point, called the "impulse point." According to Schram,[22] PC-CRASH will "in the future" allow the user to choose either a stiffness-based model that enforces conservation of energy, or the momentum-based analysis. No information is provided as to how this stiffness analysis will work.

The collision model employs a contact plane between the vehicles that contains the impulse point. At the end of the "compression phase" (loading phase) of the collisions, the relative vehicle velocity normal to the contact plane at the impulse point is assumed to be zero. For a "full collision," the tangential relative velocity at that point is assumed zero also. In a sliding contact, there can be a relative tangential velocity, which is handled by the introduction of an inter-vehicle friction coefficient. This parameter must be bounded to avoiding adding energy to the crash. Since the friction force calculation is dependent on the normal force between the vehicles, PC-CRASH cannot be expected to handle forces arising from other sources (engagement of the left front wheels in a left frontal offset vehicle-to-vehicle crash, or snagging of a wheel by a guard rail post, for example).

Tire forces occurring during the collision are ignored, so impacts with low speeds, or large mass differences between the vehicles, are subject to erroneous results.

PC-CRASH contains an interactive graphical environment, which allows a DFX-format CAD drawing or bit-mapped scale overhead photo to underlay the graphic representations of the crash. Three-dimensional animations can be created directly from the program results.

Steffan and Moser state that

> It is important for a good prediction of the collision phase to define the correct overlapping of the vehicle bodies when the forces are exchanged .... it is best to calculate this position by driving the cars from the point of first contact over the defined time distance (45 to 60 msec) with pre-impact velocity .... It is also important to control the overlapping of both vehicle bodies on the computer screen in a manner consistent with the documented residual deformations.[23] (p. 141)

To simulate a noncollision trajectory, the vehicle model in PC-CRASH has six degrees of freedom: three translations and three rotations.[23] The global (scene) coordinates form a right-handed system, but in distinction from SAE coordinates, the Z-axis is positive above the plane. The X- and Y-axes are in the ground surface plane, which can be tilted relative to Mother Earth. If the plane is not level, the gravity vector is tilted accordingly. The vehicle coordinates are also right handed, but again in contrast to SAE convention, the Y-axis is positive on the left side of the vehicle, and the Z-axis is positive "up." The vehicle rotation sequence is yaw, roll, and pitch.

The four vehicle suspensions are modeled with linear springs and dampers; each having the same characteristics in tension and compression. Their orientation and movement is assumed to be normal to the ground surface plane. The sprung "masses" are considered to be massless, meaning that the vertical tire forces are solely functions of the static vehicle weight and the compressions of the springs, and the compression rates of the dampers. The vehicle CG is considered to be in the middle of the vehicle (which conveniently decouples the angular momentum equations).

The tire model is described in a paper by Cliff and Montgomery,[24] and appears to be unique to PC-CRASH. The lateral and longitudinal forces are defined over three regions of the slip angle; how the parameters separating these regions are related to actual tire performance is not provided. The brake or acceleration forces can be specified as functions of time, for each wheel separately.

The tire model appears to stay within the confines of a "friction circle," which of course means that on a level surface, the resultant lateral acceleration in Gs cannot exceed the friction coefficient. Cliff and Montgomery present test data from step steer maneuvers that conflict with such behavior in the more severe maneuvers,[25] and suggest that perhaps the measured lateral accelerations were contaminated by vehicle body roll.

Cliff and Montgomery claim to have validated PC-CRASH against the RICSAC tests. However, difficulties are noted in dealing with the experimental data, some of which were discussed in Chapter 4. They commented that

> In cases where there was significant rotation, sometimes it was not possible to have the model match the tire marks during the initial rotational part of the post-impact trajectory and have the vehicle stop at its rest position. This may possibly be due to inaccurately-reported wheel lockup factors. For these cases, it was found to be more important to match the vehicle with the tire marks immediately post-impact, rather than match it with its rest position.[24] (p. 104)

They found that in collisions with little run-out and low separation speeds (nearly equal and opposite vehicle momenta, for example), that a crush energy analysis was needed. After all, equal momenta can arise from an almost infinite combination of vehicle impact velocities, but each such velocity combination will produce a unique closing speed, which is directly related to the total crush energy, as discussed in Chapter 19.

Cliff and Montgomery concluded that

> In cases where was little post-impact rotation and long rollout trajectories, the wheel brake factors were critical in order to assess the speeds accurately. When there was significant post-impact spin-out (more than about 90 degrees) reasonably close results could be obtained by matching the tire marks during the initial spin-out without matching the rest position.[24] (p. 111)

This is to be expected, since much more energy is expended in the initial rotation than in subsequent run-out, and getting the initial spin-out right means that the departure angles will be more accurately modeled.

Noncollision Simulations

HVOSM

HVOSM, or Highway-Vehicle-Object Simulation Model,[26] was a program developed by the U.S. Federal Government, Federal Highway Administration, in the late 1960s to describe the motion of an automobile in three-dimensional space, including interaction with roadway features such as berms, ramps, and so on. It came in two versions: one with a highway design emphasis, and another with a vehicle dynamics emphasis.

The highway design version provided for a simplified interaction with fixed objects, by modeling the vehicle with a crushable outer layer having linear crush characteristics—sort of a three-dimensional extension of the concept used in the SMAC program family. The model had 11 degrees of freedom: six for the sprung mass, one each for the unsprung masses, and one steering angle. It used a friction circle tire model, and allowed road roughness to be specified over a rigid terrain described by five input tables. The program allowed the specification of curb geometries, and modeled suspension stops by asymmetric energy absorption characteristics. Tables were used for steering and wheel torque inputs.

The vehicle dynamics version was for the evaluation of vehicle trajectories due to launch, vault, or handling maneuvers. It lacked the ability to model impacts to the body. However, the vehicle dynamics version incorporated four more degrees of freedom (15 in all) to deal with tire spins, plus provisions for aerodynamic and rolling resistance, and brake system modeling. It incorporated a detailed model of the suspension, from the tire interface through the suspension geometry, to the body mass. Of course, the more complex the model, the more input information is required. Consequently, inputs were required for such things as spring and damping rates, rear axle inertia, throttle setting, and transmission ratio. The program also included built-in models for engine torque and drag, hydraulic brake pressure versus brake torque at a given wheel, and so forth. Open-loop driver inputs could also be used. HVOSM has been used successfully to model vehicle dynamics behavior in response to complex roadway design features.

The best-known application of HVOSM is the design of ramps used for an airborne corkscrew maneuver used by a stunt driver in a James Bond movie[27] (in the days before computer-generated images). Application of either version of HVOSM to real-world rollovers was limited by the fact that the only parts of the vehicle allowed to touch the ground were the tires. Contacts with the body surface were not allowed.

Typical of programs of its era, HVOSM ran on a mainframe computer, required voluminous input data that had to be prepared with the greatest of care, and produced reams of tabular output. Pre- and post-processor programs were developed to alleviate the misery, but the user interface was meager by modern standards.

EDSVS (Engineering Dynamics Single Vehicle Simulator)

This simulation is based on the University of Michigan Transportation Research Institute (UMTRI) TBST (Tractor Braking and Steering) program,[28] and represents a nonarticulated—or unit—truck (no trailers). The truck can have tandem rear axles and dual rear tires. The vehicle travels on a flat horizontal road surface with a single friction value. (There are no friction boundaries, as in the SMAC family.) The vehicle is analyzed in the yaw plane with three degrees of freedom (translation and rotation). Roll and pitch are not modeled. Since there is no roll degree of freedom, bump steer and roll steer are ignored. As soon as a wheel load becomes negative, wheel lift is assumed, and computations are halted. There are no suspension effects; that is, the wheels remain vertical, and camber stiffness does not contribute to lateral force generation. All external forces are applied at the tires, so there are no aerodynamic effects, and no undercarriage contact with the ground.

According to program documentation,[29]

> Accident investigators can use EDSVS to determine how a driver may have lost vehicular control .... It is not intended that the user should consider the results as the only way that a particular trajectory can be obtained. That is not the purpose of EDSVS. Rather, a successful result indicates the user has found particular wheel forces and/or steering characteristics which are consistent with the manner in which the driver controlled the vehicle. (p. 13,51)

> In some cases, data which violates [sic] [program] assumptions will cause a fatal error, along with a message indicating the reason for the error. In other cases, the error is not with the data but with the use of the program under conditions which violate the assumptions inherent to the computations. EDSVS will issue results which may not be valid for the given circumstances.[29] (p. 86)

EDVTS (Engineering Dynamics Vehicle–Trailer Simulator)

EDVTS is a simulation of a vehicle–trailer system (a passenger car pulling a standard trailer, or a commercial tractor–trailer vehicle), in response to driver inputs (accelerating, braking, or steering). The response is in terms of path, velocity, acceleration, tire forces, and other data as a function of time, and is based on UMTRI TBSTT program.[30] The vehicle model consists of two rigid bodies: one for the tractor and one for the trailer. The model has four degrees of freedom; namely, the planar translation and rotation of the tractor, and the articulation angle between the tractor and trailer. There are no roll or pitch degrees of freedom. Load transfers due to braking and steering are computed quasi-statically.

In the simulation, the hitch is assumed to transmit a moment due to friction in yaw, but not pitch or roll. The normal load on each wheel is equal to its static load, plus the load transfer occurring at that time. The load transfer at the trailer wheels is based on the trailer CG height, the hitch height, the forces on the trailer at the hitch and the road, the trailer track width, and the distance between the fifth wheel and the trailer axle.

The load transfer on the tractor depends on the front and rear suspension roll stiffness, which is not modeled (there being no roll degree of freedom). Instead, the user inputs the (fixed) proportion distributed to the front and rear axles.

A simplified model for tandem axles is included. The properties of all the tires on each axle in the tandem pair are assumed equal, and treated as an equivalent single tire. Inter-axle load transfer is specified by quasi-static transfer coefficients for the tractor and trailer tandem axles. Dual tires are treated as two single tires with equal loads vertically, longitudinally, and laterally.[31]

According to program documentation,

> Since EDVTS is a 4-degree of freedom analysis, suspension effects are ignored. Therefore, the program is well suited for the study of vehicle trajectories on low-friction surfaces, such as ice or snow. It also serves as an excellent first-order approximation for normally encountered road surfaces, such as dry asphalt and concrete. Good geometrical, inertial, and tire data is [sic] essential for accurate results.[32] (p. 14)

Phase 4

As its name implies, Phase 4 grew out of the Phase I, Phase II, and Phase III computer programs. These simulations were developed at the University of Michigan Transportation Research Institute (UMTRI) to evaluate new vehicle designs that might be developed to meet stricter safety requirements imposed on the trucking industry by the U.S. Federal Government. Like HVOSM, Phase 4 is a highly complex program that runs on a mainframe computer, requires voluminous input data meticulously prepared, and produces massive amounts of tabular output data. Having no user interface, it requires an extraordinary amount of patience on the part of the user.

Phase 4[33] is a simulation of a truck/tractor, a semitrailer, and up to two full trailers. The first vehicle is the power unit and may be a truck or tractor, and may carry a payload. It is distinguished by the fact that it can have only a single front axle with single tires, and can be arbitrarily steered. All other axles on the vehicle combination can be represented as single or tandem axles with single or dual wheel sets.

The second unit is always a semitrailer. Phase 4 does not provide for a truck towing a full trailer. The third and fourth units are full trailers consisting of semitrailers on either a fixed or converter dolly. A separate payload may be specified for each trailer. Thus, Phase 4 can be used for the following configurations:

a. Straight truck, empty and loaded

b. Bobtail tractor

c. Tractor-semitrailer, 3–5 axles, empty and loaded, or

d. Tractor-semitrailer/full trailer/full trailer, 7–13 axles, empty and loaded

The mathematical model incorporates up to 71 degrees of freedom, depending on the vehicle configuration being modeled. These degrees of freedom derive from:

a. Six degrees of freedom for the truck/tractor sprung mass: three translational and three rotational.

b. Three degrees of freedom for the semitrailer, the remaining three having been removed by constraints at the hitch.

c. Five degrees of freedom for each of the two full trailers allowed.

d. Two degrees of freedom (vertical and roll) for each of the 13 axles allowed.

e. One rotational degree of freedom for each of the 26 wheels allowed.

Small-angle assumptions are made, so that Phase 4 is only valid up to the point at which wheel lift-off occurs. A simulation begins with the tow vehicle in equilibrium at the origin of the global (earth-fixed) coordinate system. The program can be operated either with open-loop steering inputs, or with closed loop steering (path following).

The fifth wheel is a rigid connection in roll (for small articulation angles) between the trailer sprung mass and the top of the tractor rear suspension. The torsional frame compliance included in the tractor links the tractor sprung mass to the fifth wheel connection. The dynamic load equalization that occurs in tandem suspensions during braking and handling is calculated continuously in order to achieve a more accurate calculation of suspension forces. Both four-spring and walking-beam tandem suspensions can be modeled.

EDVDS

EDVDS (Engineering Dynamics Vehicle Dynamics Simulator)[34] is the result of an (undoubtedly massive) effort by EDC to port the Phase 4 program to the company's HVE platform. The most significant difference in the behavior of EDVDS compared to that of Phase 4 stems from the removal of the restriction to small angles. Also, instead of the user-supplied ROAD subroutine in Phase 4 to represent irregular travel surfaces, EDVDS uses HVE's more generalized terrain model. A drive train model was also incorporated into EDVDS. Internally, Phase 4 was written in FORTRAN, whereas EDVDS was programmed in C. Also, the original input and output routines were replaced with HVE input and output interface functions.

The vehicle configurations supported by EDVDS remain the same as those in Phase 4. Suspension configurations are the same. Connection forces between "vehicles" are unchanged, including the support of both fixed and converter dollies. The HVE drive train model was added in order to model tractive effort. Brake torques are computed in the same way, but the Phase 4 antilock model is not supported in EDVDS. The Phase 4 driver model was replaced with the HVE path follower model.

The vertical tire stiffness model in EDVDS has been extended to incorporate a bilinear force–deflection characteristic having an initial and a secondary radial stiffness. In-plane tire forces are computed using one of two user-specified tire models: a linear model or semi-empirical model.

The linear model calculates longitudinal and lateral forces (relative to the tire axis system), based on the tire longitudinal slip stiffness and tire cornering stiffness, independent of load or speed. According to Day, the linear model

> should be used only for non-limit maneuvers. In fact, the simulation will terminate if the longitudinal tire slip exceeds 10 per cent or the lateral tire slip exceeds 0.10 radians (these limits are editable).[34] (p. 238)

The semi-empirical model is based on the HSRI tire model developed at the University of Michigan, and is explained in some detail in Ref. 34. The original model had a tangent function that produced infinite lateral forces at 90° slip angle; its replacement by a sine function allows proper behavior at large slip angles. Also, the longitudinal slip has been replaced with its absolute value, to allow the modeling of drive torque.

In both Phase 4 and EDVDS, inter-vehicle connections are not rigid; rather, the connections are compliant. This method allows for the ability to model the roll compliance and moment transfer between vehicles within the combination. The roll stiffness is actually the frame torsional stiffness, resisting twist about the vehicle-fixed x-axis.

The suspension is modeled by linear springs, damping, and coulomb friction on each axle. A friction null band is applied to the friction value to prevent a friction force in the absence of suspension velocity.

As a result of changes to the Phase 4 model, the suspension force due to static vehicle weight no longer drops out. In EDVDS, the total suspension force includes the portion necessary for equilibrium.

For steering inputs, the user has the option of entering the angle at the road wheels; the angles do not have to be the same for the left and right sides. Alternatively, the steering wheel angle can be specified, along with the steering gear ratio. Then, the steer angles at both road wheels are initially equal, subject to additional roll steer.

The engine is modeled using torque vs. RPM tables for wide-open and closed throttle conditions.

As of 1999, these portions of the Phase 4 model had not been implemented in EDVDS; (a) semi-empirical brake model; (b) antilock brake model; (c) table look-up brake model; (d) table look-up tire model; and (e) table look-up suspension model.

EDVDS was compared with Phase 4 for various maneuvers and vehicle configurations. In many cases, transient responses were very close in nature. In other cases, however, there was an initial oscillatory response in EDVDS, caused by the fact that Phase 4 assumes the vehicle is initially in equilibrium, while EDVDS does not, and allows the vehicle to settle into equilibrium. It is important that the user be aware of this difference.

According to Day,

> The inherent assumption of initial equilibrium in the Phase 4 vehicle model is sometimes an advantage. For example, no oscillation related

to settling occurs at the start of the simulation. However, there are also disadvantages. In particular, it is not possible to begin a simulation on a sloped surface. Also, the simulation must begin with the tow vehicle's CG at the earth-fixed origin at zero heading angle ....[34] (p. 238)

Also, tire deflections are ignored while positioning the vehicle at the beginning of the simulation:

If initial equilibrium is important, the user can over-ride the default position by manually entering a different CG elevation (i.e., by dropping the initial CG elevation by about an inch).[34]

In all, the comparison revealed differences between Phase 4 and EDVDS, ranging from negligible to significant, depending on the severity of the maneuver. Much of that difference was attributable to the elimination of the small-angle limitation from Phase 4, and the elimination of a couple of problems in the Phase 4 mathematics model.

Occupant Models

Another class of simulations involves the kinetics (kinematics plus forces and moments) of occupants in a crashing vehicle. Needless to say, these tend to be complex, representing the interactions of various body parts with each other and with seat belts, air bags, seats, and other parts of the vehicle interior, subject to vehicle crash accelerations. Among these models is CVS (crash victim simulator), its descendants ATB (articulated total body) and Dynaman, and MADYMO. These models are far beyond the scope of this book, and will not be discussed. The interested reader is encouraged to do library research among the large number of papers and books devoted to this subject.

References

1. McHenry, R.R., *User's Manual for the Crash Computer Program*, Calspan Report No. ZQ-5708-V-3, Contract DOT-HS-5-01124, National Highway Traffic Safety Administration, 1976.
2. US D.O.T., National Highway Traffic Safety Administration, *CRASH3 User's Guide and Technical Manual*, U.S. Department of Transportation, National Highway Traffic Safety Administration, Washington, DC, April 1982, p. 3–11.
3. Campbell, K., Energy basis for collision severity, K. Campbell, SAE Paper 740565, SAE International, 1974.

4. Kahane, C.J., Smith, R.A., and Tharpe, K.J., The National Crash Severity Study, *Proceedings, 6th Internal Technical Conference on Experimental Safety Vehicles*, Washington, DC, U.S. Department of Transportation, National Highway Traffic Safety Administration, 1976.

5. Sharma, D., Stern, S., Brophy, J., and Choi, E.-H., An overview of NHTSA's Crash Reconstruction Software WinSMASH, *Proceedings, 20th International Technical Conference on the Enhanced Safety of Vehicles*, U.S. Department of Transportation, National Highway Traffic Safety Administration, 2007.

6. Marquard, E., Progress in the theoretical investigation of vehicle collisions, *Automobiltechnische Zeitschrift*, 68(3), 1966.

7. Fonda, A.G., Computer implementation of momentum and energy solutions: Computer solutions, refinements and graphical extensions to the CRASH treatment, In: *Forensic Accident Investigation: Motor Vehicles*, T.L. Bohan and A.C. Damask, Eds, Michie Butterworth, Charlottesville, VA, 1995.

8. Fonda, A.G., CRASH revisited: Additions to its clarity, generality, and utility, *Transportation Research Record*, 1111, 1987.

9. Engineering Dynamics Corporation, *EDCRASH Training Manual*, Engineering Dynamics Corporation, Beaverton, OR, 1994.

10. US D.O.T., National Highway Traffic Safety Administration, *CRASH3 User's Guide and Technical Manual*, U.S. Department of Transportation, National Highway Traffic Safety Administration, Washington, DC.

11. Smith, R.A. and Noga, J.T., *Accuracy and Sensitivity of CRASH*, DOT HS-80-152, U.S. Department of Transportation, National Highway Traffic Safety Administration, Washington, DC, 1982.

12. Struble, D.E., Generalizing CRASH3 for reconstructing specific accidents, SAE Paper 870041, SAE International, 1987.

13. Woolley, R.L., Warner, C.Y., and Tagg, M.D., Inaccuracies in the CRASH3 Program, SAE Paper 850255, SAE International, 1985.

14. McHenry, R.R., *Development of a Computer Program to Aid in the Investigation of Highway Accidents*, Calspan Report No. VK-2979-V-1, 1971.

15. Solomon, P.L., *The Simulation Model of Automobile Collisions (SMAC) Operator's Manual*, U.S. Department of Transportation, National Highway Traffic Safety Administration, Washington, DC, 1974.

16. Warner, C.Y. and Perl, T.R., The accuracy and usefulness of SMAC, SAE Paper 780902, *Proceedings, 22nd Stapp Car Crash Conference*, Ann Arbor, MI, SAE International, 1978.

17. Engineering Dynamics Corporation, *EDSMAC: Simulation Model of Automobile Collisions*, Fifth Edition, Engineering Dynamics Corporation, Beaverton, OR, 1994.

18. Day, T.D. and Hargens, R.L., An overview of the way EDSMAC computes delta-V, SAE Paper 880069, SAE International, 1988.

19. McHenry, R.R., Jones, I.S., and Lynch, J.P., *Mathematical Reconstruction of Highway Accidents: Scene Measurement and Data Processing System*, U.S. Department of Transportation, National Highway Traffic Safety Administration, Buffalo, NY, 1974.

20. Jones, I.S. and Baum, A.S., *Research Input for Computer Simulation of Automobile Collisions*, Vol. IV, Staged Collision Reconstructions, Report DOT-HS-805 040, Calspan Corporation, 1978.

21. Day, T.D., An overview of the EDSMAC4 collision simulation model, SAE Paper 1999-01-0102, SAE International, 1999.

22. Schram, R., *Accident Analysis and Evaluation of PC-Crash*, Chalmers University of Technology, Eindhoven University of Technology, CA, 2005.
23. Steffan, H. and Moser, A., The collision and trajectory models f PC-CRASH, SAE Paper 960886, SAE International, 1996.
24. Cliff, W.E. and Montgomery, D.T., Validation of PC-Crash—A momentum-based accident reconstruction program, SAE Paper 960885, SAE International, 1996.
25. Dugoff, H., Segel, L., and Ervin, R.D., Measurement of vehicle response in severe braking and steering maneuvers, SAE Paper 710080, 1971.
26. Segal, D.J., *Highway-Vehicle Object Simulation Model*, Report No. FHWA-RD-76-162, U.S. Department of Transportation, Federal Highway Administration, 1976.
27. McHenry, R.R., The Astro Spiral Jump—An automobile stunt designed via computer simulation, SAE Paper 760339, SAE International, 1976.
28. Moncarz, H.T., Bernard, J.E., and Fancher, P.S., *A Simplified, Interactive Simulation for Predicting the Braking and Steering Response of Commercial Vehicles*, Report No. UM-HSRI-PF-75-8, Highway Safety Research Institute, University of Michigan, 1975.
29. Engineering Dynamics Corporation, *EDSVS: Engineering Dynamics Single Vehicle Simulation*, Version 4, Engineering Dynamics Corporation, Beaverton, OR, 1994.
30. Bernard, J.E., Winkler, C.B., and Fancher, P.S., *A Computer Based Mathematical Method for Predicting the Directional Response of Trucks and Tractor-Trailers, Phase II Technical Report, Motor Truck Braking and Handling Study*, Report No. UM-HSRI-PF-73-1, Highway Safety Research Institute, University of Michigan, 1973.
31. Woolley, R.L., Warner, C.Y., and Perl, T.R., *An Overview of Selected Computer Programs for Automotive Accident Reconstruction*, Collision Safety Engineering, Orem, UT, 1985.
32. Engineering Dynamics Corporation, *EDVTS: Engineering Dynamics Vehicle-Tractor Simulation, Version 4*, Engineering Dynamics Corporation, Beaverton, OR, 1994.
33. MacAdam, C.C., Fancher, P.S., Hu, G.T., and Gillespie, T.D., *A Computerized Model for Simulating the Braking and Steering Dynamics of Trucks, Tractor-Semitrailers, Doubles, and Triples Combinations: User's Manual—Phase 4*, MVMA Project 1197, Highway Safety Research Institute, University of Michigan, Ann Arbor, MI, 1980.
34. Day, T.D., Differences between EDVDS and Phase 4, SAE Paper 1999-01-0103, SAE International, 1999.

Index

A

A-pillar, 10, 86, 113
Abscissa, 249, 315, 335
Absolute reference operator ($), 235
AccelAvg. bas program, 173
Acceleration, 10; *see also* Velocity
  CG X-axis, 11
  target vehicle CG, 11
Acceleration integration
  crash pulse, 167
  displacement, 169
  piecewise linear acceleration, 168
Acceleration of gravity, *see* Gravity,
  acceleration of
Accelerometer channels identification
  blank lines, 161
  reports, 159
  vehicle accelerometer locations, 162
  vXXXXX.EV5, 160–161
Accelerometer location effect, 188, 189
  acceleration traces from test 5404, 188
  frontal elements, 188
  velocity–time traces from test
    5404, 189
Accelerometer mount strategy
  base of B-pillars, 156
  bi-axial mount, 155
  lumped-parameter model, 154
  rear seat cross member, 155, 156
Accelerometers, 56
Accident investigation, 103
  crush measurement, 114
  information gathering, 103–106
  scene inspection, 106–109
  vehicle inspection, 109–114
Accident phase, crash, 7–8
Accident reconstruction, 16
Accident reconstructionist, 81
  information sources, 99–100
  model run, 98
  nomenclature and terminology,
    81–90
  people sizes, 100, 101

production change-over, 98
SAE standard dimensions, 90–92
vehicle inertial properties, 97–98
vehicle specifications, 95–96
VIN, 92–95
Accident Reconstruction Journal, 399
Accident vehicle crush profiles; *see also*
  Triangular crush profile
  form factor, 323
  segment-by-segment analysis of, 322
  segment average, 324
Accuracy, 3
  crush energy calculation, 415
  of filtering process, 184–185
  of integration process, 183–184
  restitution coefficient, 296
  vehicle deceleration, 32
Ackerman geometry, *see* Ackerman
  steering
Ackerman steering, 83
Adhesion limit, 27, 28, 72
ADR, *see* Australian Design Rule (ADR)
Aerodynamic drag, 8, 23
"Air-gap" problem, 357, 411
Airbag sensor, 147
Airworthiness, 9
Aliasing
  instrumentation engineering, 197
  signal reconstructing, 195
  signals sampling, 196
Alliance of Automotive
  Manufacturers, 95
American National Standards
  Institute Specifications (ANSI
  Specifications), 394
Amplitude, 135, 136
Analog data, 167
Analog filters
  differential equations, 136
  Laplace transform properties, 137
Analytical reverse projection, 128
Angle measurements, 3, 63, 242
Angle of approach, 92, 299, 382
Angle of departure, 92

Angular momentum, 9, 388–389
    impact velocities, 365
    momentum conservation, 288, 426
    vehicle CG, 436
Angular velocity, 21, 54, 367
ANSI Specifications, *see* American
        National Standards Institute
        Specifications (ANSI
        Specifications)
Anthropometry data, 101
Approach angle, *see* Angle of approach
Approach speed, 384, 427
Architectural incompatibility, 407
Arc length (S), 37, 41, 72, 235
Arctangent function, 242
Articulated total body (ATB), 443
Articulated vehicle, 434
As-built, 89, 108, 109
At-scene photos, 111, 220
    evidence on vehicle, 207
    scene inspection, 106
ATB, *see* Articulated total body (ATB)
Attenuation, 135, 136
Australian Design Rule (ADR), 93
AutoCAD, 7
    digitizing, 122
    splines, 37
Automotive News, 96
Automotive News Data Center, 96
Axis of vision, 116, 117
Azimuth angle, 72

B

B-pillar, 10, 86, 94, 355
Back light, 82
Backlight header, 86
Backward-looking methods, 17, 32; *see
        also* Forward-looking models
Backward measure, 72
Badge engineering, 98
Barrel roll, 209, 240
Barrier-to-pole stiffness comparison, 401
Barrier crash, 14, 258
    deformation mode in underride
            crashes, 414
    high-speed, 154
    large car, 263, 264
    restitution coefficient, 16, 288
    small car, 263
    square-on flat, 323
    zero velocity, 260
Barrier equivalence, 272
Barrier Equivalent Velocity (BEV), 15, 272
Barrier force readings, 194, 195
Barrier impact speed, 15
Barrier transducers, 161, 193
Baseline, 131, 348
Base units, 2
Belt line, 85, 86, 203
Bessel filters, 139
BEV, *see* Barrier Equivalent Velocity
        (BEV)
BFE, *see* Breakaway fracture energy
        (BFE); Energy, breakaway
        fracture (BFE)
Bi-axial mount, 155
Bilinear transforms, 144, 170; *see also*
        Z-transform
Binomial expansion, 150
Bode plots
    corner frequency, 138
    filter requirements from SAE J211, 138
    logarithmic scale, 137
Body-on-frame, 82, 87
Body in white, 86
Body mount, 87
Body type, 82
Bootleg steer, 31
Bottom out, 188
Bound up, 20
    brake factors, 61
    in wreckage, 23, 25
Bowing, 349, 350, 352, 403
Brach's approach, 288, 289
Brake factors, 23, 61, 437
    effects, 24
    friction coefficient, 26
    measurements, 25
Brake force, 23, 28, 61, 71
Brake hop, 84
Braking, 9, 20
    benefit of antilock, 21
    demand, 28
    EDSMAC, 430–433
    SMAC, 61
    TBST, 438
    tests, 53

Braking, anti-lock, 21
Braking deceleration, 75
Braking demand, 20, 22, 27
Braking force, 20
Break-point frequency, 138
Breakaway fracture energy (BFE), 395
Bumper height, 394, 408
Bump stop, 84
Butt (of pole), 394
Butterworth filters, 139, 164

C

C-pillar, 86, 356
Calendar year, 82
Calibrated camera, 129, 130
Calibration coefficients, 119, 120, 121
Calibration point, 120
Calspan Reconstruction of Accident
        Speeds on the Highway
        (CRASH), 54, 422
    assumptions, 423
    SMAC users, 422
    vehicle crush damage, 423
Camber, 84, 439
Camera reverse projection methods
    advantages, 123–124, 125
    at-scene photograph, 124
    focusing screen, 123
    tire tracks, 122
    trajectory procedure, 125–127
    two-photograph, 127
Camera, single lens reflex, 123
Campbell's formulation, 308
Campbell's observation, 304, 306
CAMPOSE program, 128
Cantilever beam, 394, 397
Car and Driver, 99
Car line, 82, 95, 98, 402
Cartesian coordinates, 6, 278
Caster, 84
Catwalk, 86, 87
CDC Vital and Health Statistics, 101
CDS, *see* Crashworthiness Data System
        (CDS)
CE, *see* Crush energy (CE)
Center of gravity (CG), 10
Center of mass
    accelerometer mount, 172

    angular momentum, 389
    Cartesian coordinates, 278
    crash duration, 378
    momentum conservation, 8
    reference frame, 14
    tri-axial accelerometer pack, 178
Center of vision, 116, 117
Central collisions, 299, 301
Centripetal acceleration, 5, 57
Centroid, 284, 358, 425
CFC, *see* Channel Frequency Class
        (CFC)
CFR, *see* Code of Federal Regulations
        (CFR)
CG, *see* Center of gravity (CG)
Channel Frequency Class (CFC), 10, 167,
        195
Channel number, 159, 161
    load cells, 192, 193
    NNN, 162, 198
    v05404. 049 data file, 163
Characterizing the structure, 316
Chebyshev filter, 139
Circular arc, 72, 74
Circular frequency, 258, 268, 271
Clockwise, 6
Clones, 98, 99, 316
Closing speed, 13, 14, 426
Closing velocity, 13, 14, 387
Cloth dress maker's tape, 114
(Coeff)i, 323, 324, 336
Co-planar, 16
Co-planar analysis, 23
Coast down, 19, 20
Coasting, 23, 76
Code of Federal Regulations (CFR), 93
Collision force,
    low-speed collisions, 8
    PDOF, 425
    RICSAC run-out trajectories, 58
Collision Safety Engineering (CSE),
        288
    Brach's approach, 289
    IMPAC and PLASMO methods,
        288, 289
    linear momentum conservation, 288
Collisions, low-speed, 8
Common velocity, 364, 366, 425
Companion mass, 282, 284, 285

Compiled BASIC language, 148
    Crash3 code, 424
    EDSMAC, 430–433
    NHTSA's filtering algorithm, 169
Concentric impact, 282
Conservation laws, 8–9, 295
Conservation of angular momentum, 365
Conservation of Energy, 289, 295, 370, 386
    closing velocity to restitution
        coefficient sensitivity, 297
    crush energy, 295
    "damage-only" reconstruction, 296
    increase in crush energy, 386
    loss in kinetic energy, 385
    velocity change to restitution
        coefficient sensitivity, 297
Conservation of Momentum, 291, 363,
        386–387, 426
    angular momentum, 286
    for central collisions, 299
    coefficient of restitution, 292
    kinetic energy, 294, 295
    time-forward solution, 293
    trigonometric identities, 300
    uniaxial collision, 301
    velocities, 291
    velocity change, 293
    velocity entities, 287
Constant-stiffness crash plots, 315, 319
    absorbed energy, 325
    Campbell's original observation, 321
    crash plot ordinate, 319, 322
    data clustering, 317
    ECF, 316
    form factor, 315
    frontal crash tests, 316
    pertinent data for, 318
    for repeated impacts testing, 324
    restitution coefficients, 317
    rigid moving barrier, 320
    structural parameters, 320
Constant-velocity joint (CV joint), 88
Constant acceleration; see also Variable
        acceleration
    accident sequence, 76
    constant velocity, 71
    intermediate variables, 73
    spreadsheet, 74
    time–distance study, 72, 74

truck/bicyclist accident, 75, 77
Constant force model, 331, 333, 399
    crash plot for, 334
    crush energy, 417
    flat barrier tests, 415
    with piecewise linear crush profiles,
        335–337
    with saturation force, 419
    structural stiffness parameters,
        337–343
Constant stiffness
    Campbell's formulation, 308
    coefficients equation, 307
    conversion of kinetic energy, 303, 304
    crash plots, 337
    crush energy and force
        calculations, 332
    crush saturation, 330
    dimensionless shape function, 306
    dynamic crush, 303
    ECF, 327
    force–deflection characteristic, 304
    with force saturation, 326, 328,
        331, 354
    using piecewise linear crush
        profiles, 329
    saturation crush, 328
    time residual crush, 305
Constant velocity, 4, 71, 88
Contact force, 8, 277, 285, 299
Contact patch, 19
Controlled rest, 60
Control point, 120, 121, 122, 130
Coordinate systems
    Cartesian, 6, 278
    CG, 7, 231
    key points, 41
    mathematical, 6
    reference frame, 5, 6
    SAE, 7, 157
    small-angle assumptions, 441
    tie-in to scene, 388
    vehicle CGs, 58
    vehicle position, 36
Coplanar collision analysis, 375
    angular momentum, 388–389
    approach angles, 382
    checking results in, 375
    closing velocity, 387

configuration estimation, 380
crab angles, 381–382
crash duration, 376, 378
energy conservation, 385–386
force balance, 389–390
impact yaw rate, 379
mass positions vehicle center, 379–380
momentum and energy conservation calculations, 377
momentum conservation, 386–387
momentum vector direction, 387
PDOF, 384, 385
restitution coefficient, 383–384
root choice, 376
sample spreadsheet calculations, 375–376
trajectory analysis, 379
vehicle headings, 381
vehicle inputs, 390
vehicle selection, 378
yaw rate degradation, 378
Coplanar collisions reconstruction, 363
angular momentum, 365
conservation of momentum, 363
energy, 364
governing equations development, 366–370
impact and separation conditions, 370
impact velocity vectors, 366
sample reconstruction, 372
Coplanar crashes
crush measurement issues in, 357
crush profile, 359
reconstruction procedures, 358
stiffness properties, 358, 359
testing agencies, 360
Core support, 87, 357
Corner frequency, 138
barrier load cell channels, 197
filter coefficients, 164
Cornering coefficient, *see* Cornering stiffness
Cornering force, 31
Cornering stiffness, 30, 67, 432
Coulomb's friction law, 21, 442
Cowl, 88
Crab angle
deceleration, 36

drag factor, 49
effects, 32–33
nose-leading, 209
vehicle drag factor, 33
vehicle scratch angles, 240
yaw angles, 65
Cramer's Rule, 137
CRASH, *see* Calspan Reconstruction of Accident Speeds on the Highway (CRASH)
Crash3 user, 301, 422, 423
calculation procedure, 426
reconstruction procedures, 358, 385
source code for, 365
stiffness coefficients, 427
Crash duration, 376, 378
changes in, 380
vehicle headings, 371
velocity–time curve, 15
Crash phase, 8
backward-looking methods, 17
linear force–deflection characteristic, 277
momentum conservation, 387
post-crash phase, 58
reference frame, 281
run-out phase, 31
stiffness, 278
Crash plot, 316, 410
constant-force model, 333–335
constant-stiffness, 315
crash tests, 376
NEI, 343
ordinate, 319
regression line, 327
spreadsheet formulas, 322
with uniform crush and force saturation, 328
Crash pulse data analyzing, 167, 177, 432
acceleration data, 177
acceleration signal, 185, 186
accelerometer location effect, 188–189
CG x-axis acceleration, 179
CG x-axis displacement, 180
CG x-axis velocity, 179
digitizing hardcopy plots, 181
filtering process accuracy, 184–185
integration process accuracy, 183–184
plotted curve quality effects, 182–183

Crash pulse data analyzing *(Continued)*
 Test 5683 analyzing, 180
 Test 5883 analyzing, 180
 velocity, 178
Crash severity measures
 closing speed, 13, 14
 closing velocity, 13, 14
 intrusion, 275
 separation velocity, 14
 velocity change, 274
Crash test data files, 194, 390
 data filtering, 164
 from NHTSA, 159
 structural characterization, 322
Crash test dummies, 101, 156
Crash test report, 10
 acceleration data, 177
 data plots in, 164
 data presentation, 157
 NHTSA, 167
 right-side accelerometer data from, 182
Crash victim simulator (CVS), 443
Crashworthiness, 9, 14, 113
Crashworthiness Data System (CDS),
 350, 423
Crown, 90, 121
Crush analyzing
 in full-width tests, 409
 offset override tests, 409
Crush, average, 256
 crush energy calculation, 322
 crush profile, 308
 NEI's calculation of, 340
 piecewise-linear crush profiles, 311
 triangular crush profile, 314
Crush documentation, 417
Crush, dynamic, 262, 264, 305
Crush energy (CE), 304
 closing velocity, 295
 crush saturation, 330
 energy absorption, 252–254
 equating coefficients of, 307
 flat barrier tests, 412
 force–deflection characteristic, 198
 lower fraction, 257
 restitution coefficient, 304
Crush energy assessment, 303
 accident vehicle crush profiles,
 322–324

calculation, 398
constant-force model, 333–335
constant-stiffness crash plots,
 315–316
constant-stiffness models, 303–308
constant stiffness with force
 saturation, 326–328
force–deflection characteristic, 198
half-sine wave crush profile, 308–309
half-sine wave squared crush profile,
 309–311
lower fraction, 257
piecewise-linear crush profiles,
 311–314
structural stiffness parameters,
 337–343
triangular crush profile, 314–315
Crush measurement, 114
 bowing constant, 350
 coplanar collisions, 204
 crash test, 346
 jig, 348
 Neptune, 357
 for rollover accidents, 360
 upper-structure, 410, 411
Crush profile, 284
 crush force, 415
 de facto assumption, 348
 force balance, 389
 form factors for piecewise-linear, 311
 half-sine wave, 308–309
 NASS protocol, 356
 pre-and post-test shape
 measurements, 410
 segment-by-segment calculations, 322
 triangular, 314–315
Crush, residual, 250
 crash-tested vehicles, 303
 depths and separation velocities, 430
 linear relationship, 261
Crush width, 304
 NEI's, 340
 overall width, 360
 pre-crash spacing, 400
 stiffness coefficients, 398
Crush zone, 9–10, 153–154
 accelerometers, 155
 B-pillars, 156
 crash phase duration, 277

CSE approach, 288
momentum calculation, 425
C_s, *see* Saturation crush (C_s)
CSE, *see* Collision Safety Engineering
(CSE)
Curb trips, 243, 244
Curb weight, 88
MVMA specifications, 96
Neptune, 97
specification, 100
Cutoff frequency, 139, 174
CV joint, *see* Constant-velocity joint
(CV joint)
CVS, *see* Crash victim simulator (CVS)

D

D-pillar, 86
Damage analysis, 426, 427
Damage, contact (direct), 59, 345, 358, 389
Damage, induced, 204, 345, 358
Damage onset, 317, 399
Damper, viscous, 252, 254
Damping, Coulomb, 254
Damping, friction, 254
Damping, hysteresis, 254
Damping, structural, 254
Dash panel, 86
DPLC, 162
frontal elements, 188
velocity–time curve, 189
Data acquisition
RICSAC results, 58
separation speeds comparison, 57
separation velocity, 56
translational and rotational
accelerations, 57
Data clustering, 317, 319
Data file parsing, 163–164, 171
Data filtering, 164–165, 169
Data plots, 158, 162, 194
Data recorders, 16, 428
Daylight opening (DLO), 85, 86
dB, *see* decibel (dB)
DBD crush deformation jig, 348
DDW, *see* Direct damage width (DDW)
Debeading, 216
Debris, 95, 106, 115
Decade, 135, 138, 403

Deceleration, translational, 424
decibel (dB), 137
Deck lid, 86, 95, 110
Define name feature, 39
Degree of freedom, 278
car-to-car crash, 429
EDVTS, 440
in lumped-mass model, 154
ΔV, *see* Velocity change delta-V (ΔV)
Demarcation line, 412
Department of Transportation (DOT), 157
Dependent variable, 38
array value, 164
spline, 39
Difference equation, 140
in compiled BASIC, 148
IIR filter, 141
from transfer function, 143
Differential equations of motion, 17,
35, 421
Digital filters, 139, 147
in airbag sensor, 147
difference equation in BASIC,
148, 149
filter output comparison, 149
filter's responses, 147
fourth-order bandpass filter, 149, 150
half-sine pulses, 148
Digitizing hardcopy plots, 181
direct, *see* Damage, contact (direct)
Direct damage width (DDW), 340
Direction of travel
lock fraction, 33
tire, 30
vehicle scratches, 240
vehicle's heading, 32
X-axis points in, 7
Direct linear transformation method
(DLT method), 118
Displacement, 10, 156
AccelAvg. bas program, 173
acceleration data, 177
barrier force readings, 194
CG X-axis, 13
change in, 12
demarcation line, 412
force–deflection characteristic, 199
linear and angular, 425
maximum, 183

Displacement (*Continued*)
  row force, 198
  SAE coordinate system, 157
  smoothing, 186
DLO, *see* Daylight opening (DLO)
DLT method, *see* Direct linear
      transformation method (DLT
      method)
Dog that didn't bark, 107
DOT, *see* Department of Transportation
      (DOT)
DOT number, 157, 159, 162, 193, 198
Double-filter procedure, 164, 165
Downloading test data, 158, 191
Drag factor, 33, 49
  on crab angle, 49
  rollover test data, 206
  speed at initiation, 228
  yaw angles, 65
Drag factor, overall, 228, 230
  inverse relationship, 228, 229
  rollover reconstructions, 228
  speed at roll initiation, 229
  speed *vs.* roll distance, 229
Drag force
  brake factor, 26
  friction coefficient, 27
  longitudinal, 22
Drag sled, 21, 23
Drip rail, 88, 208
Driven axle, 23
Driven wheel, 61
Dry pavement, 60
Dummy weights and dimensions, 101
Duration, 8
  crash duration, 15
  crash phase, 277
  elastic collision, 261
  pulse duration, 15
  time, 73
  vehicle-to-vehicle collision, 278

E

Eccentric collisions, 282, 371
  companion mass, 285, 286
  particle-type analysis, 285
Eccentric impact, 282, 285, 385
ECF, *see* Energy of crush factor (ECF)

EDC, *see* Engineering Dynamics
      Corporation (EDC)
EDCRASH program, 423, 424, 430
EDSMAC, *see* Engineering Dynamics,
      Inc.'s version of SMAC
      (EDSMAC)
EDSMAC4 program, 433–434
EDSVS, *see* Engineering Dynamics
      Single Vehicle Simulator
      (EDSVS)
EDVDS, *see* Engineering Dynamics
      Vehicle Dynamics Simulator
      (EDVDS)
EDVTS, *see* Engineering Dynamics
      Vehicle–Trailer Simulator
      (EDVTS)
EES, *see* Energy equivalent speed (EES);
      Equivalent energy speed (EES)
Effective mass, 282, 283
  Crash3, 385
  eccentric collisions, 301, 371
  yaw moment of inertia, 285
Effective weight, 320
Elapsed distance, 72, 73, 75
Elapsed time, 72, 73, 75, 236
Elastic crash, 260
Elevation, 8, 109, 443
Emblem, 82, 111, 208, 209
End over end, 215, 240
Energy absorption, 252, 253
  crush phase, 271
  damping mechanism, 254
  energy recovery, 253
  kinetic energy, 252
Energy, breakaway fracture (BFE), *see*
      Breakaway fracture energy
      (BFE)
Energy, conservation of, *see*
      Conservation of energy
Energy dissipated, 32, 53, 254, 367
Energy equivalent speed (EES), 272, 416
Energy, fracture, 394, 396
Energy of crush factor (ECF), 316, 327
Energy recovery, 253, 272
Energy, shear
  in underride crashes, 413–414
  weighting factors, 418
Engine cradle, 88
Engine drag, 20, 22

Engineering Dynamics Corporation
(EDC), 424, 430
Engineering Dynamics, Inc.'s version of
SMAC (EDSMAC), 31
crash pulse, 432
crush patterns, 431
EDCRASH, 430
initial-value problem, 430
tire forces, 432
Engineering Dynamics Single Vehicle
Simulator (EDSVS), 438
program documentation, 439
single friction value, 438
Engineering Dynamics Vehicle
Dynamics Simulator
(EDVDS), 441
roll stiffness, 442
tire deflections, 443
vehicle configurations, 441
vertical tire stiffness model in, 441
Engineering Dynamics Vehicle–Trailer
Simulator (EDVTS), 439
suspension effects, 440
vehicle–trailer system, 439
Equivalence
barrier, 272
barrier crash, 14
EES, 16
effective mass, 285
velocity–time trace, 15
Equivalent energy speed (EES), 15
Event data recorder, 16, 159
Evidence at scene inspection
asphalt deposits in wheel rim, 220
brush marks, 222
classic crescent moon mark, 219
classic rim-down mark, 219
crescent mark visible, 220
ejection portal evidence, 223
glass deposit long post-accident, 224
pavement edge effects, 221, 222
plant materials in wheel rim, 221
tarp marks, 222
window glazing, 223
Evidence on vehicle
black tire marks on roadway, 218
direction of roll, 207
fractured wheel with impacts, 218
in-plane compressive stresses, 215

material flow on plastic bumper
cover, 209
nose-leading, 209
pitch-over, 216, 217
rear lift gate damage, 217
roll direction, 208
scratch pattern demarcation, 213
trailing side, 208
vehicle inspection sketch, 210, 211, 212
wood fibers, 216
yaw motions, 212
Excel
Goal Seek feature, 27, 380
"name manager" feature, 40
spline operation in, 42
Exemplar vehicle, 25
crush measurement, 114
equivalent profiles on, 347
VIN masks, 95
Expert AutoStats, 96
Extrapolation, 38, 39
effort of, 360
point coordinates, 122
Extrication tools, 104

F

Failure Analysis Associates, 243
Fastback, 82
FE, *see* Fracture energy (FE)
Federal Highway Administration, 437
Federal Motor Vehicle Safety Standard
(FMVSS), 85, 101, 156
Standard 105, 85
Standard 204, 156
Standard 208, 101
Fictitious test, 343
Fifth wheel, 10, 85, 441
Film plane, 116
Filter(j) algorithm, 170
Filter, band-pass, 147
fourth-order, 149
60 ms half-sine wave, 151
10 ms sine wave, 150
25 ms sine wave, 150
Filter, band-stop, 147
Filter, Bessel, *see* Bessel filter
Filter, Butterworth, *see* Butterworth filter
Filter, Chebyshev, *see* Chebyshev filter

Filter class, 12, 170, 177
Filter coefficient, 140, 164, 170
Filtering, 135
    analog filters, 136–137
    bilinear transforms, 144
    bode plots, 137–138
    digital filter, 139
    filter order, 137
    filter types, 139
    FIR filter, 140
    frequency, 135
    IIR filters, 140, 141
    Z-transform, 141–143
Filtering effects
    acceleration signal, 185, 186
    channel 280 accelerations, 186
    channel 280 velocity data, 187
    velocities from channel 280, 187
Filtering process accuracy; *see also*
        Integration process accuracy
    CFC 60 filters for CG x-axis
        accelerations, 185
    Channel 280 of Test 5683, 184
Filter, low-pass, 139
Filter order, 137, 139, 141
Filter subroutine, 170–171
Finite element model, 247
Finite impulse response filter (FIR
        filter), 140; *see also* Infinite
        impulse response (IIR filters)
Firewall, 86
FIR filter, *see* Finite impulse response
        filter (FIR filter)
Fixed objects, 53, 204, 393
Flat barrier tests
    frontal and rear, 360
    Marine Taurus tests, 417
    override tests, 412
    Plymouth reliant underride
        crash, 419
    stiffness coefficients, 398
FMVSS, *see* Federal Motor Vehicle
        Safety Standard (FMVSS)
Focal length, 116
    calibrated camera, 129, 130
    rough match, 125
Focal plane, *see* Film plane
Fog line, 90, 108
FOH, *see* Front overhang (FOH)

Force-deflection characteristics
    constant-force model, 333
    constant-stiffness, 433
    crush energy, 198
    EDVDS, 441
    harmonic oscillator, 277
Force, 308
    balance, 389–390
    contact, 8, 285, 389
    dynamic, 252
    external, 8, 19, 305, 429, 439
    friction, 9, 31, 254, 285
    static, 252
    tire, 30, 31
Force, lateral, *see* Lateral force
Force, longitudinal, *see* Longitudinal
        force
Force saturation
    constant stiffness with, 326, 333, 337
    crash plot, 328
    structural stiffness parameters,
        337–343
Ford taurus underride crash
        reconstruction
    crush force, 415, 416
    EES, 416
    flat barrier tests, 415
    Marine Taurus tests, 417
    NHTSA Taurus tests, 416
FOR loop, 149
Form factor, 306, 309, 310, 322, 335
    for constant-stiffness crash plots,
        315–316
    for half-sine wave crush profile,
        308–309
    for half-sine wave squared crush
        profile, 309–311
    for piecewise-linear crush profiles,
        311–314
    in segment-by-segment analysis,
        322–324
    for triangular crush profile, 314–315
FORTRAN language, 164, 169, 441
Forward-looking models, 17; *see also*
        Backward-looking methods
Forward measure, 72
FotoGram program, 120, 121, 128
FotoIn3.bas program, 122
Fourier Transform, 143

Fourth-order bandpass filter, 149, 150
  60 ms half-sine wave, 151
  10 ms sine wave, 150
  25 ms sine wave, 150
Fracture energy (FE), 396, 397
Free to roll, 20, 24, 29
Frequency
  bode plot, 137
  circular, 258
  cutoff frequency, 139
  digital filters, 147
  maximum, 197
  nonlinear function, 141
  ringing, 10
Frequency domain, 137, 148
Frequency response curve, *see* Bode
    plots
Friction
  available, 24
  circle, 31, 429, 436
  ellipse, 31
  sliding, 19, 23, 30, 31
  static, 21
Friction coefficient, 27, 28
  reverse trajectory calculation, 50
  tangential forces, 430
  tire/road interface, 26
  weighted average, 27
Frontal pole impact tests, 398, 399, 400
Front bumper height, 91
Front overhang (FOH), 91
Front track width, 92, 410
Full-scale mapping
  accident vehicle, 351, 352
  C-pillar striker, 356
  correction factor, 355
  crush energy calculation, 354
  crush measurements, 353
  crush profiles, 352, 353
  crush vectors, 354, 355
  damage map, 351
  match-up, 352
Full-width barrier tests, 410
Furrowing, 243, 244

G

Geometric incompatibility, 407
Gerber Variable Scale, 62, 63

Glazing, 85, 94, 223
Global coordinates, 6
Goal Seek feature, 27, 242
Governing equations development
  for coplanar collision analysis, 371
  energy equation, 368
  momentum conservation, 367
  quadratic equation, 369
  separation velocity components, 366
Gravity, acceleration of, 8, 73, 258, 280
Greenhouse, 85
  crush measurement, 204
  deformation, 204
  demarcation, 203
  structure, 409
Gross axle weight rating (GAWR), 88
Gross vehicle weight rating (GVWR), 88
Grouping load cell data channels, 194–195
Guard crash testing, 408, 410

H

H-point, 85
  couple distance, 92
  couple, 85, 92
  machine, 85
  SgRP, 85
  travel, 85
H-point template, machine, 85
Half-sine wave crush profile
  average crush, 308
  dimensionless shape function, 309
Half-sine wave squared crush profile,
    309
  crush shapes and form factors, 310
  narrow-object impact, 309
  trigonometric identity, 310–311
Half sine wave
  bandpass filter response, 151
  in graphic form, 147
Hatchback, 82, 95
Haversine, *see* Half-sine wave squared
    crush profile
Heading angle
  for accidents, 36
  crab angle, 33
  run-out motions, 55
  transition, 301
  vehicle, 242

Head room, 92
Height, seated, 100
Height, standing, 101, 121
Hemlock poles, 396
Highway-Vehicle-Object Simulation
        Model (HVOSM), 437
    design of ramps, 438
    roadway features, 437
Honda accord underride crash
        reconstruction
    crush documentation, 417
    NHTSA civic tests, 418
    structural characterization, 418
Horizon line, 117
Horizontal curvature, 89, 109
HSRI, *see* University of Michigan
        Highway Safety Research
        Institute (HSRI)
Human-Vehicle-Environment (HVE),
        338, 441
HVE, *see* Human-Vehicle-Environment
        (HVE)
HVOSM, *see* Highway-Vehicle-Object
        Simulation Model (HVOSM)
Hysteresis, 19, 254

I

IF() function, 26
IIHS, *see* Insurance Institute for
        Highway Safety (IIHS)
IIR filters, *see* Infinite impulse response
        (IIR filters)
Image space, 118
IMPAC program, 288, 289, 383
Impact mechanics, 277
    conservation of energy, 289
    crash phase duration, 277–278
    CSE, 288, 289
    degrees of freedom, 278
    eccentric collisions and effective
            mass, 282–285
    impulse–momentum-based, 280–281
    mass, 279
    momentum, 279
    momentum conservation, 286–287
    PIM, 287, 288
    yaw moment of inertia, 280
Impact speed, velocity, 54, 411

Impact yaw rate, 379
Impulse, 279, 282
    center, 284, 285, 288, 289
    normal, 287
    ratio coefficient, 288
    response, 140
    tangential, 287, 435
    vector, 16, 284
Impulse–momentum-based impact
        mechanics; *see also* Planar
        impact mechanics (PIM)
    collision forces, 281
    Newton's Second Law application, 280
    Zamboni machines, 280, 281
Impulse–momentum methods, 284
Inches, 2, 3
Independent variable, 36
    ArrayofD, 39
    circle in parametric form, 37
    filter coefficients, 164
    GivenS, 39
    image frame, 36
    Key S values in spline, 43
    reverse trajectory analysis, 37
    selection, 36
    spline function, 38
Induced damage, 204, 345, 427
Inertial properties, 83, 96
    NHTSA, 427
    people sizes, 100
    vehicle, 97–98
Inertial reference frame, 5, 269
Infinite impulse response (IIR filters),
            140–141; *see also* Finite impulse
            response filter (FIR filter)
Information gathering
    at-scene photographs, 104
    police investigation, 105–106
    scene visit rule, 103
Initial value problem, 17, 72
    CSE approach, 288
    displacements and velocities, 269
    EDSMAC, 430
    simulations, 421
    six-equation system, 289
    SMAC program, 422
    solution to, 268
Initiation, 204
    cessation of tire marks at, 205

end of furrow at, 205
  speed, 205, 206
Injury exposure, 14
  occupant's, 178
  vehicles, 260
Instantaneous center, 240
  by crash, 285–286
  polynomials, 425
Instrument cluster, 88
Instrument panel, 88
Insurance Institute for Highway Safety
    (IIHS), 157, 275
Integration, 186
Integration process accuracy
  Channel 280 of Test 5683, 183, 184
  TRC's crash report, 184
Intermediate point, 50
  accident vehicle, 382
  trajectory coordinates, 424
  vehicle coordinates for, 64
Intermediate position, 62, 63
International Organization for
    Standardization (ISO), 93
Internet, 99
Intrusion, 155
  IIHS, 275
  mechanism, 265
  NHTSA, 174
  side impacts, 403
Inverse camera, 130, 131
Investigation, at scene, 122–123

J

JATO Dynamics, 96
Jig, crush measurement, 348
Jounce, 84
Jounce stop, *see* Bump stop

K

KE, *see* Kinetic energy (KE)
Key frames, 36, 37, 52
Key points, 72
Key positions, 36, 74
  CG locations at, 232
  scratch angle measurements, 242
  spreadsheet, 239
  vehicles in, 231

Key values, 36
Kick-up, 87–88
Kinetic energy (KE), 9, 15–16, 294
  crush energies, 415
  dissipation of, 301
  energy, 252
  energy conservation, 385–386
  in NHTSA Taurus tests, 416
  in vehicle collision, 295
Kirchhoff's Laws, 136
Known conditions, 35, 421

L

Ladder frame, 87
Laminated glass, 85
Lamp inspection, 112
Laplace s space, 144
Laplace Transform, 136
  frequency-domain solution, 137
  unilateral Z-transform, 142
Large car
  barrier crashes, 263–264
  comparison, 264
  dynamic crush in, 264
  hits small car, 270–272
  small car *vs.*, 263–264
Laser gun, 10
Latch, 88, 155
Latency, 139
Lateral acceleration, 54, 436, 437
Lateral force, 72
  components, 30
  generation, 29, 30
  semi-empirical model, 442
  tire forces, 31, 432
  weight transfer effects, 278
Lateral force generation; *see also*
    Longitudinal force generation
  slip angle, 29
  tire cornering characteristic, 30
LCAnal. bas, 198
LCBs, *see* Load cell barriers (LCBs)
LCCD equation, *see* Linear constant-
    coefficient difference equation
    (LCCD equation)
Leading side in rollovers, 208
Least significant digit, 4
LF, *see* Lock fraction (LF)

Linear constant-coefficient difference
    equation (LCCD equation), 140
Linear model, 442
Line of alignment, 117
Line of sight, 69, 74, 76, 108
Load cell, 191
  analysis application, 174
  array, 156
  barrier information, 412–413
  barrier transducers, 161
  channels, 197
  in NHTSA, 4
  LC for, 193
  NCAP, 156
Load cell barriers (LCBs), 413
  demarcation line, 412
  override tests, 412
  Volpe Tests, 413
Load cell data analysis, 195
  barrier force *vs.* time, 200
  crush energy, 199
  displacement file, 199
  force–displacement characteristic,
    200
  NHTSA CFC 60 filter, 198
Load–deflection curves
  load cell data, 273
  load–unload cycle, 250
  lumped-mass model, 249
  negative forces, 274
  quasi-static test, 252
  second loading cycle, 251
  for test DOT 5404, 273
  unloading from different
    deflection, 251
Load paths, 154, 247–248
  crush in, 269
  to dynamic tests, 252
  load–deflection curves, 249
Local coordinates, 6; *see also* Global
    coordinates
Lock fraction (LF), 27
  impacts with fixed objects, 53
  in RICSAC reconstructions, 66
  for various trajectory segments, 28
  vehicle inputs, 390
Lock up, 21, 27, 31
Longitudinal acceleration, 29, 54, 412
Longitudinal force, 30

Longitudinal force generation
  brake factor measurements, 25
  free-to-roll, 29
  friction coefficients, 27
  lock fraction calculation, 28
  spread sheet, 26
  tire slip characteristic, 21
  tire's rotation, 22
  vertical loading effect, 20
  weight transfer effects, 23
Loose parts, 357
Loose vehicle parts, 111, 357
Lumped-mass model, 154, 155, 249, 252
Lumped-parameter model, *see* Lumped-
    mass model

M

Mag nails, 107
Make, 82
Mapped point, 120
Mapping, full scale, 351, 352, 353
Marine Taurus tests, 417
Market data, 96
Marquard's procedure, 425
Mass, 279, 367
  accelerometer mounts, 155
  incompatibility, 407
  lumped-mass model, 154
  slug, 2
  sprung and unsprung, 83, 84
  stationary system center, 178
  units, 2
  vehicle's center of, 31
Mass moment of inertia, 279
Matched z-transform method, 148
Materials for review, 105
Maximum crush, 402
Mechanics of Materials, *see* Strength of
    Materials
MIN() function, 26
mkfilter program, 147
Model run, 98
Model year, 82, 92
  CFR format, 93
  VIN codes, 94
Moiré interference pattern, 195, 196
Moment of inertia, 3, 279
  hemlock poles, 396

mass, 279
pitch, 216
yaw, 279, 280
Momentum, 279
Momentum, angular, 9, 388–389
  conservation, 365
  linear momentum conservation, 288
  vehicle CG, 436
Momentum, conservation of, 365
Momentum vector direction, 387
Monotonic, 235–236
Most significant digit, 4
Motion table, 52
Motor vehicle, 9
  crashes, 1
  parking brake system, 85
  SAE standard dimensions, 90–92
Motor Vehicle Manufacturing
    Association (MVMA), 95
Mount, accelerometer, 10
  at center of mass, 172
  strategy, 154–156
Moving average, 140, 149
Multi-valued, 39, 40
Musical pitch, 135
MVMA, *see* Motor Vehicle
    Manufacturing Association
    (MVMA)

N

Narrow fixed-object collisions, 393
  crash pulses, 265
  crush profiles and vehicle crush
      energy, 398–402
  frontal collisions, 264
  impact speed, 402, 403
  maximum crush, 402, 403
  side impacts, 403–404
  wooden utility poles, 394–397
Narrow fixed object impact, 393
National Accident Sampling System
    (NASS), 348, 423
  bowing, 349
  DBD crush deformation jig, 348
  deflection points, 349
  inspection procedures, 350
National Crash Severity Study (NCSS),
    274, 423

National Highway Safety Bureau, 7
National Highway Traffic Safety
    Administration (NHTSA), 4, 92
  acceleration channels, 172–173
  acceleration integration, 167–169
  automated signal analysis
      package, 201
  civic tests, 418
  data file parsing, 171
  data filtering, 169
  EV5 ASCII X-Y, 160
  filter algorithm, 165
  filter subroutine, 170–171
  NHTFiltr.bas program output, 172
  signal analysis software, 174
  signal browser, 174
  tests, 411, 416
National Oceanographic and
    Atmospheric Administration
    (NOAA), 106
NCAP, *see* New Car Assessment
    Program (NCAP)
NCSS, *see* National Crash Severity Study
    (NCSS)
Neptune, 357, 411
Neptune Engineering (NEI), 99, 337
Neutral steer, 83
New Car Assessment Program (NCAP),
    156, 191
Newspaper photographs, 115
Newton-meters (NWM), 194
Newtons (NWT), 194
Newton's First Law, 281
Newton's laws of motion, 4–5
Newton's Second Law, 2, 17
  accelerometer mount strategy, 154
  application, 280
  differential equations, 17
  net force on particle, 4
  summation, 429
Newton's Third Law, 5, 287, 389
NHTFiltr.bas, 169, 172
NHTSA, *see* National Highway Traffic
    Safety Administration
    (NHTSA)
NHTSA crash test data, 153
  accelerometer channels
      identification, 159–162
  accelerometer data, 157–158

NHTSA crash test data *(Continued)*
  accelerometer mount strategy,
      154–156
  crush zone, 153–154
  data file parsing, 163–164
  data filtering, 164–165
  desired channels downloading, 162
  measurement parameters and
      transducers, 156
  sign convention, 157
  vehicle crashes, 153
NHTSA load cell barrier data, 191
  aliasing, 195–197
  configuration, 192
  crash test data files, 194
  downloading, 192–194
  grouping load cell data channels,
      194–195
  load cell, 191
  load cell data analysis, 195
NHTSA web site
  crash test data files, 194
  downloading data from, 158–159
  signal browser, 174
Nikon F3, F4, 123
NOAA, *see* National Oceanographic and
      Atmospheric Administration
      (NOAA)
Noise, 10, 197
Noise, vibration, and harshness
      (NVH), 87
Nomenclature, terminology, 81–90
Non-driven axle, 23, 24
Non-driven wheel, 24
  brake factors, 25
  FADriven, 26
Non-Newtonian reference frame, 6, 56
Noncollision simulations; *see also*
      Simulations
  EDSVS, 438–439
  EDVDS, 441–443
  EDVTS, 439–440
  HVOSM, 437–438
  Phase 4, 440–441
Noncollision trajectories with splines, 35
  collision phase, 35, 36
  independent variable selection,
      36–37
  key values, 36

smoothing function, 37–38
  splines properties, 38–40
Normal force, 21, 23
  brake factor, 26
  car in a car-to-car crash, 429
  friction force calculation, 435
  lateral force generation, 29
  longitudinal and lateral force
      components, 30
Northwestern University Traffic
      Institute, 62, 63
Nose-leading, 209
Nose leading roll, 212, 241
Notchback, 82
NVH, *see* Noise, vibration, and
      harshness (NVH)
NWM, *see* Newton-meters (NWM)
NWT, *see* Newtons (NWT)

O

OAL, *see* Overall length (OAL)
Object point, 120, 128
Object space, 118, 119, 130
Occupant kinematics, 16, 105, 113
Occupant model, 227, 443
Occupant protection, restraints, 113
Octave, 135, 138, 197
Off-tracking, 83
Ordinate (vertical axis), 315
  calculation, 315
  crash plot, 319, 321, 325
  in parameter calculation, 322
Origin of coordinates, 107
Overall length (OAL), 91, 280
Overall vehicle width (OAW), 427
Overall width (OAW), 92, 280
Overhanging barrier, 407, 411
Overlay tracing, 125
Oversteer, 83
Overtone, 135

P

Parking brake, 31
Parking brake system, 85
Parsing the data file, 163–164, 171, 172
Part 565, 49 CFR, 93, 94
Particle mass, 285–286

Pascal's Triangle, 150
Pavement gouges, damage, 36, 107
PC-CRASH program
  collision model, 435
  equal momenta, 437
  pre-and post-impact trajectories, 435
  tire model, 436
PDOF, *see* Principal Direction of Force
    (PDOF)
PE, *see* Potential energy (PE)
Pendulum, 394, 395, 396
Perception, 70–71
Perception, reaction, 70–71, 73
Perfectly elastic, 259, 292, 295
Perimeter frame, 87
Perspective, 116, 118
Perspective, mathematical basis, 117
Phase shift, 141
  double-filter procedure, 164
  NHTSA CFC 60 filter, 198
Photogrammetry
  DLT method, 118
  imposing planarity process, 119
  two-dimensional, 119–122
Photographic analysis
  center of vision, 116
  film plane, 116
  geometry, 118
  vanishing point, 116, 117
  vertical line, 117
Photographs
  analytical reverse projection, 128
  camera reverse projection methods,
    122–127
  information from, 115
  photogrammetry, 118–119
  photographic analysis, 116–118
  police, 74, 115
  3D multiple-image photogrammetry,
    128–132
  2D photogrammetry, 119–122
  two-photograph camera reverse
    projection, 127
Photography, vehicle, 110, 112
PhotoModeler, 129
Picture point, 120
Piecewise-linear crush profiles
  crush measurements, 313
  segment of, 311

smooth crush profiles, 311
  weighted average, 312
PIM, *see* Planar impact mechanics (PIM)
Pin-hole camera, 116, 118
Pinch weld, 351, 356
Pitch-over, 216, 217
Pitch
  angle, 7
  moment, 216
  musical, 135
Planar impact mechanics (PIM), 287
  impulse ratio coefficient, 288
  relative velocity, 287
  restitution coefficient, 287, 288
PLANTRAN, 120, 121, 128
PLASMO methods, 288, 289, 383
Platform, 82, 99
Plenum, plenum chamber, 88
Plotted curve quality effects
  crash pulse, 183
  digitizing poor-quality hard copy, 183
  right-side accelerometer data, 182
Plymouth reliant underride crash
    reconstruction, 419
Point of tangency, point on tangent, 89
Poisson's ratio, 414
Polaroid photographs, 110, 111
Pole, 137
  class, 394
  fracture energies, 397
  NP-1 in Wolfe study, 396
Pole impact
  crush stiffness, 400
  paucity, 402
  vehicle-to-pole impact, 395
Polynomial, 137
  second-order, 396
  in simulations, 425
  transfer function, 142
Polyvinyl butyral (PVB), 85
Position variables, 49, 52
Positive driveline torque, 22
Post-collision phase, 35
Post-crash trajectories analyzing, 53–54
Post-test photos, 61
Potential energy (PE), 9, 228
Pre-crash hydroplaning, 112
Pre-crash motions, 47
Precision, 3

Primary latched position, fully latched
    position, 89
Principal Direction of Force (PDOF), 284
    crash duration, 384
    scene coordinates, 385
Principal point, 116, 129, 130
Prism, 90, 131, 280
Production change-over, 98
Production run, 98
    Honda Civic Structure, 418
    Internet, 99
    for vehicle, 316
Profile, 89
    CRASH programs, 423
    crush, 305, 308
    green, 351
    half-sine wave squared, 309
    piecewise linear crush, 314
    at regular intervals, 109
    smooth, 308
Pseudo-ellipse, 33
Pseudo arc length variable, 40
Pseudo force, 397, 398
PulseInt. bas, 181
PVB, *see* Polyvinyl butyral (PVB)
Pythagorean Theorem, 40, 232

Q

Quasi-static, 10, 248
    load transfer, 434
    test, 252

R

Radians, 3, 242
Radius of gyration, 98, 280
Ramp breakover angle, 92
Rate of change, 4, 12, 257
Reaction, 70–71
Rear bumper height, 91–92
Rear crashes, 408
Rear overhang (ROH), 91
Rear track width, 92
Rebound, 84
Rebound stop, 84
Rebound velocity, 15
Reconstructionist, 16
Redundant, 247

Reference frame, 5, 6
Reference frame, inertial or Newtonian, 4
Reference point, 48, 85
Regression, 57, 415
Repair of structural separations,
    326, 350
Repeatability, 203
    crash pulse, 183
    digitizing hardcopy plots, 181
Repeated impacts, 251
    CE, 416
    constant-stiffness crash plots for,
        324–326
    Prasad's tests, 359
Reports, 159
Research Input for Computer
        Simulation of Automobile
        Collisions (RICSAC), 54, 63
    collision configuration, 58
    impact configurations, 55
    impact velocities, 59
    key point and segment
        information, 64
    test report, 60–61
    Tests, 56, 62
    tri-axial accelerations, 54
Residual crush, *see* Static crush
Restitution
    crush energy, 255
    dynamic crush, 262
    load–deflection characteristic, 256
    partial restitution, 262
    relative exit velocity, 254
    restitution coefficients, 257, 261
    rigid barrier, 256
Restitution coefficient, 14
    rate of separation, 383
    run-out analysis, 384
Reverse-trajectory technique, 229
Reverse trajectory analysis, 36
    application of, 64
    need for, 37
    yaw and roll rates, 236, 237, 238, 239
Reverse trajectory calculation using
        splines, 47
    brake factors, 61
    calculations, 52
    controlled rest, 60
    data acquisition, 56–58

friction coefficient, 51
post-crash trajectories analyzing,
    53–54
reconstruction technique, 65
RICSAC crash tests, 54–55
run-out motions, 55–56
run-out trajectory, 50
secondary impacts with fixed
    objects, 53
side slap impacts, 59
spline results, 51
surface friction, 60–61
test 4 separation velocities in, 66
time-forward simulations, 67
validation run, 62–65
variables, 49
velocity–time histories, 47–48
Reverse trajectory spreadsheet
differences, 231–232
input section of, 233
key point coordinates, 232
roll rate, 236
spline points, 233, 234, 235
RICSAC, *see* Research Input for
    Computer Simulation of
    Automobile Collisions
    (RICSAC)
Right-of-way, 90, 109
Right hand rule, 6
Rigid moving barrier, 320–322
Ringing, 10
Road loads, 87
Roadway evidence, 232, 380
Roadway striping, 108
Rocker panel, 86
crush profile points, 347
near-side, 156
passenger vehicles, 351
ROH, *see* Rear overhang (ROH)
Roll-off characteristic, 138, 197
Roll angle, 7, 232
center of mass, 278
initiation, 204
rollover calculations, 239
valuable insights, 243
Roll direction, 16
drawings, 209
flow direction, 208
longitudinal axis, 241

Rollei-metric MR2 Close Range
    Photogrammetry System, 129
Rolling radius, 21, 84
Rolling resistance
energy dissipation, 19
test plan, 19, 20
Roll moment, 23
center of mass, 278
inertia, 216
Rollover, 203
on-road, 204, 238
roof deformation measurements, 360
tripped, 204, 243
Rollover analysis, 227
deformation, 227
overall drag factor, 228–229
reverse trajectory spreadsheet,
    231–236
rollover trajectory, 229–231
scratch angle directions, 239–242
soil and curb trips, 243–244
yaw and roll rates, 236–239
Rollover forensics, 203
belt line, 203
evidence at scene, 218–224
evidence on vehicle, 207–218
greenhouse crush measurement, 204
severity measurements, 204–207
Rollover trajectory
CG coordinates, 231
scene measurements, 229
vehicle placements, 230, 231
Roll rate, 236, 237, 238, 239
curve, 238
units, 3
Roof deformation, 275, 360
Roof rail, 86
Root (of quadratic), 370
Root choice, 376
Rotational acceleration, 57, 279
Run-out
backward-looking analysis, 48
crab angle, 379
motions, 55–56
post-collision phase, 35
post-crash, 271
reverse trajectory reconstruction, 50
RICSAC results, 58, 59
rollover calculations, 239

Run-out (*Continued*)
    separation speed, 228
    SMAC TRAJ simulation, 425
    splining operation, 378
    system momentum vector, 384
    tire forces in, 31
    trajectory, 50
    wreckage, 112

S

SAE, *see* Society of Automotive
        Engineers (SAE)
Safety Standard numbering, 7
Sample rate, 147
Sampling period, 142
Saturation crush (C_s), 326, 333
    constant-force model, 337, 419
    for crush values, 327
Saturation force (F_s), 327
    calculation, 337
    constant-force model, 333
    segment in crush saturation, 330
Scene drawing, 109, 230, 380
Scene inspection; *see also* Vehicle
        inspection
    debris, 106
    official survey benchmarks, 106, 107
    reconstruction calculations, 107
    roadway striping, 108
    sight lines, 109
Scene investigation, 122
Scene topography, 109
Science, 1, 2
Science general principles, 1, 2
    accident phases, 7–8
    accident reconstruction, 16
    base units, 2–4
    conservation laws, 8–9
    coordinate systems, 5–7
    crash severity measures, 12–14
    crush zones, 9–10
    equivalence, 14–16
    Newton's laws of motion, 4–5
Scratch angle, 241, 242
Scratch angle directions
    Arctangent function, 242
    Goal Seek of Excel, 242
    roll axis, 240

translational and tangential
        velocity, 241
Scratch angle progression, 212
Scratch marks, 221
Scratch pattern, 209
    angles on vehicle, 212
    demarcation, 213
    highlighted in red, 213, 215
Scratch pattern demarcation, 213,
        214, 215
Scuff marks, 62
Seating Reference Point (SgRP), 85
Seaworthiness, 9
Secondary impacts
    and controlled rest, 59–60
    with fixed objects, 53
Secondary latched position, 88, 89
Section, 89, 109
Section modulus, 414
Segment-by-segment analysis, 49,
        322–324
Self-aligning torque, 84
Self-equilibrating, 5, 363
Semi-empirical model, 442
Semitrailer, 89
Senate filibuster rule, 366
Separation conditions, 300, 422
Separation position
    intermediate positions, 63–64
    for RICSAC run-out trajectories,
        58–59
    velocity directions, 387
    x-direction accelerometer trace, 60
Separation speed, velocity, 57
Separation velocity, 14, 56
Series, 82
Service brakes, 84
Severity, 204
Severity measurements
    initiation, 204
    initiation speed, 205, 206
    rollover distance, 207
    rollover severity, 206
SgRP, *see* Seating Reference Point (SgRP)
Shape factor, 274, 398
Shear deformation, 414
Shear energy
    shear modulus, 414
    in underride crashes, 413

Shear force, 414, 416
Shear modulus, 414
Shutting the door, 356
Side slap, 59, 432
Side slap impacts, 59, 432
Side sway, end shift, 360
Sight lines, 109
Sign convention, 157
Significant figures, 2, 4
Sill, 86
"Sill averaging" procedure, 349
Sill averaging, 349, 350
Simulation Model of Automobile
     Collisions program (SMAC
     program), 31, 54, 429
  EDSMAC, 430–433
  model crash behavior, 429
  RHO vector, 430
  roadway forces, 429
Simulations, 421
  CRASH program, 422–428
  PC-CRASH program, 435–437
  SMAC program, 429–434
Sisters & Clones List, *see* Vehicle Year &
     Model Interchange List (Sisters
     & Clones List)
Sisters, 98, 99
Skid resistance value, 60
Slack adjustor, 89
Slip angle, 29
Slip, longitudinal, 22, 442
Slope, 90
Slug (unit of mass), 2
Slug, 2
SMAC program, *see* Simulation Model
     of Automobile Collisions
     program (SMAC program)
SMAC Traj routine, 425
Small angle assumption, 441
Small car, 264
  barrier crashes, 263
  comparisons, 264
  large car hits, 270–272
Small car barrier crashes, 263
Smoothing function
  animation software, 37
  electronic spreadsheets, 38
Smoothing process, 12
Snagging, 366, 435

Society of Automotive Engineers
     (SAE), 6, 7
  coordinate system, 157
  coordinates, 6, 7, 62
  J1100, 91
  J211, 138, 167, 177, 178
  J656, 89
  J826, 85
  Paper 940569, 89
  Paper, 338
  Recommended Practice J1100, 138
  Recommended Practice J670e, 83
  Soil trips, 228, 244
  Standard Dimensions, 90–92
Southwest Research Institute, 394
Speed units, 204
Spin2, 424
Spin axis, 83
Splines, 37
  absolute cell reference indicators, 39
  Excel's name manager feature, 40
  independent variable, 43
  key points, 38, 40, 41
  operation in Excel, 42, 45
  piecewise linear function, 42, 43
  post-crash trajectories, 40
  segmented curve, 43
  spline curve, 41
  spreadsheet, 44
  trajectory, 44
Splining operation, 49
Spread sheet, 26
Spreadsheet calculations, 375–376
Sprung mass, 83
Sprung weight, 83
Standard weight, 343
Statically indeterminate, 247
Static crush, 303
Static loaded radius, 84, 92
Station, 106
Steering inputs, 29
Stereoplotter, 129
Stiffness, 278
Stiffness incompatibility, 407
Stimuli, 70
Stimulus, response, 70
Strain rate, 248, 252, 325
Strength of Materials, 247, 248
Striker, 88, 356

String pot, 10
Striping, 108, 220
Structural damping, 254
Structural dynamics
    barrier crash time response, 259
    circular frequency, 258
    elastic crash, 260
    linear load–deflection characteristic,
        261
    simple barrier crash model, 258
    in vehicle crashes, 257
    velocity–time plot, 259, 260
Structural parameter, characteristic,
        305, 316
Structural stiffness parameters
    coefficients, 338
    for constant-force models, 337
    crash plot numbers, 338
    NEI crash plot results, 340, 342, 343
    Neptune stiffness parameters, 341
    restitution coefficient, 342
    Toyota Siennas calculation, 339
Structure crush, upper, lower, 409, 411, 412
Sub-frame, 87
Subdivision, 36
Summagraphics MM Series, 122
Superelevation, 90
Surface friction, 60–61
SURFINT program, 128
Suspension rate, 84
Synectics Road Safety Research
        Corporation, 409
System center of mass, 9
    angular momentum, 389
    crashes, 14
    stationary, 178
    system momentum, 281
Système International d-Unitès, 2
System of Units, Technical English, 2

T

Tail leading roll, 209, 241
Tangential acceleration, 57
Tangential velocity, 240, 435
Taurus tests, 416
TBST program, *see* Tractor Braking
        and Steering program (TBST
        program)

Tempered glass, 86, 106
Terrain topology, 49, 51, 432
Theft prevention standard, 95
Theodolite, 90
Three-dimensional (3D), 38
Three-dimensional multiple-image
        photogrammetry; *see also* Two-
        dimensional photogrammetry
    crush on vehicles measurement, 128
    inverse camera, 130
    photogrammetric measurements,
        131, 132
    PhotoModeler, 129
    total station, 131
Three parameter model, 328
Time-backwards, 421
Time-forward, 293, 435
Time-forward simulations, 67
Time base, 232
Time–distance study, 69
    constant acceleration, 71–74
    perception, 70–71
    reaction, 70–71
    variable acceleration, 77–80
Time line, 69
    for bicyclist, 76, 77
    crash duration, 378
    for truck, 75
Time remaining, 72
Tire conditions, 112
Tire deflation, 22
    additive effect, 27
    at-scene photographs, 24
Tire drag, 21
Tire forces, 31, 432
    lateral, 424
    longitudinal, 442
Tire inspection, 24
Tire marks, 55
    cars, 115
    on road, 84
    on roadway, 218
Tire models
    backward-looking approach, 32
    crab angle effects, 32–33
    lateral force generation, 29–30
    longitudinal force generation, 20–29
    rolling resistance, 19–20
Tire patch (V_H), 21, 24

Tire pressure, 20
Tire saturation
Tire's rotation, 22
Toe-in, 84
Toe-in, toe-out, 84
Toe-out, 84
Toe board, 86
Topography, 109
Torque, 279
   box, 87
   drive line, 22
   retardation, 23
   yaw, 32, 53
Total station method, 90, 356
Track width, 92
Traction, *see* Positive driveline torque
Tractor, 89
Tractor Braking and Steering program
   (TBST program), 438
Trailing side in rollovers, 208, 424
Trajectory, 26
Trajectory segment, 22, 27
   sample calculation, 28
   segment-by-segment analysis, 49
   using spreadsheets, 239
TRANS4, 121
Transaxle, 88
Transducer, 10, 156
Transfer function, 137
   IIR filter designs, 142
   Z-transform, 141
Translational acceleration, 57
Translational velocity, 55, 56, 206
Transmission gear, 20, 25
Transportation Research Center of Ohio
   (TRC), 178
Transport Canada, 157, 340
Travel speed, 14, 16
TRC, *see* Transportation Research
   Center of Ohio (TRC)
Tread, 19, 92
Trees, 393, 397
Tri-axial mount, 155
Triangular crush profile, 314–315; *see
   also* Piecewise-linear crush
   profiles
Triangulation, 127
Triaxial, 56
Trim level, 82

Truck's acceleration, 75
Tumblehome, 206
Tunnel, 86
Turn radius, 364, 379
Two-dimensional photogrammetry
   calibration coefficients, 120
   control points selection, 121–122
   FotoIn3.bas, 122
   object space, 119
Two-Photograph Camera Reverse
   Projection, 127
Two parameter model, 308, 333

U

UMTRI, *see* University of Michigan
   Transportation Research
   Institute (UMTRI)
Underride guard, 408, 410
Underride or override collisions, 407
   crash tests, 409
   crush energy, 408
   Ford Taurus underride crash
     reconstruction, 415–417
   Honda accord underride crash
     reconstruction, 417–418
   irregular crush profiles, 412
   load cell barrier information, 412–413
   plymouth reliant underride crash
     reconstruction, 419
   shear energy in, 413–414
   synectics bumper underride crash
     tests, 409
Underride/override, 272, 414
Understeer, 83
Uniaxial collisions, 291, 301
   conservation of energy, 295–297
   conservation of momentum, 291–295
Unibody, unitized, 87
Uninflated tire, 24
unit of mass, *see* Slug (unit of mass)
University of Michigan Highway Safety
   Research Institute (HSRI), 442
University of Michigan Transportation
   Research Institute (UMTRI),
   438, 440
Unloading, 248, 262
   from different deflection, 251
   strain rate, 325

Unspring weight, mass, 83
Unsprung masses, 83
Unsprung weight, 83
Upper longitudinal member, 86–87
Utility poles, wooden, 394–397

V

v05404.049 data file, 163–164
Vanishing point, 116–117
Variable acceleration; *see also* Constant
        acceleration
    acceleration performance test curves,
        78, 79
    time–distance problems, 77, 80
    time–distance study with, 80
VDS, *see* Vehicle Description Section
        (VDS)
Vehicle-to-vehicle collisions
    barrier collisions, 265, 270
    barrier crashes, 268
    crash model, 266
    crush force, 269
    load–deflection equation, 266
    motion of two-mass system, 267
Vehicle
    coordinates, 64
    crashes, 7, 153
    deceleration, 32, 47
    headings, 49, 50, 381
    inertial properties, 97–98
    inputs, 390
    interior inspection, 113, 178
    position, 32, 36, 380
    schematics, 58
    selection, 378
    yaw rotation, 285
Vehicle Crash Test Database, 159
Vehicle crush measurement, 345
    contact damage, 345
    coplanar crashes, 357–360
    exemplar vehicle, 347
    full-scale mapping, 351–356
    inch tape marking, 347
    loose parts, 357
    NASS protocol, 348–350
    rollover roof deformation
        measurements, 360
    tape line, 346

    total station method, 356
Vehicle Description Section (VDS), 93
Vehicle Identification Number (VIN), 92
    codes, 94
    decoding importance, 95
    formats, 92, 93
    reasons, 93, 94
Vehicle Identification Section (VIS), 93
Vehicle inspection; *see also* Scene
        inspection
    case materials, 109
    cloth dress maker's tape, 114
    loose parts, 111
    sequence for, 110
    tire conditions, 112
    vehicle identity, 111
    vehicle interior, 113
Vehicle specifications
    MVMA specifications, 96
    reconstructionist, 95
Vehicle structure crash mechanics, 247
    barrier equivalence, 272
    crash severity measures, 274–275
    energy absorption, 252–254
    large car barrier crashes, 263–264
    large car hits small car, 270–272
    limitations and assumptions, 248
    load–deflection curves, 249–252
    load paths, 247–248
    narrow fixed object collisions,
        264–265
    restitution, 254–257
    small and large car comparisons, 264
    small car barrier crashes, 263
    structural dynamics, 257–261
    vehicle-to-vehicle collisions, 265–270
Vehicle Year & Model Interchange List
        (Sisters & Clones List), 99, 316
Velocity-time (V-t) curve, history, 178,
        189, 260
Velocity, 2, 10
    acceleration by integration
        process, 178
    CG X-axis, 12
    change in, 12
    common, 364, 366
    components at separation, 56
    curvature, 89
    IMPAC and PLASMO, 288–289

vector direction, 381, 387
vector, 14, 241
wheel's tangential, 21
Velocity change delta-V (ΔV), 385
Velocity–time histories, 47–48
Velocity–time trace, 15
Vertical curvature, 89
Vertical load, 28, 423
VIN, *see* Vehicle Identification Number
    (VIN)
VIN DeCoder, 95, 96
VINLink, 95
VINPower, 95
VIS, *see* Vehicle Identification Section
    (VIS)
Volpe Tests, 413
vXXXXX. EV5 file, 160, 193
vXXXXX file, 160, 193

W

WardsAuto, 96
Water line, 85
WaveFront, 37
WB, *see* Wheelbase (WB)
WeCARE, 348
Weight distribution, 20, 27
Weighted average, 27, 33
    crush measurements, 312
    filtering data, 164
    in oblique impacts, 355
Weighting factor, 389, 390, 418
Weight transfer, 20
Weight transfer effects, 23, 278
Weight units, 101
Weld inspection, 113
What-if, 71, 379
Wheelbase (WB), 91
Wheel bound up, 61
Wheel hop, 84
Wheel lift-off, 441
Wheel rate, *see* Suspension rate
Wheel well, 87, 112
Wind direction, 105, 109
Window glazing, 223
Windows XP, 3
Windshield header, 86
WinSMASH program
    damage width, 427

enhancements, 426, 427
NHTSA, 428
PDOF, 428
stiffness parameters, 427
WinZip, 160
WMI, *see* World Manufacturer Identifier
    (WMI)
Wooden utility poles
    BFE for, 395
    butt, 394
    energy accounting, 397
    hemlock poles, 396
    reconstruction procedure, 395
    speed estimation, 398
Work done
    deflection, 253
    energy absorption, 252, 253–254
    on vehicle in segment, 47
World Manufacturer Identifier (WMI),
    93
Wreckage, 112

X

XlXtrFun, 38

Y

Yaw angle, 7, 56
Yaw moment
    of inertia, 279, 280
    mass distribution, 285
Yaw moment of inertia, 279, 280
Yaw rate degradation, 378
Yaw rates, 49, 53, 237
Yellow pine pole, 396
Yielding (of the soil), 397

Z

Z-transform
    bilateral or two-sided, 141
    inverse discrete-time Fourier
        transform, 143
    region of convergence, 142, 143
    transfer function, 142
Zamboni machines, 280, 281
Zero acceleration, *see* Constant velocity
z space, 144

Printed in the United States
by Baker & Taylor Publisher Services